ORDERED
VECTOR SPACES
AND
LINEAR OPERATORS

ROMULUS CRISTESCU

ORDERED VECTOR SPACES AND LINEAR OPERATORS

EDITURA ACADEMIEI
BUCUREŞTI
ROMÂNIA

ABACUS PRESS
TUNBRIDGE WELLS, KENT
ENGLAND
1976

Revised and updated translation of
Spaţii lineare ordonate şi operatori lineari
EDITURA ACADEMIEI, Bucharest, 1970

The English edition appears under the joint imprints of
Editura Academiei
Calea Victoriei 125, Bucureşti, Romania
and
Abacus Press
Abacus House, Speldhurst Road, Tunbridge Wells, Kent, England

Translated from the Romanian by Silviu Teleman
Institute of Atomic Physics, Bucharest
Translation Editor: J. C. Hammel, Ph. D.

ISBN 0 85626 090 8

Printed in Romania

Contents

Foreword

At the International Congress of Mathematicians, held in Bologna in 1928, F. Riesz considered for the first time the set of all linear functionals defined on the space of continuous real functions, as an ordered vector space.

The theory of ordered vector spaces developed with the papers of L. V. Kantorovitch, written between 1935 and 1937, and with that of H. Freudenthal, which was independently published in 1936. This theory was subsequently enriched by the results of H. Nakano, published after 1941, as well as by those of B. Z. Vulich, A. G. Pinsker and others.

We must also mention the role played by the investigations made by G. Birkhoff on the lattice theory, as well as on the theory of ordered groups and spaces.

Birkhoff's results have been systematically expounded in his famous monograph Lattice theory, *first published in New York, in 1940.*

Research by M. G. Krein and his students on the theory of normed spaces with cones is also noteworthy.

Normed ordered vector spaces have also been considered by L. V. Kantorovitch and H. Nakano.

During the last two decades the theory of locally convex ordered vector spaces has also been developed in some papers written by H. Schaefer, I. Namioka, I. Kawai, A. Peressini and others.

The theory of ordered vector spaces is now one of the most important branches of functional analysis.

The first monographs on this subject appeared in 1950. These are Functional analysis in partially ordered spaces, *by L. V. Kantorovich, B. Z. Vulich and A.G. Pinsker, published in Moscow, in Russian, and two books written in English,* Modulared semi-ordered linear spaces *and* Modern spectral theory *by H. Nakano, which appeared the same year in Tokyo.*

In 1959 the author's book Ordered vector spaces *(in Romanian) appeared in Bucharest, whereas in 1961 the book* Introduction to the theory of partially ordered spaces, *by B. Z. Vulich (in Russian) appeared in Moscow.*

The present book differs from the author's book published in 1959 by the new form of exposition of the material, as well as by the inclusion of some new results.

For the understanding of this book it is assumed that the reader is acquainted with some general facts from Functional Analysis, as, for instance, those contained in the author's book Functional Analysis, *which was published in Romanian, in 1970. For the English-speaking reader, the treatise* Linear operators *by N. Dunford and J. Schwartz (vol. I) is an excellent reference.*

1

Basic facts about ordered sets

The aim of this chapter is to present some results from the theory of ordered sets, which will be constantly used in the study of ordered vector spaces.

1.1 General facts about ordered sets

1.1.1 The notion of ordered set

A relation $x \text{ o } y$, defined for some pairs x, y of elements of a set E, is called an *order relation* in E if the following conditions are satisfied.

(i) $x \text{ o } x$, for any $x \in E$ (reflexivity);

(ii) "$x \text{ o } y$ and $y \text{ o } z$" implies $x \text{ o } z$ (transitivity);

(iii) "$x \text{ o } y$ and $y \text{ o } x$" implies $x = y$ (antisymmetry).

A set E, in which an order relation has been given, is called an *ordered set**.

If A is a subset of an ordered set E, then the relation $x \text{ o } y$, considered for elements x, y belonging to A, is an order relation in A. This order relation is said to be *induced* by the order relation given in E.

A reflexive and transitive relation is called a pre-order relation and a set in which a pre-order relation has been given is called a pre-ordered set.

If E is a pre-ordered set with the pre-order relation $x \text{ p } y$, then the relation $x \sim y$, defined by the condition "$x \text{ p } y$ and $y \text{ p } x$", is an equivalence relation in E; in the quotient set (which in this case is denoted by E/p) we get an order relation by defining

$$\dot{x} \text{ o } \dot{y} \Leftrightarrow x \text{ p } y,$$

where $\dot{x}, \dot{y}$ are the equivalence classes determined by x and y, respectively.

The symbol $\leqslant$ is generally used for an order relation, that is, we write $x \leqslant y$ (instead of $x \text{ o } y$) and read "x is smaller or equal to y". We also write $y \geqslant x$ and

* The terms *semi-ordered set* and *partially ordered* set are also used.

read "y is greater or equal to x". The notation $x < y$ is used for : $x \leqslant y$ and $x \neq y$; $y > x$ means that: $y \geqslant x$ and $y \neq x$.

If E is an ordered set and if for a given pair of elements $x, y \in E$ the relation $x \leqslant y$ does not hold, we write $x \nleqslant y$. If $x \nleqslant y$ and $y \nleqslant x$, then the elements x, y are called *incomparable*.

An ordered set E is called *increasingly directed* (or directed to the right) if for any pair of elements $x, y \in E$ there exists an element $z \in E$, such that $x \leqslant z$ and $y \leqslant z$. An ordered set E is called *decreasingly directed* (or directed to the left) if for any two elements $x, y \in E$ there exists an element $z \in E$, such that $z \leqslant x$ and $z \leqslant y$.

An ordered set E is called *directed on both sides* if it is directed in both ways.

We shall make the convention to call simply a *directed set* any set which is increasingly directed.

An ordered set E is called a *totally ordered set* if it has no incomparable elements.

Two ordered sets E and E' are said to be *isomorphic* if there exists a one-to-one mapping f of E onto E' such that

$$x \leqslant y \Leftrightarrow f(x) \leqslant f(y).$$

The mapping f is said to be an isomorphism (for the order structure).

1.1.2 Least upper bounds and greatest lower bounds

Let E be an ordered set. A subset A of E is said to be *bounded from below* or *minorized* if there exists an element $y \in E$ such that $y \leqslant x$, for any $x \in A$. The element y is called a *lower bound* for A. If there exists a lower bound a for the set A, which belongs to A, then a is called the *smallest element* of A.

In a decreasingly directed set any finite subset is bounded from below.

A subset B of E is said to be *bounded from above*, or *majorized*, if there exists an element $z \in E$, such that $x \leqslant z$, for any $x \in B$. The element z is called an *upper bound* for B. If there exists an upper bound b, which belongs to B, then b is called the *greatest element* of B.

In an increasingly directed set any finite subset is bounded from above.

A subset of E, which is at the same time bounded from below and from above, is simply said to be *bounded*.

If a and b are two elements of E, such that $a \leqslant b$, then the set

$$[a, b] = \{x \in E; a \leqslant x \leqslant b\}$$

is called an *interval*, and a and b are its *extremities*.

A subset C of E is bounded iff*) there exists an interval $[a, b]$, such that $C \subset [a, b]$. In a set which is directed on both sides any finite subset is bounded.

One says that a subset A of E has a *greatest lower bound* if it is bounded from below and if the set of its lower bounds has a greatest element a. The element

*) 'Iff' stands for 'if and only if'.

a is called the *greatest lower bound* of A and it is denoted by inf A or glb A. The greatest lower bound, if it exists, is unique.

A subset B of E is said to have a *least upper bound* if it is bounded from above and if the set of its upper bounds has a smallest element b. The element b is called the *least upper bound* of B and it is denoted by sup B or lub B.

The least upper bound, if it exists, is unique.

One can also speak about a bounded from below, or from above, *system* of elements of E, about the greatest lower bound or the least upper bound of a system $\{x_j\}_{j\in J}$ of elements of the set E. In this case the notation $\bigwedge_{j\in J} x_j$ is used for the greatest lower bound and $\bigvee_{j\in J} x_j$ for the least upper bound. If $J=\{1, 2, \ldots, n\}$, one also writes $\bigwedge_{j=1}^{n} x_j$ and $\bigvee_{j=1}^{n} x_j$, respectively for the two bounds. They can also be written as follows: $x_1 \wedge x_2 \wedge \ldots \wedge x_n$, and $x_1 \vee x_2 \ldots \vee x_n$, respectively. For any two elements x, y we shall always denote by $x \wedge y$ and $x \vee y$ the two bounds (if they exist).

If E and E' are two isomorphically ordered sets and if f is an isomorphism between them, if C is a subset of E such that $a = \inf C$ (or $b = \sup C$) exists, then $f(a) = \inf f(C)$ ($f(b) = \sup f(C)$ respectively).

It is easily verified that in an arbitrary ordered set E, the greatest lower bound and the least upper bound of its subsets have the following properties:

(i) if $A_1 \subset A_2$ and if $a_1 = \inf A_1$ and $a_2 = \inf A_2$ exist, then $a_1 \geqslant a_2$;

(ii) if $B_1 \subset B_2$ and if $b_1 = \sup B_1$ and $b_2 = \sup B_2$ exist, then $b_1 \leqslant b_2$;

(iii) if $a = \sup A$ and $b = \inf B$ exist and if $x \leqslant y$ for any $x \in A$ and $y \in B$, then $a \leqslant b$.

It is also easily verified that if $\{A_j\}_{j\in J}$ is a family of subsets of the ordered set E, if $A = \bigcup_{j\in J} A_j$ and if $a=\inf A$ and $a_j=\inf A_j$ exist, for any $j \in J$, then $\bigwedge_{j\in J} a_j$ also exists and $a = \bigwedge_{j\in J} a_j$.

A similar proposition holds for the least upper bounds.

1.1.3 Minimal elements and maximal elements

Let E be an ordered set.

An element $a \in E$ is said to be *minimal* if $x \leqslant a$ implies $x = a$; an element $b \in E$ is said to be *maximal* if $b \leqslant x$ implies $x = b$.

An ordered set may have no minimal (or maximal) elements, or it can have one or more minimal elements (or maximal elements, respectively).

If an ordered set has the smallest element a (or the greatest element b), then this element is the only minimal (the only maximal respectively) element of the set.

The following proposition is known as Zorn's theorem [1]) (also Zorn's lemma).

Let E be an ordered set, such that any totally ordered subset of E has an upper bound. Then for any element $x_0 \in E$ there exists a maximal element $y \in E$, such that $x_0 \leqslant y$.

---

[1]) See M. Zorn [1].

1.2 Lattices

1.2.1 The notion of lattice

Definition 1. An ordered set E is said to be a *lattice*, if for any pair of its elements x, y, elements $x \wedge y$ and $x \vee y$ exist.

Obviously, any lattice is an ordered set, directed on both sides.

If E is a lattice, then for any finite subset of E there exist the greatest lower bound and the least upper bound.

If A is a subset of a lattice E, and if A is a lattice for the induced order relation, then the greatest lower bound (or the least upper bound) in A of any two elements $x, y \in A$ generally do not coincide with the corresponding greatest lower bound (or the least upper bound) in E.

Definition 2. A subset A of a lattice E is said to be a *sublattice* if from $x \in A$ and $y \in A$ it follows that $x \wedge y \in A$ and $x \vee y \in A$.

Any sublattice of a lattice is obviously a lattice itself (with respect to the induced order relation).

Examples. (1) Let T be any set and let $F(T)$ be the set of all real functions defined on T. We shall put $x \leqslant y (x, y \in F(T))$ if $x(t) \leqslant y(t)$, for any $t \in T$. The set $F(T)$ is then a lattice in which $(x \wedge y)(t) = \min\{x(t), y(t)\}$, $(x \vee y)(t) = \max\{x(t), y(t)\}$.

(2) Let $C([0,1])$ be the subset of the set $F([0,1])$ consisting of all the continuous functions. The set $C([0,1])$ is a sublattice of $F([0,1])$.

1.2.2 Completeness

Definition 1. A lattice E is said to be *complete* if any subset of E admits of a greatest lower bound and a least upper bound. The lattice E is said to be *relatively complete* if any bounded subset of E admits a greatest lower bound and a least upper bound.

If in definition 1 the subsets of E are assumed to be *countable*, then we get the definition of the *σ-complete lattice* (*relatively σ-complete, respectively*).

Proposition 1. *Any relatively complete lattice E has the following properties:*

(i) *for any non-empty subset $A \subset E$, bounded from below, there exists the element* $\inf A$:

(ii) *for any non-empty subset $B \subset E$, bounded from above, there exists the element $\sup B$.*

Conversely, if a lattice E has one of the properties (i), (ii), *then E is relatively complete.*

Proof. Let E be a relatively complete lattice and A a bounded from below subset of E. Let us consider an arbitrary $a_0 \in A$ and let

$$A_0 = \{a_0 \wedge x | x \in A\}.$$

The set A_0 is non-empty and bounded, and therefore there exists $a = \inf A_0$; we obviously have $a = \inf A$.

One can similarly prove that if B is bounded from above, then $\sup B$ exists.

Now let E be a lattice having property (i).

Let us show that if B is a non-empty bounded from above subset of E, then $\sup B$ exists.

Let M be the set of all the upper bounds of B. By hypothesis, $M \neq \varnothing$. Since the set M is bounded from below (by any element of B), it follows that the element $b = \inf M$ exists. Since $x \leqslant b \leqslant y$ for any $x \in B$ and any $y \in M$, it follows that $b = \sup B$.

One can show similarly that if the lattice has property (ii), then E is relatively complete.

Remark. A relatively σ-complete lattice has properties (i), (ii) from proposition 1 if the subsets A and B are, moreover, assumed to be countable.

In the next proposition we shall use the following two definitions:

Definition 2. A family $\{E_j\}_{j\in J}$ of sets has the *binary intersection property* if for any subset $J_0 \subset J$, such that $E_{j_1} \cap E_{j_2} \neq \varnothing$ for $j_1, j_2 \in J_0$, one has

$$\bigcap_{j\in J_0} E_j \neq \varnothing.$$

Definition 3. An ordered set E has the *interpolation property* if for any elements $x_1, x_2, y_1, y_2 \in E$, such that $x_i \leqslant y_j$ $(i, j = 1, 2)$, there exists $z \in E$, such that

$$x_i \leqslant z \leqslant y_j, \qquad i, j = 1,2.$$

A lattice has the interpolation property because one can take $z = x_1 \vee x_2$ or $z = y_1 \wedge y_2$.

Proposition 2. *Let E be an ordered set which is directed on both sides. Then the set E is a relatively complete lattice, if and only if, it has the interpolation property and the family of its intervals has the binary intersection property.*

Proof. Let E be a relatively complete lattice. Let $\{[x_j, y_j]\}_{j\in J_0}$ be any family of intervals of E, such that

$$[x_h, y_h] \cap [x_k, y_k] \neq \varnothing,$$

for any $h, k \in J_0$. For any $h, k \in J_0$ there then exists a $z \in E$, such that $x_h \leqslant z \leqslant y_h$ and $x_k \leqslant z \leqslant y_k$. It follows that $x_h \leqslant y_k$, for any $h, k \in J_0$ and, therefore, there exist

$$x = \bigvee_{j\in J_0} x_j, \quad y = \bigwedge_{j\in J_0} y_j;$$

it follows that $x \leqslant y$. If $a \in [x, y]$, then $a \in [x_j, y_j]$, for any $j \in J_0$.

It follows that the family of all the intervals of E has the binary intersection property.

Conversely, let E be an ordered set, directed on both sides. Let A be subset of E, bounded from above, and let B be the set of all its upper bounds. If E has the interpolation property, then for any $x_1, x_2 \in A$, and $y_1, y_2 \in B$ the intervals $[x_1, y_1]$, $[x_2, y_2]$ have a common point. If the family of all the intervals of the set E has the binary intersection property, then the set of all intervals $[x, y]$, with $x \in A$ and $y \in B$, have a common point $z \in E$. It is then obvious that $z = \sup A$. One can similarly show that from the interpolation property of E and the binary intersection property of the family of all the intervals of E it follows that any non-empty bounded from below subset of E admits of a greatest lower bound.

Examples (1). The set $\mathscr{L}$ of all the Lebesgue measurable subsets of the interval $[0,1]$ is a σ-complete lattice, if the order relation is taken to be the inclusion. For a finite or a countable system of elements from $\mathscr{L}$, the greatest lower bound coincides with the intersection, whereas the least upper bound coincides with the union of the given elements.

(2) The lattice $F(T)$ (Section 1.2.1, Example 1), is relatively complete. In this set, $a = \inf A$ means $a(t) = \inf\limits_{x \in A} x(t)$, and $b = \sup B$ means $b(t) = \sup\limits_{x \in B} x(t)$.

(3) The lattice $C([0,1])$ (Section 1.2.1, Example 2) is not relatively σ-complete.

Definition 4. Let E and E' be two ordered sets. The set E' is called an *extension* of E, and E is said to *embed* in the set E', if E is isomorphic to a subset of E'.

Theorem. *Any ordered set E, which is directed on both sides, may be embedded in a relatively complete lattice.*

Proof. For any bounded from above subset A of E we shall denote by $s(A)$ the set of all the upper bounds of A. If B is a bounded from below subset of E, we shall denote by $i(B)$ the set of all its lower bounds.

We shall denote by $\tilde{E}$ the set of all non-empty, bounded from above subsets $A \subset E$ satisfying the condition $A = i(s(A))$. Let us note that if A is any non-empty bounded from above subset of E, then the set $A_1 = i(s(A))$ is the smallest element of $\tilde{E}$ which contains A.

Indeed, it is obvious that $A \subset A_1$. We then have

$$s(i(s(A))) = s(A),$$

whence $i(s(A_1)) = A_1$ and, therefore, $A_1 \in \tilde{E}$. If $B \in \tilde{E}$ and $A \subset B$, then $s(A) \supset s(B)$, whence

$$A_1 = i(s(A)) \subset i(s(B)) = B.$$

One can also state that

$$i(s(x)) = i(x),$$

for any $x \in E$.

We shall consider in $\tilde{E}$ the inclusion as an order relation.

The set $\tilde{E}$ is directed on both sides.

Indeed, let $A_1, A_2 \in \tilde{E}$. Let $x_1 \in A_1$ and $x_2 \in A_2$. Since set E is directed to the left, there exists an $x \in E$, such that $x \leqslant x_1$ and $x \leqslant x_2$. It follows that $i(x) \subset A_1$ and $i(x) \subset A_2$; consequently, $\tilde{E}$ is directed to the left. In the same way one can show that the set $\tilde{E}$ is directed to the right, by taking into account the fact that if A_1, $A_2 \in \tilde{E}$, then the set $A_1 \cup A_2$ is bounded from above.

Now let $\{A_j\}_{j \in J}$ be an arbitrary, bounded from below system of elements of $\tilde{E}$. If we put

$$A_0 = \bigcap_{j \in J} A_j,$$

the set A_0 is non-empty and $A_0 \in \tilde{E}$. Indeed, we have $A_0 \subset A_j$, for any $j \in J$; this implies

$$i(s(A_0)) \subset i(s(A_j)) = A_j,$$

for any $j \in J$, and this shows that

$$i(s(A_0)) \subset A_0.$$

Since the opposite inclusion always holds, it follows that $A_0 \in \tilde{E}$. It is therefore obvious that A_0 is the greatest lower bound of the system $\{A_j\}_{j \in J}$ in the ordered set $\tilde{E}$.

Similarly, if the system $\{A_j\}_{j \in J}$ is bounded from above and if we put $A' = \bigcup_{j \in J} A_j$ and $A^0 = i(s(A'))$, then A^0 is the least upper bound of the system $\{A_j\}_{j \in J}$ in $\tilde{E}$.

We have thus shown that $\tilde{E}$ is a relatively complete lattice.

If we put

$$\tilde{E}_0 = \{i(x);\ x \in E\},$$

then the mapping $x \mapsto i(x)$ is an isomorphism between the ordered sets E and $\tilde{E}_0$ as one can easily establish. The theorem is proved.

The method by which $\tilde{E}$ has been constructed is called *Dedekind's method* and it is fully analogous to the method used in order to construct the set of real numbers. The set $\tilde{E}$ is called the *Dedekind extension of E.*

Remarks. (1) If $A \subset E$ and $y = \inf A$, then

$$i(y) = \bigwedge_{x \in A} i(x),$$

whereas if $z = \sup A$, one has

$$i(z) = \bigvee_{x \in A} i(x).$$

The verification can be easily made.

The embedding of E in $\tilde{E}$ is therefore performed "with the conservation of the bounds".

(2) If $A \in \tilde{E}$, then

$$A = \sup_{x \in A} i(x) = \inf_{y \in s(A)} i(y).$$

Indeed, it is sufficient to state that

$$A = \bigcup_{x \in A} i(x) = \bigcap_{y \in s(A)} i(y).$$

1.2.3 Distributivity laws

Let E be a lattice.

The following formulae are called the *finite distributivity laws*

$$x \wedge \left(\bigvee_{j=1}^{n} y_j \right) = \bigvee_{j=1}^{n} (x \wedge y_j), \tag{1}$$

$$x \vee \left(\bigwedge_{j=1}^{n} y_j \right) = \bigwedge_{j=1}^{n} (x \vee y_j), \tag{2}$$

where $x \in E$, $y_j \in E$ $(j = 1, 2, \ldots, n)$, and the number n is arbitrary.

The finite distributivity laws do not hold in any lattice. In order that they be satisfied it is sufficient that for any elements x, y, z the following inequality hold

$$x \wedge (y \vee z) \leqslant (x \wedge y) \vee (x \wedge z). \tag{3}$$

Indeed, it is not difficult to see that in any lattice the opposite inequality holds, which, in conjunction with (3), leads to the equality

$$x \wedge (y \vee z) = (x \wedge y) \vee (x \wedge z). \tag{4}$$

From (4) we also get

$$x \vee (y \wedge z) = (x \vee y) \wedge (x \vee z). \tag{5}$$

From (4) and (5) we get, by induction, (1) and (2).

The same conclusion can be reached if we assume that for any elements x, y, z of the considered lattice the inequality

$$x \vee (y \wedge z) \geqslant (x \vee y) \wedge (x \vee z)$$

holds.

A lattice in which the finite distributivity laws hold is usually referred to as a *distributive lattice.*

By definition, in a lattice E the *infinite distributivity laws* hold, if E satisfies the following conditions:

(i) If $\{y_j\}_{j\in J}$ is any system of elements of E and if $\bigvee_{j\in J} y_j$ exists, then $\bigvee_{j\in J} (x \wedge y_j)$ also exists for any $x \in E$ and the following equality holds:

$$x \wedge (\bigvee_{j\in J} y_j) = \bigvee_{j\in J} (x \wedge y_j). \tag{6}$$

(ii) If $\{y_j\}_{j\in J}$ is any system of elements of E and if $\bigwedge_{j\in J} y_j$ exists, then $\bigwedge_{j\in J} (x \vee y_j)$ also exists for any $x \in E$ and the following equality holds.

$$x \vee (\bigwedge_{j\in J} y_j) = \bigwedge_{j\in J} (x \vee y_j). \tag{7}$$

If in a lattice E the infinite distributivity laws hold, then the equalities

$$(\bigvee_{i\in I} x_i) \wedge (\bigvee_{j\in J} y_j) = \bigvee_{\substack{i\in I\\ j\in J}} (x_i \wedge y_j), \tag{8}$$

$$(\bigwedge_{i\in I} x_i) \vee (\bigwedge_{j\in J} y_j) = \bigwedge_{\substack{i\in I\\ j\in J}} (x_i \vee y_j), \tag{9}$$

hold, as soon as the left-hand members exist.

In other words, if $\bigvee_{i\in I} x_i$ and $\bigvee_{i\in I} y_j$ exist, then $\bigvee_{i,j} (x_i \wedge y_j)$ exists too and equality (8) holds. Similarly for equality (9). The proof can be easily made.

If in (6) and (7) one assumes that the set J is at most countable, then one says that E satisfies the countable distributivity laws (or, equivalently, the σ-distributivity laws). In this case, in (8) and (9), the sets I and J will be countable by assumption.

1.2.4 Boolean lattices

Definition 1. A *Boolean lattice* (or Boolean algebra) is any lattice $\mathscr{B}$ which satisfies the following conditions:

(i) the set $\mathscr{B}$ has the smallest element $\mathbf{0}$ and the greatest element u;

(ii) in the set $\mathscr{B}$ the finite distributivity laws hold;

(iii) for any element $e \in \mathscr{B}$ there exists an element $e^* \in \mathscr{B}$ such that $e \wedge e^* = \mathbf{0}$ and $e \vee e^* = u$.

The element e^* from (iii) is called the *complement* of e. It is unique, because from $e_1 \wedge e = \mathbf{0}$ and $e_1 \vee e = u$ it follows that

$$e_1 = e_1 \vee 0 = e_1 \vee (e \wedge e^*) = (e_1 \vee e) \wedge (e_1 \vee e^*) = e_1 \vee e^*,$$

and this implies that $e_1 \geqslant e^*$. By exchanging the rôles played by the two elements e_1, e^*, it follows that $e^* \geqslant e_1$ and, therefore, $e_1 = e^*$.

In particular, we get $(e^*)^* = e$.

We have also

Proposition 1. *In a Boolean lattice the complement has the following properties*

(1) *if* $e \wedge e_1 = \mathbf{0}$ *then* $e_1 \leqslant e^*$;

(2) *if* $e_1 \leqslant e_2$, *then* $e_1^* \geqslant e_2^*$;

(3) *if* $\bigvee_{j \in J} e_j$ *exists, then* $\bigwedge_{j \in J} e_j^*$ *exists too and the equality*

$$(\bigvee_{j \in J} e_j)^* = \bigwedge_{j \in J} e_j^*$$

holds.

Proof (1) If we put $e_2 = e_1 \vee e^*$, we obtain

$$e_2 \wedge e = (e_1 \vee e^*) \wedge e = (e_1 \wedge e) \vee (e^* \wedge e) = \mathbf{0}$$

$$e_2 \vee e = (e_1 \vee e^*) \vee e = e_1 \vee (e^* \vee e) = e_1 \vee u = u,$$

and this implies $e_2 = e^*$, whence $e_1 \leqslant e^*$.

(2) If $e_1 \leqslant e_2$, then

$$e_1 \wedge e_2^* \leqslant e_2 \wedge e_2^* = \mathbf{0};$$

hence $e_1 \wedge e_2^* = \mathbf{0}$ and this shows that $e_2^* \leqslant e_1^*$, by virtue of property (1).

(3) Let

$$e = \bigvee_{j \in J} e_j;$$

then we have $e_j \leqslant e$ and, with property (2) we get $e^* \leqslant e_j^*$ for any $j \in J$. If $f \leqslant e_j^*$ for any $j \in J$, then $e_j \leqslant f^*$ for any j, and this shows that $e \leqslant f^*$. It follows that $e^* \geqslant f$ and, therefore

$$e^* = \bigwedge_{j \in J} e_j^*.$$

Remark. From property (3) one infers that

$$(e_1 \vee e_2)^* = e_1^* \wedge e_2^*,$$

$$(e_1 \wedge e_2)^* = e_1^* \vee e_2^*,$$

for any elements e_1, e_2.

Proposition 2. *In any Boolean lattice the infinite distributivity laws hold.*

Proof. Let us suppose that the system $\{f_j\}_{j \in J}$ has the least upper bound f. Let e be an arbitrary element and let us put $g = e \wedge f$. It is obvious that

$$e \wedge f_j \leqslant g, \quad j \in J.$$

Let now $e \wedge f_j \leqslant h$, for any $j \in J$. We have

$$h \geqslant e \wedge h = (e \wedge h) \vee (e \wedge e^*) = e \wedge (h \vee e^*).$$

On the other hand, we have

$$h \vee e^* \geqslant (e \wedge f_j) \vee e^* = (e \vee e^*) \wedge (f_j \vee e^*) = f_j \vee e^* \geqslant f_j$$

for any $j \in J$, and this shows that

$$h \geqslant e \wedge (h \vee e^*) \geqslant e \wedge f = g.$$

It follows that

$$g = \bigvee_{j \in J} (e \wedge f_j).$$

In a similar way one can prove the formula

$$e \vee (\bigwedge_{j \in J} f_j) = \bigwedge_{j \in J} (e \vee f_j),$$

if $\bigwedge_{j \in J} f_j$ exists.

Definition 2. *A Boolean sub-lattice of a Boolean lattice* $\mathscr{B}$ is any sub-lattice $\mathscr{B}_1$ of $\mathscr{B}$ such that

$$e \in \mathscr{B}_1 \Rightarrow e^* \in \mathscr{B}_1.$$

A complete (σ-complete) Boolean lattice is any Boolean lattice which is complete (σ-complete, respectively) as a lattice.

Examples. (1) The set $\mathscr{P}(E)$ of all the subsets of a set E, ordered by inclusion, is a complete Boolean lattice, in which $\mathbf{0} = \varnothing, u = E$, whereas $A^* = \complement A \; (A \in \mathscr{P}(E))$.

(2) Any subset $\mathscr{A}$ of $\mathscr{P}(E)$, which has the properties

(i) $A_1, A_2 \in \mathscr{A} \Rightarrow A_1 \cup A_2 \in \mathscr{A}$

(ii) $A \in \mathscr{A} \Rightarrow \complement A \in \mathscr{A}$,

is a Boolean lattice, namely a Boolean sublattice of $\mathscr{P}(E)$.

In the following part of this section, $\mathscr{B}$ will stand for an arbitrary Boolean lattice.

Definition 3. A subset E of the Boolean lattice $\mathscr{B}$ is called a *chain* if for any n elements $e_1, e_2, \ldots, e_n$ of E we have $\bigwedge_{j=1}^{n} e_j \neq \mathbf{0}$. A chain E is said to be *maximal* if it is contained in no other chain.

Lemma 1. *Any maximal chain E has the following properties:*

(1) *if* $e \in E$ *and* $e \leqslant f$, $(f \in \mathscr{B})$, *then* $f \in E$;

(2) *if* $e_j \in E, j = 1, 2, \ldots, n$, *then* $\bigwedge_{j=1}^{n} e_j \in E$;

(3) *if* $e \notin E$, *then there exists* $e_1 \in E$ such that $e \wedge e_1 = \mathbf{0}$.

Proof. (1) If $e \in E$, then the set

$$E_1 = E \cup \{f; e \leqslant f \in \mathscr{B}\}$$

is a chain. Since E is maximal, it follows that $E_1 = E$.

(2) The set

$$E_1 = E \cup \left\{\bigwedge_{j=1}^{n} e_j;\; e_j \in E,\; n = 1,2,3, \ldots\right\}$$

is a chain; therefore $E_1 = E$.

(3) If there exists no element $e_1 \in E$, such that $e \wedge e_1 = \mathbf{0}$, then the set $E_1 = E \cup \{e\}$ would be a chain which contains E but differing from E.

Lemma 2. *Any chain is contained in a maximal chain. In particular, any element* $e \neq \mathbf{0}$ *belongs to a maximal chain.*

Proof. Let us denote by $\mathscr{E}$ the set of all the chains contained in $\mathscr{B}$. By ordering this set by inclusion, the hypothesis of Zorn's lemma is satisfied. Indeed, if $\mathscr{E}_0 = \{E_j\}_{j \in J}$ is a totally ordered system of elements of $\mathscr{E}$, then the set $E_0 = \bigcup_{j \in J} E_j$ is a chain and E_0 is an upper bound of $\mathscr{E}_0$.

The lemma then follows by application of Zorn's lemma.

Let us now denote by $\mathscr{F}$ the set of all the maximal chains, contained in $\mathscr{B}$. For any $e \in \mathscr{B}$ let A_e be the set of all the maximal chains which contain the element e.

Let us now define

$$\mathscr{A} = \{A_e;\ e \in \mathscr{B}\}.$$

Proposition 3. *The system $\mathscr{A}$ of sets is a Boolean lattice with respect to inclusion and it is isomorphic to the Boolean lattice $\mathscr{B}$.*

Proof. We shall associate the element A_e to any element $e \in \mathscr{B}$. This correspondence is one-to-one. Indeed, let $e_1 \neq e_2$ and let us write $e_3 = e_1 \wedge e_2$. At least for one of the elements e_1, e_2, let it be e_1, we have the strict inequality $e_3 < e_1$. By taking into account proposition 1, we have

$$e_0 = e_1 \wedge e_3^* \neq \mathbf{0}.$$

Let E be a maximal chain which contains e_0 (lemma 2). From $e_0 \leqslant e_1$, by taking into account lemma 1, we get $e_1 \in E$. Since $e_2 \wedge e_0 = \mathbf{0}$, we have $e_2 \notin E$. It follows that $E \in A_{e_1}$; from the fact that $E \notin A_{e_2}$ we get $A_{e_1} \neq A_{e_2}$.

Let us now prove the equality

$$A_{e_1 \vee e_2} = A_{e_1} \cup A_{e_2}, \tag{1}$$

for any two elements $e_1, e_2 \in \mathscr{B}$.

Let us now write $e_3 = e_1 \vee e_2$. If E is a maximal chain containing e_1, then $e_3 \in E$. This means that if $E \in A_{e_1}$, then $E \in A_{e_3}$ and this shows that $A_{e_1} \subset A_{e_3}$. By analogy we get that $A_{e_2} \subset A_{e_3}$, and this shows that

$$A_{e_1} \cup A_{e_2} \subset A_{e_3}.$$

Let E be a maximal chain which contains e_3. If $e_1 \notin E$, then there would exist $f_1 \in E$, such that $e_1 \wedge f_1 = \mathbf{0}$ (lemma 1). If $e_2 \notin E$, then one could find an element $f_2 \in E$ such that $e_2 \wedge f_2 = \mathbf{0}$. By defining

$$f_3 = f_1 \wedge f_2$$

we would infer that $e_3 \wedge f_3 = \mathbf{0}$. But $e_3 \in E$ and $f_3 \in E$, and this is a contradiction. It follows that

$$A_{e_3} \subset A_{e_1} \cup A_{e_2},$$

and we thus obtain equality (1).

In view of equality (1) the mapping $e \mapsto A_e$ is an isomorphism and the proposition is proved.

Remark. Besides equality (1), the following equalities also hold

$$A_{e_1 \wedge e_2} = A_{e_1} \cap A_{e_2}, \tag{2}$$

$$A_{e^*} = \complement A_e. \tag{3}$$

Indeed, in order to prove (3), let us state that if $e \in E \in \mathscr{F}$, then $e^* \notin E$, and this implies that

$$A_e \cap A_{e^*} = \varnothing. \tag{4}$$

On the other hand, from $e \vee e^* = u$, by taking into account equality (1) we get

$$A_e \cup A_{e^*} = \mathscr{F}. \tag{5}$$

From (4) and (5) relation (3) immediately follows.

Equality (2) now immediately follows from (1) and (3).

From equalities (1), (2) and (3) we get, in particular, that $\mathscr{A}$ is an algebra of sets [2]).

Let us now introduce a topology in the set $\mathscr{F}$.

A subset of $\mathscr{F}$ will be considered to be open if it is a union of sets of $\mathscr{A}$. From $\complement A_e = A_{e^*}$ it follows that any set A_e is simultaneously closed and open.

Lemma 3. *The set $\mathscr{F}$ is a totally disconnected compact*[3]) *topological space, in which $\mathscr{A}$ is the set of all its closed and open subsets.*

Proof. The axioms for the open sets are easily verified. Let us check the Hausdorff axiom. Let $E_1, E_2 \in \mathscr{F}$, such that $E_1 \neq E_2$. There then exists an $e \in \mathscr{B}$, such that $e \in E_1$ and $e \notin E_2$ (or inversely). It follows that $E_1 \in A_e$ and $E_2 \in \complement A_e = A_{e^*}$ (or inversely), and hence the space $\mathscr{F}$ is a Hausdorff space.

Let us now check the compactness condition. Let us first consider a system $\{A_{e_j}\}_{j \in J}$ of sets from $\mathscr{A}$, which we assume to be a chain, i.e.

$$\bigcap_{i=1}^{n} A_{e_{j_i}} \neq \varnothing,$$

for any indices $j_1, j_2, \ldots, j_n$. From the equality

$$A_{\bigwedge_{i=1}^{n} e_{j_i}} = \bigcap_{i=1}^{n} A_{e_{j_i}}$$

one can easily infer, due to the existing isomorphism of $\mathscr{A}$ with $\mathscr{B}$, that

$$\bigwedge_{i=1}^{n} e_{j_i} \neq \varnothing.$$

It follows that the system $\{e_j\}_{j \in J}$ is a chain of elements of $\mathscr{B}$.

---

[2]) See P. Halmos [1].

[3]) We use here the term compact in the sense of N. Bourbaki [1], by virtue of which it is a Hausdorff space in which any open cover contains a finite cover (or, equivalently, from any family of closed sets whose intersection is empty one can choose a finite subfamily having the same property). A compact space $\mathscr{F}$ is said to be *totally disconnected* if any open set is a union of open and closed sets.

By taking into account lemma 2, there exists a maximal chain E, such that $e_j \in E$ for any $j \in J$. It follows that $E \in A_{e_j}$, for any $j \in J$, and this shows that

$$\bigcap_{j \in J} A_{e_j} \neq \emptyset.$$

Let us consider a chain $\{F_j\}_{j \in J}$ of closed sets of $\mathscr{F}$. For any $j \in J$ there exists a system $\{A_{e_{ji}}\}_{i \in I_j}$, such that

$$F_j = \bigcap_{i \in I_j} A_{e_{ji}}.$$

The system of all the sets $A_{e_{ji}}$ is a chain having the same intersection as the system $\{F_j\}_{j \in J}$. It follows that

$$\bigcap_{j \in J} F_j \neq \emptyset,$$

and this shows that $\mathscr{F}$ is a compact space.

Let now A be a subset of $\mathscr{F}$ which we assume to be open and closed. It is of the form

$$A = \bigcup_{j \in J} A_{e_j}.$$

On the other hand, since the space is compact, and since A is closed, hence compact, one can write

$$A = \bigcup_{i=1}^{n} A_{e_{j_i}} = A_{\bigvee_{i=1}^{n} e_{ji}},$$

which shows that $A \in \mathscr{A}$. The lemma is proved.

From this lemma and from proposition 2 we get the following theorem.

Theorem. *Any Boolean lattice is isomorphic to the set of all clopen sets of a totally disconnected compact topological space.*

Remark. The word *clopen* is a contraction, meaning both closed and open.

1.3 Convergence with respect to the order relation

1.3.1 The notion of (*o*)-convergence

Let E be an ordered set. A sequence [4] $\{x_n\}_{n \in \mathbb{N}}$ of elements of E is said to be *increasing* if $m < n$ implies $x_m \leqslant x_n$. In this case one writes $x_n \underset{n \in \mathbb{N}}{\uparrow}$ or simply $x_n \uparrow$. If, moreover, the element $x = \bigvee_{n \in \mathbb{N}} x_n$ exists, one writes $x_n \underset{n \in \mathbb{N}}{\uparrow} x$ or simply $x_n \uparrow x$.

[4]) By $\mathbb{N}$ we denote the set of natural numbers.

Analogously, one can define the *decreasing* sequence and one uses the notations $x_n \underset{n\in\mathbb{N}}{\downarrow}$ or $x_n \downarrow$. If moreover, the element $x = \bigwedge_{n\in\mathbb{N}} x_n$ exists, then one writes $x_n \underset{n\in\mathbb{N}}{\downarrow} x$ or $x_n \downarrow x$.

Definition 1. The sequence $\{x_n\}_{n\in\mathbb{N}}$ of elements of E is said to *converge with respect to the order relation* to x (abbreviated, (o)-converges to x) if there exist the sequences $\{a_n\}_{n\in\mathbb{N}}$, $\{b_n\}_{n\in\mathbb{N}}$ of elements of E, such that

(i) $a_n \leqslant x_n \leqslant b_n \quad , \quad n \in \mathbb{N}$;
(ii) $a_n \uparrow x$ and $b_n \downarrow x$.

In this case one writes $x_n \xrightarrow{o} x$. We also say that x is the limit with respect to the order relation (in an abbreviated form, the (o)-limit of the sequence $\{x_n\}_{n\in\mathbb{N}}$) and we write $x = (o)\text{-}\lim\limits_{n} x_n$.

Proposition 1. *The convergence with respect to the order relation has the following properties.*

(1) *If* $x_n \uparrow$ (*or* $x_n \downarrow$), *then* $x_n \xrightarrow{o} x$ *iff* $x_n \uparrow x$ ($x_n \downarrow x$, *respectively*);
(2) *Any* (o)-*convergent sequence is bounded*;
(3) *If* $x_n \leqslant y_n$ *and* $x_n \xrightarrow{o} x$, $y_n \xrightarrow{o} y$, *then* $x \leqslant y$.
(4) *If* $x_n \xrightarrow{o} x$ *and* $x_n \xrightarrow{o} y$, *then* $x = y$.
(5) *If* $x_n \xrightarrow{o} x$, *then any subsequence of the sequence* $\{x_n\}_{n\in\mathbb{N}}$ *is* (o)-*convergent to the same limit.*
(6) *If* $y_n \leqslant x_n \leqslant z_n$ *for any* $n \in \mathbb{N}$, *and if* $y_n \xrightarrow{o} x$, $z_n \xrightarrow{o} x$, *then* $x_n \xrightarrow{o} x$.

Proof. (1) Let $x_n \xrightarrow{o} x$ and let $\{a_n\}_{n\in\mathbb{N}}$ and $\{b_n\}_{n\in\mathbb{N}}$ as in definition 1. If $x_n \uparrow$, then $x_m \leqslant b_n$, for any m and n. It follows that $x_m \leqslant x$, for any $m \in \mathbb{N}$. If for an element $y \in E$ we have $x_m \leqslant y$ for any $m \in \mathbb{N}$, then $a_m \leqslant y$, for any $m \in \mathbb{N}$, whence $x \leqslant y$. It follows that $x = \bigvee x_n$. Conversely, if $x_n \uparrow x$, then by taking $a_n = x_n$ and $b_n = x$, the conditions (i) and (ii) are satisfied and, therefore, $x_n \xrightarrow{o} x$.

The case of the decreasing sequence may be similarly treated.

(2) From the conditions of definition 1 it follows that $a_1 \leqslant x_n \leqslant b_1$, for any $n \in \mathbb{N}$.

(3) Let $a_n \leqslant x_n$ be such that $a_n \uparrow x$ and $b_n \geqslant y_n$ be such that $b_n \downarrow y$. Then

$$a_n \leqslant x_n \leqslant y_n \leqslant b_n;$$

since $a_n \uparrow$ and $b_n \downarrow$, it follows that $a_n \leqslant b_m$, for any m, $n \in \mathbb{N}$. Taking into account the remarks made in Section 1.1.2, it further follows that $x \leqslant y$.

(4) This is an immediate consequence of (3).

(5) This is an immediate consequence of the definition.

(6) Since $y_n \xrightarrow{o} x$, there exists a sequence $\{a_n\}_{n\in\mathbb{N}}$, such that $a_n \leqslant y_n$ and $a_n \uparrow x$. There also exists a sequence $\{b_n\}_{n\in\mathbb{N}}$, such that $z_n \leqslant b_n$ and $b_n \downarrow x$.
Since

$$a_n \leqslant y_n \leqslant x_n \leqslant z_n \leqslant b_n,$$

the conditions (i), (ii) from definition 1 are satisfied by the sequence $\{x_n\}_{n\in\mathbb{N}}$.

Proposition 2. *If E is a lattice in which the countable distributivity laws hold and if* $x_n \xrightarrow{o} x$, $y_n \xrightarrow{o} y$, *then*

$$x_n \vee y_n \xrightarrow{o} x \vee y, \; x_n \wedge y_n \xrightarrow{o} x \wedge y.$$

Proof. Let $a_n \leqslant x_n \leqslant b_n$, such that $a_n \uparrow x$, $b_n \downarrow x$ and $a'_n \leqslant y_n \leqslant b'_n$, such that $a'_n \uparrow y$, $b'_n \downarrow y$. We obviously have

$$a_n \vee a'_n \leqslant x_n \vee y_n \leqslant b_n \vee b'_n.$$

By taking into account the fact that the sequences $\{a_n\}_{n\in\mathbb{N}}$, $\{a'_n\}_{n\in\mathbb{N}}$ are increasing sequences, we get

$$x \vee y = (\bigvee_{m\in\mathbb{N}} a_m) \vee (\bigvee_{n\in\mathbb{N}} a'_n) = \bigvee_{m,\, n\in\mathbb{N}} (a_m \vee a'_n) = \bigvee_{n\in\mathbb{N}} (a_n \vee a'_n),$$

hence

$$a_n \vee a'_n \uparrow x \vee y.$$

In a similar way, by taking into account the countable distributivity and the fact that the sequence $\{b_n\}_{n\in\mathbb{N}}$, $\{b'_n\}_{n\in\mathbb{N}}$ are decreasing sequences, we get

$$x \vee y = (\bigwedge_{m\in\mathbb{N}} b_m) \vee (\bigwedge_{n\in\mathbb{N}} b'_n) = \bigwedge_{m,\, n\in\mathbb{N}} (b_m \vee b'_n) = \bigwedge_{n\in\mathbb{N}} (b_n \vee b'_n),$$

hence

$$b_n \vee b'_n \downarrow x \vee y.$$

It follows that $x_n \vee y_n \xrightarrow{o} x \vee y$.

In the same manner one shows that $x_n \wedge y_n \xrightarrow{o} x \wedge y$.

Definition 2. Let $\{x_n\}_{n\in\mathbb{N}}$ be a sequence of elements belonging to an ordered set E. Then the sequence is said to be (o^*)-convergent to an element x (and x is called the (o^*)-limit of the sequence) if from any subsequence $\{x_{j_n}\}_{n\in\mathbb{N}}$ of the given set one may choose another subsequence $\{x_{j_{i_n}}\}_{n\in\mathbb{N}}$ which is (o)-convergent to x.

One writes $x_n \xrightarrow{o^*} x$ or $x = o^* \lim\limits_n x_n$.

One also used the name of (t)-convergence for the (o^*)-convergence.

If $x_n \xrightarrow{o} x$, then $x_n \xrightarrow{o^*} x$, but the converse is not true, in general.

It is easy to check that for monotone sequences the (o^*)-convergence coincides with the (o)-convergence.

1.3.2 Superior limit and inferior limit

Convergence with respect to order for the sequences whose members belong to a relatively σ-complete lattice may also be introduced with the help of the superior and inferior limits, which are defined in this section.

Definition. Let E be a relatively σ-complete lattice and $\{x_n\}_{n\in\mathbb{N}}$ a sequence of elements of E. If this sequence is bounded from above, and if the sequence $\{\bigvee_{n\geqslant m} x_n\}_{m\in\mathbb{N}}$ is bounded from below, then the element of E, defined by the formula

$$\overline{\lim_n}\, x_n = \bigwedge_{m\in\mathbb{N}} (\bigvee_{n\geqslant m} x_n)$$

is called the *superior limit* of the given sequence.

Similarly, if the sequence $\{x_n\}_{n\in\mathbb{N}}$ is bounded from below and if the sequence $\{\bigwedge_{n\geqslant m} x_n\}_{m\in\mathbb{N}}$ is bounded from above, then the element of E given by

$$\underline{\lim_n}\, x_n = \bigvee_{m\in\mathbb{N}} (\bigwedge_{n\geqslant m} x_n),$$

is called the *inferior limit* of the given sequence.

Proposition 1. *If the sequence $\{x_n\}_{n\in\mathbb{N}}$ is bounded, then $\overline{\lim}\, x_n$ and $\underline{\lim}\, x_n$ exist and the following inequalities hold*

$$\bigwedge_{n\in\mathbb{N}} x_n \leqslant \underline{\lim}\, x_n \leqslant \overline{\lim}\, x_n \leqslant \bigvee_{n\in\mathbb{N}} x_n.$$

Proof. For $j > m$ $(j, m \in \mathbb{N})$ we obviously have

$$\bigwedge_{n\geqslant 1} x_n \leqslant \bigwedge_{n\geqslant m} x_n \leqslant \bigwedge_{n\geqslant j} x_n \leqslant \bigvee_{n\geqslant j} x_n \leqslant \bigvee_{n\geqslant m} x_n \leqslant \bigvee_{n\geqslant 1} x_n.$$

From the first inequality we obviously get

$$\bigwedge_{n\geqslant 1} x_n \leqslant \underline{\lim}\, x_n,$$

whereas from the last one

$$\overline{\lim}\, x_n \leqslant \bigvee_{n\geqslant 1} x_n.$$

From the other inequalities we infer that

$$\bigwedge_{n\geqslant m_1} x_n \leqslant \bigvee_{n\geqslant m_2} x_n$$

for any $m_1, m_2 \in \mathbb{N}$; this implies that

$$\bigvee_{m\geqslant 1} \bigwedge_{n\geqslant m} x_n \leqslant \bigwedge_{m\geqslant 1} \bigvee_{n\geqslant m} x_n.$$

Proposition 2. *In order that* $x_n \xrightarrow{o} x$, *in a relatively* σ-*complete lattice* E, *it is necessary and sufficient that the sequence* $\{x_n\}_{n\in\mathbb{N}}$ *be bounded and that*

$$\underline{\lim}_n\, x_n = \overline{\lim}_n\, x_n = x. \tag{1}$$

Proof. If $x_n \xrightarrow{o} x$, then the sequence $\{x_n\}_{n\in\mathbb{N}}$ is bounded and, therefore, the elements $\underline{\lim}\, x_n$ and $\overline{\lim}\, x_n$ exist. Let the sequences $a_n \uparrow x$ and $b_n \downarrow x$ be such that $a_n \leqslant x_n \leqslant b_n$. It is obvious that

$$x = \overline{\lim}\, a_n = \overline{\lim}\, b_n$$

and we have

$$x = \overline{\lim}\, a_n \leqslant \overline{\lim}\, x_n \leqslant \overline{\lim}\, b_n = x,$$

whence $\overline{\lim}\, x_n = x$. In a similar way one shows that $\underline{\lim}\, x_n = x$, and this shows that equalities (1) are satisfied.

Conversely, let us assume that equalities (1) hold. Let us denote

$$a_n = \bigwedge_{j\geqslant n} x_j \quad , \quad b_n = \bigvee_{j\geqslant n} x_j.$$

Then $a_n \uparrow x$ and $b_n \downarrow x$, and we obviously have $a_n \leqslant x_n \leqslant b_n$; this implies that $x_n \xrightarrow{o} x$.

Proposition 3. *Let* E *be a relatively* σ-*complete lattice in which the* σ-*distributivity laws hold. Let* $\{x_n\}_{n\in\mathbb{N}}$ *be a sequence of elements of* E. *If* $\overline{\lim}\, x_n$ *exists, then* $\overline{\lim}\, (x_n \vee y)$ *and* $\overline{\lim}\, (x_n \wedge y)$ *exist too, and the following equalities hold*

$$\overline{\lim}\, (x_n \vee y) = (\overline{\lim}\, x_n) \vee y, \tag{2}$$

$$\overline{\lim}\, (x_n \wedge y) = (\overline{\lim}\, x_n) \wedge y. \tag{3}$$

Proof. If $\overline{\lim}\, x_n$ exists, then, by assumption, the sequence $\{x_n\}_{n\in\mathbb{N}}$ is bounded from above and, therefore, the sequence $\{x_n \vee y\}_{n\in\mathbb{N}}$ is bounded from above too. Since

$$\bigvee_{n\geqslant m} (x_n \vee y) = (\bigvee_{n\geqslant m} x_n) \vee y,$$

and because the sequence $\{\bigvee_{n\geqslant m} x_n\}_{m\in\mathbb{N}}$ is bounded from below, it follows that the sequence $\{\bigvee_{n\geqslant m} (x_n \vee y)\}_{m\in\mathbb{N}}$ is bounded from below. Therefore there exists $\overline{\lim}\, (x_n \vee y)$ and we have

$$\bigwedge_{m\in\mathbb{N}} \bigvee_{n\geqslant m} (x_n \vee y) = \bigwedge_{m\in\mathbb{N}} ((\bigvee_{n\geqslant m} x_n) \vee y) = (\bigwedge_{m\in\mathbb{N}} \bigvee_{n\geqslant m} x_n) \vee y.$$

We have thus established equality (2). The existence of the element $\overline{\lim}\,(x_n \wedge y)$, as well as equality (3), may be established in a similar manner.

Remark. A proposition similar to that of proposition 3 holds for the limit inferior. The proof is similar.

Proposition 4. *Let E be a relatively σ-complete lattice in which the σ-distributivity laws hold. If $\{x_n\}_{n\in\mathbb{N}}$, $\{y_n\}_{n\in\mathbb{N}}$ are bounded sequences of elements of E, then*

$$\overline{\lim}\,(x_n \vee y_n) = (\overline{\lim}\, x_n) \vee (\overline{\lim}\, y_n), \tag{4}$$

$$\underline{\lim}\,(x_n \wedge y_n) = (\underline{\lim}\, x_n) \wedge (\underline{\lim}\, y_n). \tag{5}$$

If $y_n \overset{o}{\to} y$, then

$$\overline{\lim}\,(x_n \wedge y_n) = (\overline{\lim}\, x_n) \wedge y, \tag{6}$$

$$\underline{\lim}\,(x_n \vee y_n) = (\underline{\lim}\, x_n) \vee y. \tag{7}$$

Proof. From $x_n, y_n \leqslant x_n \vee y_n$ it immediately follows that

$$\overline{\lim}\, x_n,\ \overline{\lim}\, y_n \leqslant \overline{\lim}\,(x_n \vee y_n);$$

this shows that

$$(\overline{\lim}\, x_n) \vee (\overline{\lim}\, y_n) \leqslant \overline{\lim}\,(x_n \vee y_n).$$

On the other hand, for any $m \in \mathbb{N}$ we have

$$\overline{\lim}\,(x_n \vee y_n) \leqslant \bigvee_{n\geqslant m} (x_n \vee y_n) = (\bigvee_{n\geqslant m} x_n) \vee (\bigvee_{n\geqslant m} y_n),$$

whence we get

$$\overline{\lim}\,(x_n \vee y_n) \leqslant (\bigvee_{n\geqslant j} x_n) \vee (\bigvee_{n\geqslant m} y_n),$$

for any $j, m \in \mathbb{N}$. It follows that

$$\overline{\lim}\,(x_n \vee y_n) \leqslant \bigwedge_{j,\, m\in\mathbb{N}} (\bigvee_{n\geqslant j} x_n) \vee (\bigvee_{n\geqslant m} y_n)) = (\overline{\lim}\, x_n) \vee (\overline{\lim}\, y_n).$$

Equality (4) is therefore established. Equality (5) may be established in a similar way.

Let now $y_n \overset{o}{\to} y$. From $x_n \wedge y_n \leqslant x_n, y_n$ we get

$$\overline{\lim}\,(x_n \wedge y_n) \leqslant (\overline{\lim}\, x_n) \wedge (\overline{\lim}\, y_n) = (\overline{\lim}\, x_n) \wedge y. \tag{8}$$

On the other hand, if we denote

$$z_m = \bigwedge_{n \geqslant m} y_n,$$

we get

$$\bigvee_{n \geqslant m} (x_n \wedge y_n) \geqslant \bigvee_{n \geqslant m} (x_n \wedge z_m) = (\bigvee_{n \geqslant m} x_n) \wedge (\bigwedge_{n \geqslant m} y_n) \geqslant (\overline{\lim}\, x_n) \wedge (\bigwedge_{n \geqslant m} y_n).$$

Since

$$\bigvee_{n \geqslant m} (x_n \wedge y_n) \downarrow_{m \in \mathbb{N}} \overline{\lim}\, (x_n \wedge y_n)$$

and

$$\bigwedge_{n \geqslant m} y_n \uparrow_{m \in \mathbb{N}} y,$$

by taking into account propositions 1,2 from Section 1.3.1, we infer that

$$\overline{\lim}\, (x_n \wedge y_n) \geqslant (\overline{\lim}\, x_n) \wedge y. \tag{9}$$

From (8) and (9) we get equality (6). Equality (7) may be established in a similar way.

1.3.3 The convergence of generalized sequences

Let E be an ordered set.

A generalized sequence $\{x_\delta\}_{\delta \in \Delta}$ of elements of E is said to be an increasing generalized sequence if $\delta_1 \leqslant \delta_2$ implies $x_{\delta_1} \leqslant x_{\delta_2}$. In this case we shall write $x_\delta \uparrow_{\delta \in \Delta}$ If, moreover, the element $x = \bigvee_{\delta \in \Delta} x_\delta$ exists, we shall write $x_\delta \uparrow_{\delta \in \Delta} x$.

The *decreasing* generalized sequence can be defined in a similar way and one uses the notation $x_\delta \downarrow_{\delta \in \Delta}$ in order to express the fact that the generalized sequence $\{x_\delta\}_{\delta \in \Delta}$ is decreasing. The notation $x_\delta \downarrow_{\delta \in \Delta} x$ has an obvious meaning.

For the generalized sequences one can introduce several types of convergence with respect to the order relation.

In this section we shall limit our considerations to some remarks concerning the case of generalized sequences whose members belong to a relatively complete lattice[5]).

Let E be a relatively complete lattice. Let $\{x_\delta\}_{\delta \in \Delta}$ be a generalized sequence, whose members belong to E. If this sequence is bounded, one can define the limit inferior and the limit superior, as in the case of the usual sequences, by the formulae

$$\overline{\lim_{\delta \in \Delta}}\, x_\delta = \bigwedge_{\delta \in \Delta} \bigvee_{\delta' \geqslant \delta} x_{\delta'},$$

$$\underline{\lim_{\delta \in \Delta}}\, x_\delta = \bigvee_{\delta \in \Delta} \bigwedge_{\delta' \geqslant \delta} x_{\delta'}.$$

[5]) In Chapter III we shall introduce, for the generalized sequences whose members belong to a vector lattice, a type of convergence which is more general than that introduced in Section 1.3.1.

If

$$\underline{\lim}_{\delta \in \Delta} x_\delta = \overline{\lim}_{\delta \in \Delta} x_\delta = x, \tag{1}$$

and if we denote

$$a_\delta = \bigwedge_{\delta' \geqslant \delta} x_{\delta'} \quad , \quad b_\delta = \bigvee_{\delta' \geqslant \delta} x_{\delta'},$$

then

(i) $$a_\delta \leqslant x_\delta \leqslant b_\delta \ (\forall \delta \in \Delta):$$

(ii) $$a_\delta \underset{\delta \in \Delta}{\uparrow} x \quad \text{and} \quad b_\delta \underset{\delta \in \Delta}{\downarrow} x.$$

Conversely, if conditions (i) and (ii) are satisfied, relation (1) holds.

Bibliographical comments

In his work *The Mathematical Analysis of Logic*, G. Boole [1] introduces the lattices which bear his name. This work is the first one in which abstract ordered sets are considered. Later, C.S. Peirce [1] makes an axiomatic presentation of the ordered sets.

The theory of lattices, as well as some of its numerous applications, is developed in the well-known book by G. Birkhoff [7].

The extensions of the ordered sets have been considered by H. M. Mac Neil [1], to whom the application of the Dedekind method to general ordered sets is to be ascribed.

The theory of the representation of the Boole lattices belongs to M. H. Stone [1]. (See also H. Wallman [1]).

For the theory of Boole lattices, see also D. A. Kappos [1] and R. Shikorski [1].

Convergence with respect to the order relation, defined with the help of the inferior and the superior limits, has been introduced by L. V. Kantorovitch [1] and by G. Birkhoff [2].

Several types of topologies for the ordered sets have been introduced by O. Frink [1] and by other authors.

In the book of L. V. Kantorovitch, B. Z. Vulich and A. G. Pinsker [1] the following topology is introduced in an ordered set E: a subset A of E is closed iff from $x_n \in A$ $(n \in \mathbb{N})$ and $x_n \overset{o}{\rightarrow} x$ it follows that $x \in A$. The convergence of the (usual) sequences in this topology coincides with the (t)-convergence. In the book by B. Z. Vulich [1] a topology is introduced in a similar way, but the usual sequences are replaced by generalized sequences. For the usual sequences the convergence with respect to this topology is, generally speaking, weaker than the (t)-convergence. This topology generally is not a Hausdorff topology.

2

Elementary facts from the theory of topological vector spaces

This chapter contains some results from the theory of topological vector spaces, which are used later in this book.

2.1 Topological vector spaces

2.1.1 General facts [1])

Let $\mathscr{X}$ be a (real [2])) vector space. Let τ be a topology in the space $\mathscr{X}$, such that: (i) the mapping $(x, y) \mapsto x+y$ of $\mathscr{X} \times \mathscr{X}$ into $\mathscr{X}$ is continuous (if in $\mathscr{X} \times \mathscr{X}$ the product topology is considered): (ii) the mapping $(\alpha, x) \mapsto \alpha x$ of $\mathbb{R} \times \mathscr{X}$ into $\mathscr{X}$ is continuous (if in $\mathbb{R} \times \mathscr{X}$ we consider the product topology). In this case one says that the topology τ is *compatible* with the structure of vector space of $\mathscr{X}$.

Any vector space, in which a topology compatible with the structure of vector space has been defined, is called a *topological vector space.*

If $\mathscr{X}$ is a topological vector space, then for any element $a \in \mathscr{X}$ and any real non-zero number α, the mappings $x \mapsto x + a$ and $x \mapsto \alpha x$ (of $\mathscr{X}$ onto $\mathscr{X}$) are homeomorphisms. In particular, if W is a neighbourhood of the origin, then the set [3]) $W + a$ is a neighbourhood of a, whereas αW is a neighbourhood of the origin, for any $\alpha \neq 0$ $(\alpha \in \mathbb{R})$.

If $\mathscr{W}$ is a fundamental system of neighbourhoods of the origin, then by putting

$$\mathscr{W}_a = \{W + a;\ W \in \mathscr{W}\}$$

we get a fundamental system of neighbourhoods for a.

In order that a topological vector space be a Hausdorff space it is necessary and sufficient that for any $x \neq \mathbf{0}$ $(x \in \mathscr{X})$ a neighbourhood W of the origin should exist, that does not contain x.

---

[1]) For the proofs of the facts discussed in this section the reader may consult, for example, the book [31], by R. Cristescu.

[2]) In this chapter we are interested only in real vector spaces. We denote by $\mathbb{R}$ the field of real numbers.

[3]) We denote $W + a = \{x + a; x \in W\}$; $\alpha W = \{\alpha x; x \in W\}$.

In a topological vector space any neighbourhood of the origin is an absorbent set [4]).

In any topological vector space there exists a fundamental system of neighbourhoods of the origin consisting of equilibrated sets.

If W is a neighbourhood of the origin, then there exists a neighbourhood W_1 of the origin, such that $W_1 + W_1 \subset W$.

Let $\mathscr{X}$ be a vector space. Let $\mathscr{W}$ be a set of equilibrated and absorbent sets of $\mathscr{X}$, such that

(i) if $W_1, W_2 \in \mathscr{W}$, then there exists a $W_3 \in \mathscr{W}$, such that $W_3 \subset W_1 \cap W_2$;

(ii) for any $W \in \mathscr{W}$, there exists a $W_1 \in \mathscr{W}$, such that $W_1 + W_1 \subset W$;

(iii) for any $W \in \mathscr{W}$ and any $\lambda \neq 0$ ($\lambda \in \mathbb{R}$) one has $\lambda W \in \mathscr{W}$.

In this case, there exists a unique topology τ, which is compatible with the structure of vector space of $\mathscr{X}$, such that $\mathscr{W}$ is a fundamental system of neighbourhoods of the origin.

One can show that if $\mathscr{X}$ is a topological vector space and if A is an equilibrated subset of $\mathscr{X}$, then its closure $\bar{A}$ is equilibrated too; if $\mathscr{X}_0$ is a vector subspace of $\mathscr{X}$, then its closure $\overline{\mathscr{X}}_0$ is also a vector subspace.

Let $\mathscr{X}$ be a topological vector space. A generalized sequence $\{x_\delta\}_{\delta \in \Delta}$ whose members belong to $\mathscr{X}$, is called a *generalized Cauchy sequence* if for any neighbourhood W of the origin of $\mathscr{X}$, there exists an index $\delta_0 \in \Delta$, such that $\delta', \delta'' \geqslant \delta_0$ imply $x_{\delta'} - x_{\delta''} \in W$.

It is easy to show that any convergent generalized sequence is a generalized Cauchy sequence.

The topological vector space $\mathscr{X}$ is said to be *complete* if any generalized Cauchy sequence converges to an element of $\mathscr{X}$ The space $\mathscr{X}$ is said to be *sequentially complete* if any Cauchy sequence $\{x_n\}_{n \in \mathbb{N}}$, whose members belong to $\mathscr{X}$, converges to an element of $\mathscr{X}$. A subset A of a topological vector space $\mathscr{X}$ is said to be *complete* if any generalized Cauchy sequence, whose members belong to A, converges to an element of A.

In a topological vector space $\mathscr{X}$ a subset A is said to be *bounded* if for any neighbourhood W of the origin there exists a number $\lambda > 0$ such that $\lambda A \subset W$.

In order that a subset A of a topological vector space $\mathscr{X}$ be bounded it is necessary and sufficient that for any sequence $\{x_n\}_{n \in \mathbb{N}}$ whose members belong to A, and for any sequence $\{\lambda_n\}_{n \in \mathbb{N}}$ of positive numbers, which converges to zero, the sequence $\{\lambda_n x_n\}_{n \in \mathbb{N}}$ converges to zero.

2.1.2 Quotient topological vector spaces

Let $\mathscr{X}$ be a topological space and ρ an equivalence relation in $\mathscr{X}$. Let $\mathscr{X}/\rho$ be the quotient set and let l be the "natural mapping" given by $l(x) = \dot{x}$, where $\dot{x}$ is the class which contains x. The *quotient topology* in $\mathscr{X}/\rho$ is the strongest topology such that the mapping l is continuous. The set $\mathscr{X}/\rho$, endowed with the quotient topology, is called the *quotient topological space*.

---

[4]) We recall that a set A is said to be *absorbent* iff for any $x \in \mathscr{X}$ there exists a number $\varepsilon > 0$, such that $\lambda x \in A$ if $0 < |\lambda| \leqslant \varepsilon$. A set A is said to be *equilibrated* iff $|\alpha| \leqslant 1 (\alpha \in \mathbb{R})$ implies $\alpha A \subset A$.

The quotient topology has as open sets all the sets $B \subset \mathscr{X}/\rho$ such that $\overset{-1}{l}(B)$ is an open set in $\mathscr{X}$.

An equivalence relation ρ in $\mathscr{X}$ is said to be *open* if for any open set $A \subset \mathscr{X}$ the set $l(A)$ is open in the quotient topological space $\mathscr{X}/\rho$.

One can show that an equivalence relation ρ is open, iff for any $x \in \mathscr{X}$, the set of all neighbourhoods of $\dot{x}$ in $\mathscr{X}/\rho$ is equal to the set of all sets of the form $l(W)$, where W is an arbitrary neighbourhood of x.

Lemma. *If the (quotient) topological space $\mathscr{X}/\rho$ is a Hausdorff space, then the set $G = \{(x, y);\ x \equiv y\,(\rho)\}$ is closed in the product topological space $\mathscr{X} \times \mathscr{X}$. If the relation ρ is open, then the converse is also true.*

Proof. Let us assume that the space $\mathscr{X}/\rho$ is a Hausdorff space. Let $x, y \in \mathscr{X}$ be such that $(x, y) \notin G$. Since $\dot{x} \neq \dot{y}$, there exists an open neighbourhood F_1 of $\dot{x}$ and an open neighbourhood F_2 of $\dot{y}$, such that $F_1 \cap F_2 = \varnothing$. If we put $E_1 = \overset{-1}{l}(F_1)$ and $E_2 = \overset{-1}{l}(F_2)$, then $E_1 \times E_2$ is a neighbourhood of (x, y) in $\mathscr{X} \times \mathscr{X}$, whereas $E_1 \times E_2$ and G are disjoint. It follows that G is closed.

Conversely, if the relation ρ is open and if the set G is closed, let $\dot{x} \neq \dot{y}$. Let E_1 be an open neighbourhood of x in $\mathscr{X}$ and E_2 be an open neighbourhood of y in $\mathscr{X}$, such that $E_1 \times E_2$ and G be disjoint. If we put $F' = l(E_1)$ and $F'' = l(E_2)$, then F' and F'' are open, disjoint, and $\dot{x} \in F'$, $\dot{y} \in F''$. It follows that $\mathscr{X}/\rho$ is a Hausdorff space.

The lemma is proved.

Let $\mathscr{X}$ be a topological vector space, $\mathscr{X}_0$ a vector subspace and $\mathscr{X}/\mathscr{X}_0$ the quotient vector space. The relation ρ will be, in this case, the following: $x \equiv y\,(\rho) \Leftrightarrow$ $\Leftrightarrow x - y \in \mathscr{X}_0$. It is an open relation. Indeed, if A is an open set in $\mathscr{X}$, then the set $A + \mathscr{X}_0$ is also an open set. By putting $\dot{A} = l(A)$, we have $\overset{-1}{l}(\dot{A}) = A + \mathscr{X}_0$, and this shows that $\dot{A}$ is open in virtue of the definition of the quotient topology.

Proposition 1. *The quotient topology in $\mathscr{X}/\mathscr{X}_0$ is compatible with the structure of vector space.*

Proof. Since the equivalence relation is open, the topological spaces $(\mathscr{X}/\mathscr{X}_0) \times$ $\times (\mathscr{X}/\mathscr{X}_0)$ and $(\mathscr{X} \times \mathscr{X})/(\mathscr{X}_0 \times \mathscr{X}_0)$ are homeomorphic by the canonical mapping $(\dot{x}, \dot{y}) \mapsto \dot{z}$, where $z = (x, y)$. In order to establish the continuity of the mapping $(\dot{x}, \dot{y}) \mapsto \dot{x} + \dot{y}$, it is sufficient to show that the mapping $(x, y) \mapsto \dot{x} + \dot{y}$ of $\mathscr{X} \times \mathscr{X}$ in $\mathscr{X}/\mathscr{X}_0$ is continuous.

Indeed, from here we shall deduce that (by taking into account the definition of the continuity and of the quotient topology) the mapping $\dot{z} \mapsto \dot{x} + \dot{y}$ (where $z = (x, y)$) is continuous. Let us then remark that by putting $f((x, y)) = x + y$, $g((x, y)) = \dot{x} + \dot{y}$ and by denoting by l the natural mapping of $\mathscr{X}$ onto $\mathscr{X}/\mathscr{X}_0$, we have $g = l \circ f$. Since l and f are continuous, it follows that g is continuous.

Let us now show that the mapping $(\alpha, \dot{x}) \mapsto \alpha\dot{x}$ of $\mathbb{R} \times (\mathscr{X}/\mathscr{X}_0)$ into $\mathscr{X}/\mathscr{X}_0$ is continuous. Since the topological spaces $\mathbb{R} \times \mathscr{X}/\mathscr{X}_0$ and $(\mathbb{R} \times \mathscr{X})/(\{0\} \times \mathscr{X}_0)$ are homeomorphic, it is sufficient to prove the continuity of the mapping $(\alpha, x) \mapsto \alpha\dot{x}$ (of $\mathbb{R} \times \mathscr{X}$ into $\mathscr{X}/\mathscr{X}_0$).

As in the case of the addition, the continuity of the mapping $(\alpha, x) \mapsto \alpha\dot{x}$ follows from the fact that this mapping may be split as a composition of two continuous mappings.

Proposition 2. *The quotient topology in $\mathscr{X}/\mathscr{X}_0$ is a Hausdorff topology iff $\mathscr{X}_0$ is a closed subspace.*

Proof. If the quotient topology in $\mathscr{X}/\mathscr{X}_0$ is Hausdorff, then the set $\{\dot{0}\}$ is closed and, therefore, the set

$$\mathscr{X}_0 = \overset{-1}{l}(\{\dot{0}\})$$

is closed in $\mathscr{X}$.

Conversely, let us assume that $\mathscr{X}_0$ is a closed set in $\mathscr{X}$. The set

$$G = \{(x, y);\ x - y \in \mathscr{X}_0\}$$

is closed in $\mathscr{X} \times \mathscr{X}$, because by putting $f((x, y)) = x - y$ we get a continuous function f, defined on $\mathscr{X} \times \mathscr{X}$ and taking values in $\mathscr{X}$, and

$$G = \overset{-1}{f}(\mathscr{X}_0).$$

Since the equivalence relation we have considered in $\mathscr{X}$ is open, the space $\mathscr{X}/\mathscr{X}_0$ is a Hausdorff space, by the preceding lemma.

2.1.3 Products of topological vector spaces

Let $\{\mathscr{X}_j\}_{j \in J}$ be a family of topological vector spaces and let

$$\mathscr{X} = \prod_{j \in J} \mathscr{X}_j,$$

with the structure of a product vector space. We shall consider in $\mathscr{X}$ the product topology. *The product topology is compatible with the structure of vector space on $\mathscr{X}$.*

Indeed, let us define

$$f(\{x_j\}_{j \in J}, \{y_j\}_{j \in J}) = \{x_j + y_j\}_{j \in J},$$

$$g(\{x_j\}_{j \in J}, \{y_j\}_{j \in J}) = \{(x_j, y_j)\}_{j \in J},$$

$$h(\{(x_j, y_j)\}_{j \in J}) = \{x_j + y_j\}_{j \in J}.$$

The mapping $f: \mathscr{X} \times \mathscr{X} \to \mathscr{X}$ defines the addition in the product vector space $\mathscr{X}$.

The function $g: \mathscr{X} \times \mathscr{X} \to \mathscr{Z}$, where

$$\mathscr{Z} = \prod_{j \in J} (\mathscr{X}_j \times \mathscr{X}_j)$$

is endowed with the structure of a product vector space, is continuous if $\mathscr{Z}$ is endowed with the product topology (it is easy to show that $\mathscr{Z}$ is homeomorphic to $\mathscr{X} \times \mathscr{X}$, g being a homeomorphism).

The function $h: \mathscr{Z} \to \mathscr{X}$ is also continuous (since each mapping $(x_j, y_j) \mapsto x_j + y_j$ of $\mathscr{X}_j \times \mathscr{X}_j$ in $\mathscr{X}_j$ is continuous).

Since $f = h \circ g$, the function f is continuous.

Let us now define

$$f_0((\alpha, \{x_j\}_{j \in J})) = \{\alpha x_j\}_{j \in J}$$

$$f_i((\alpha, \{x_j\}_{j \in J})) = \alpha x_i, \quad i \in J,$$

$$g_i((\alpha, \{x_j\}_{j \in J}) = (\alpha, x_i), \quad i \in J,$$

$$h_i((\alpha, x_i)) = \alpha x_i, \quad i \in J.$$

For each $i \in J$, the function g_i, defined on $\mathbb{R} \times \mathscr{X}$ and taking values in $\mathbb{R} \times \mathscr{X}_i$, is continuous; the function h_i: $\mathbb{R} \times \mathscr{X}_i \to \mathscr{X}_i$ is also continuous. Since $f_i = h_i \circ g_i$, the function f_i (defined on $\mathbb{R} \times \mathscr{X}$ and taking values in $\mathscr{X}_i$) is continuous. It therefore follows that the function f_0, that is, the multiplication by scalars in $\mathscr{X}$, is continuous too.

The product vector space endowed with the product topology is called the *product topological space*.

Remark. If $\{\mathscr{X}_j\}_{j \in J}$ is a family of vector spaces, the (algebraic) *direct sum* of the family is, by definition, the subspace $\mathscr{Y}$ of the product linear space $\mathscr{X} = \prod_{j \in J} \mathscr{X}_j$ consisting of all the elements $x = \{x_j\}_{j \in J}$, such that $x_j \neq \mathbf{0}$ only for a finite number of indices.

If $\{\mathscr{X}_j\}_{j \in J}$ is a family of topological vector spaces and if $\mathscr{X}$ is the product topological vector space, then the (algebraic) direct sum $\mathscr{Y}$ of the family is a dense subset of $\mathscr{X}$.

2.1.4 Convex sets

As it is well known, a subset E of a vector space $\mathscr{X}$ is said to be *convex* if $x, y \in E$ implies that $\lambda x + (1 - \lambda) y \in E$, for any $\lambda \in [0,1]$.

The convex hull of a set A, i.e. the smallest convex set which contains A, will be denoted by $\operatorname{co}(A)$.

Proposition 1. *In a topological vector space $\mathscr{X}$ the closure and the interior of a convex set are convex sets.*

Proof. Let A be a convex subset of $\mathscr{X}$ and $\bar{A}$ its closure. For any $\lambda \in [0,1]$ let us put

$$f_\lambda(x, y) = \lambda x + (1 - \lambda) y.$$

The function f_λ, defined on $\mathscr{X} \times \mathscr{X}$, is continuous and maps the set $A \times A$ into A. It follows that f_λ maps the set $\bar{A} \times \bar{A}$ into $\bar{A}$ and this shows that $\bar{A}$ is convex.

Now let $x, y \in \text{int } A$ and

$$z = \lambda x + (1 - \lambda) y,$$

where $\lambda \in [0,1]$. Then there exists a neighbourhood W of the origin, such that

$$y + W \subset A.$$

If $\lambda \neq 1$, then the set

$$W_1 = (1 - \lambda) W$$

is a neighbourhood of the origin and

$$z + W_1 = \lambda x + (1 - \lambda)(y + W) \subset A;$$

hence $z \in \text{int } A$.

If $\lambda = 1$, then $z = x \in \text{int } A$.

The proposition is proved.

Remark. In the second part of the proof of proposition 1 we have shown that if $x \in A$ and $y \in \text{int } A$, then for any $\lambda \in [0,1)$ we have $z = \lambda x + (1 - \lambda) y \in \text{int } A$.

Proposition 2. *If $A_1, A_2, \ldots, A_n$ are convex compact subsets of a Hausdorff topological vector space $\mathscr{X}$, then the set* $\text{co}\left(\bigcup_{i=1}^{n} A_i\right)$ *is compact.*

Proof. Let $A = \prod_{i=1}^{n} A_i$ and

$$\Lambda = \left\{(\lambda_1, \ldots, \lambda_n) \in \mathbb{R}^n;\ \lambda_i \geqslant 0,\ i = 1, 2, \ldots, n,\ \sum_{i=1}^{n} \lambda_i = 1\right\}.$$

Let us consider the function f defined on the subset $\Lambda \times A$ of $\mathbb{R}^n \times \mathscr{X}^n$, taking values in $\mathscr{X}$, given by the formula

$$f(\lambda_1, \ldots, \lambda_n, x_1, \ldots, x_n) = \sum_{i=1}^{n} \lambda_i x_i.$$

The function f is obviously continuous. By putting

$$E = \text{co}\left(\bigcup_{i=1}^{n} A_i\right),$$

it is easy to observe that the set E is equal to the set $f(\Lambda \times A)$. Since the set $\Lambda \times A$ is obviously compact and since f is continuous, it follows that the set E is compact.

The proposition is proved.

If A is a subset of a topological vector space $\mathscr{X}$, its closed convex hull, that is, the smallest closed convex set containing A, will be denoted by $\overline{\text{co}}\,(A)$. It obviously coincides with the intersection of all convex, closed sets which contain A. It is also obvious that

$$\overline{\text{co}}\,(A) = \overline{\text{co}\,(A)},$$

for any set A (see proposition 1).

2.1.5 Continuous linear functionals

Let $\mathscr{X}$ be a (real) topological vector space.

Proposition 1. *A linear functional φ, defined on $\mathscr{X}$, is continuous iff there exists an open set $A \subset \mathscr{X}$, such that $\varphi(A)$ be bounded.*

Proof. If φ is continuous, then the set $A = \overset{-1}{\varphi}((-1,1))$ is open and $|\varphi(x)| < 1$ for any $x \in A$.

Conversely, let us assume that there exists an open set $A \subset \mathscr{X}$ and a number $\lambda > 0$, such that $|\varphi(x)| \leqslant \lambda$, for any $x \in A$. If $x_0 \in A$, then

$$|\varphi(x - x_0)| \leqslant 2\lambda, \; x \in A.$$

The set $E = A - x_0$ is a neighbourhood of the origin. Since the mapping $x \mapsto \alpha x$ (of $\mathscr{X}$ into $\mathscr{X}$) is continuous, for any number $\mu > 0$, there exists a neighbourhood W of the origin, such that $\mu W \subset E$. It follows that

$$|\varphi(x)| \leqslant \frac{2\lambda}{\mu}, \; (x \in W),$$

and this shows that φ is continuous at the origin, hence at each point.

Proposition 2. *Let H be a hyperplane in the space $\mathscr{X}$:*

$$H = \{x; \varphi(x) = \lambda\}, \tag{1}$$

where φ is a (non-zero) linear functional on $\mathscr{X}$. Then H is a closed set iff φ is continuous.

Proof. Assume that the hyperplane H is closed. As it is well known, the hyperplane H may be written as $H = \mathscr{X}_0 + a$, where $\mathscr{X}_0$ is a maximal vector subspace, and a is a certain element of $\mathscr{X}$[5]). Without lessening the generality, we can assume

[5]) See R. Cristescu [31].

λ to be zero and hence, that H is a maximal vector subspace. The quotient space $\mathscr{X}/H$ will then be a one-dimensional topological vector space. This space is, therefore, isomorphic to $\mathbb{R}$. If we denote by l the natural mapping of $\mathscr{X}$ onto $\mathscr{X}/H$, and if we define $\psi(\dot{x}) = \varphi(x)$, where $\dot{x} = l(x)$, we get a linear continuous functional defined on $\mathscr{X}/H$. From the fact that $\varphi = \psi \circ l$ we infer that φ is continuous.

Conversely, if φ is a linear continuous functional, then the hyperplane H is obviously a closed set.

The proposition is proved.

A hyperplane H, which is closed for the corresponding topology, is said to be a *closed hyperplane.*

If H is a closed hyperplane (of the form (1)), then the sets

$$H_1 = \{x;\, \varphi(x) \leqslant \lambda\},\ H_2 = \{x;\, \varphi(x) \geqslant \lambda\}$$

are closed and convex. They are called the *closed semi-spaces* determined by the hyperplane H.

The sets

$$H' = \{x;\, \varphi(x) < \lambda\},\ H'' = \{x;\, \varphi(x) > \lambda\}$$

are open and convex. They are called the *open semi-spaces* determined by H.

If A and B are two sets, such that $A \subset H_1$ and $B \subset H_2$, then one says that H *separates* A from B. If $A \subset H'$ and $B \subset H''$, then one says that H *strictly separates* A from B.

2.2 Locally convex spaces

2.2.1 General properties[6])

Let $\mathscr{X}$ be a topological vector space and let τ be its topology. If there exists a fundamental system of neighbourhoods of the origin consisting of convex sets, then topology τ is said to be *locally convex* and $\mathscr{X}$ is said to be a *locally convex space.*

If in the topological vector space $\mathscr{X}$ there exists a fundamental system of neighbourhoods of the origin, consisting of convex sets, then there also exists a fundamental system consisting of equilibrated and convex sets.

Indeed, if W is a convex neighbourhood of the origin, then the set $W \cap (-W)$ is convex and equilibrated.

If $\mathscr{P} = \{p_j\}_{j \in J}$ is a family of semi-norms on a vector space $\mathscr{X}$, then $\mathscr{X}$ becomes a locally convex space if we endow it with the topology τ for which a fundamental system of neighbourhoods of the origin is the system of all sets of the form

$$E(j_1, j_2, \ldots, j_n;\, \varepsilon) = \{x;\, p_{j_i}(x) < \varepsilon,\ i = 1, 2, \ldots, n\}.$$

One says that the topology τ is determined by the family $\mathscr{P}$.

---

[6]) For the proofs of the results discussed in this section the reader can consult R. Cristescu [31]. As in Section 2.1, we shall consider only real spaces.

The topology τ is Hausdorff iff the family $\mathscr{P}$ is *sufficient,* i.e., for any $x \neq \mathbf{0}$ there exists an index $j \in J$, such that $p_j(x) \neq 0$.

If $\mathscr{X}$ is any locally convex space, then there exists a family of semi-norms which determines the topology of the space. More precisely, if $\mathscr{W}$ is a fundamental system of neighbourhoods of the origin, consisting of equilibrated and convex sets, then the topology τ of $\mathscr{X}$ is determined by the family $\{p_W\}_{W \subset \mathscr{W}}$ of the "Minkowski semi-norms", i.e., of the semi-norms given by

$$p_W(x) = \inf \{\mu;\ x \in \mu W,\ \mu > 0\}. \tag{1}$$

If p_1 and p_2 are two semi-norms on a vector space $\mathscr{X}$, then one writes $p_1 \leqslant p_2$ for the relation

$$p_1(x) \leqslant p_2(x), \qquad \text{for any } x \in \mathscr{X}.$$

If $\mathscr{P}$ is a directed set of semi-norms on a vector space $\mathscr{X}$, then for the topology determined by $\mathscr{P}$, the system of all sets of the form

$$E(p;\ \varepsilon) = \{x;\ p(x) < \varepsilon\},$$

where $p \in \mathscr{P}$, is a fundamental system of neighbourhoods of the origin.

Any set $\mathscr{P}$ of semi-norms may be replaced by a directed set $\mathscr{P}'$ of semi-norms, which determines the same topology. If $\mathscr{P} = \{p_j\}_{j \in J}$, one can take as a directed system $\mathscr{P}'$ of semi-norms the system of all the semi-norms given by

$$p_{(j_1, \ldots, j_n)}(x) = \max_{1 \leqslant i \leqslant n} p_{j_i}(x),$$

where $\{j_1, \ldots, j_n\}$ is any finite system of indices of J.

Let $\mathscr{P}'$ and $\mathscr{P}''$ be two directed sets of semi-norms on the vector space $\mathscr{X}$ and let τ', τ'' be the topologies determined by $\mathscr{P}'$ and $\mathscr{P}''$, respectively.

One says that $\mathscr{P}'$ is *dominated* by $\mathscr{P}''$ if for any $p' \in \mathscr{P}'$ there exists a $p'' \in \mathscr{P}''$ and a number $\lambda > 0$, such that [7] $p' \leqslant \lambda p''$.

In order that $\tau' \leqslant \tau''$ it is necessary and sufficient that $\mathscr{P}'$ be dominated by $\mathscr{P}''$.

If $\mathscr{P}'$ is dominated by $\mathscr{P}''$, whereas $\mathscr{P}''$ is dominated by $\mathscr{P}'$, then the families $\mathscr{P}'$ and $\mathscr{P}''$ are said to be *equivalent.*

In order that $\tau' = \tau''$ it is necessary and sufficient that $\mathscr{P}'$ and $\mathscr{P}''$ be equivalent.

If $\mathscr{X}$ is a locally convex space and if $\mathscr{X}_0$ is a vector subspace of $\mathscr{X}$, then $\mathscr{X}_0$ is a locally convex space for the induced topology. If $\mathscr{P}$ is a set of semi-norms which determines the topology of $\mathscr{X}$, and if $\mathscr{P}_0$ is the set of restrictions to $\mathscr{X}_0$ of the semi-norms belonging to $\mathscr{P}$, then the topology of $\mathscr{X}_0$ is determined by $\mathscr{P}_0$.

★

Now let $\mathscr{X}$ be a locally convex space and E be an equilibrated, convex, absorbent subset of $\mathscr{X}$. Let p_E be the Minkowski semi-norm associated with E(i.e., the functional given by formula (1), in which $W = E$).

---

[7]) If p is a semi-norm and α is a real number, then one writes $(\alpha p)(x) = \alpha p(x)$.

The semi-norm p_E is continuous iff the set E is a neighbourhood of the origin. The proof may be easily made.

If E is an equilibrated and convex neighbourhood of the origin, then the interior and the closure of E are given by the formulae

$$\text{int } E = \{x; p_E(x) < 1\},$$

$$\bar{E} = \{x; p_E(x) \leqslant 1\}.$$

Indeed, let us denote

$$A = \{x; p_E(x) < 1\},$$

$$B = \{x; p_E(x) \leqslant 1\}.$$

We have $A \subset E \subset B$.

Since p_E is continuous, the set A is open, whereas B is closed. Let $x \in B$ and W be neighbourhoods of the origin. Then there exists an $\alpha \in \mathbb{R}$, such that $-\alpha x \in W$. One can obviously assume that $0 < \alpha < 1$. From $x \in B$ we infer that

$$p_E((1-\alpha)x) = (1-\alpha)p_E(x) < 1,$$

and this shows that $(1-\alpha)x \in A$. On the other hand, we have

$$(1-\alpha)x \in x + W,$$

and this implies that

$$A \cap (x + W \neq) \varnothing.$$

It, therefore, follows that $x \in \bar{A}$ and this shows that $B \subset \bar{A}$. Since $A \subset B$ and since B is closed, we get $B = \bar{A}$.

Now let $y \in \text{int } B$. Then there exists a neighbourhood W of the origin, such that $y + W \subset B$. Let $0 < \alpha < 1$ be such that $\alpha y \in W$. It follows that

$$(1+\alpha)y \in y + W \subset B,$$

and, consequently, we have

$$p_E((1+\alpha)y) \leqslant 1.$$

Hence $p_E(y) < 1$ and this shows that $y \in A$. We have thus shown that int $B \subset A$ and this implies that $A = \text{int } B$.

Remark. From what we have shown in this section, the following properties immediately follow.

If $\mathscr{X}$ is a locally convex space and if $\mathscr{P}$ is a set of semi-norms, which determines the topology of the space, then any semi-norm from $\mathscr{P}$ is continuous.

If $\mathscr{P}$ is a set of semi-norms, which determines the topology τ of the locally convex space $\mathscr{X}$, and if $\mathscr{P}_0$ is any set of continuous semi-norms defined on $\mathscr{X}$, then the set $\mathscr{P}_1 = \mathscr{P} \cup \mathscr{P}_0$ is equivalent to $\mathscr{P}$, hence $\mathscr{P}_1$ determines the same topology τ.

The topology of a locally convex space $\mathscr{X}$ is determined by the set of all the continuous semi-norms defined on $\mathscr{X}$.

2.2.2 Product spaces and quotient spaces

Let $\{\mathscr{X}_j\}_{j \in J}$ be a family of locally convex spaces, $\mathscr{X}$ a vector space and assume that for any $j \in J$ a linear mapping $f_j: \mathscr{X} \to \mathscr{X}_j$ is given.

The weakest topology τ in $\mathscr{X}$, such that all mappings f_j are continuous, is locally convex.

Indeed, if $\mathscr{W}_j$ is a fundamental system of neighbourhoods of the origin in the space $\mathscr{X}_j$, then a fundamental system of neighbourhoods of the origin for the topology τ is given by the system of all finite intersections of the form

$$W = \bigcap_{i=1}^{n} \overset{-1}{f_{j_i}}(W_{j_i}), \tag{1}$$

where $W_{j_i} \in \mathscr{W}_{j_i}$.

The topology τ is compatible with the structure of vector space[8]) of $\mathscr{X}$. On the other hand, $\mathscr{W}_j$ may be considered as being formed of convex sets and this will imply that any set W of the form (1) is convex.

It follows that the topology τ is locally convex.

As a particular case, we get:

If $\{\mathscr{X}_j\}_{j \in J}$ is a family of locally convex spaces, then the product topological vector space $\mathscr{X} = \prod_{j \in J} \mathscr{X}_j$ is locally convex.

If $\mathscr{P}_j$ is a set of semi-norms on $\mathscr{X}_j$, which determines the topology of the space $\mathscr{X}_j$, and if for any $p_j \in \mathscr{P}_j$ and any element $x = \{x_j\}_{j \in J}$ of the product space $\mathscr{X}$, we put

$$q_j(x) = p_j(x_j),$$

then the set of all semi-norms q_j determines the product topology. This can easily be proved by taking into account the definition of the neighbourhoods of the product space.

Now let $\mathscr{X}$ be any locally convex space and $\mathscr{X}_0$ a vector subspace of $\mathscr{X}$.

The quotient space $\mathscr{X}/\mathscr{X}_0$ is a locally convex space.

---

[8]) Let $\mathscr{X}$ be a vector space, $\mathscr{Y}$ a topological vector space, f a linear mapping of $\mathscr{X}$ into $\mathscr{Y}$, and let τ_0 be the weakest topology in $\mathscr{X}$ such that f be continuous. Any neighbourhood of the origin in the topology τ_0 contains a set of the form $\overset{-1}{f}(W)$, where W is a neighbourhood of the origin in $\mathscr{Y}$. From here one easily infers that the topology τ_0 is compatible with the structure of vector space.

The case of the topology τ (the weakest-topology such that all mappings f_j are continuous) can then be obviously derived.

Indeed, if l is the natural mapping of $\mathscr{X}$ onto $\mathscr{X}/\mathscr{X}_0$, then for any convex neighbourhood W of the origin of $\mathscr{X}$, the set $l(W)$ is a convex neighbourhood of the origin of $\mathscr{X}/\mathscr{X}_0$. Since the system of all sets of the form $l(W)$ is a fundamental system of neighbourhoods of the origin in the quotient topology, the space $\mathscr{X}/\mathscr{X}_0$ is locally convex.

Now let E be a neighbourhood of the origin in the space $\mathscr{X}$; we shall assume that E is an open, convex, equilibrated set. If p_E is the Minkowski semi-norm associated to E, then

$$E = \{x; p_E(x) < 1\}.$$

If $y \in \mathscr{X}/\mathscr{X}_0$ and if μ is a positive number, such that

$$y \in \mu l(E) = l(\mu E),$$

then there exists an $x \in \mathscr{X}$, such that $x \in \mu E$ and $l(x) = y$. Conversely, if $x \in \mu E$, then $l(x) \in \mu l(E)$. It follows that

$$\{\mu; y \in \mu l(E),\ \mu > 0\} = \{\mu; x \in \mu E,\ \mu > 0, l(x) = y\}.$$

It follows that the Minkowski semi-norm $\dot{p}_{l(E)}$ associated to $l(E)$ is given by the formula

$$\dot{p}_{l(E)}(y) = \inf_{l(x)=y} p_E(x).$$

If $\mathscr{P}$ is a directed set of semi-norms, which determines the topology of the space $\mathscr{X}$, then the topology of the space $\mathscr{X}/\mathscr{X}_0$ is determined by the family of all semi-norms of the form

$$\dot{p}(y) = \inf_{l(x)=y} p(x), \quad y \in \mathscr{X}/\mathscr{X}_0,$$

where $p \in \mathscr{P}$.

Let $\mathscr{X}$ be a (Hausdorff) locally convex space and $\mathscr{P} = \{p_j\}_{j\in J}$ be a family of semi-norms, which determines the topology of $\mathscr{X}$. The set

$$G_j = \overset{-1}{p_j}(0)$$

is a vector subspace of $\mathscr{X}$. If $\dot{x} \in \mathscr{X}/G$, then for any elements $x_1, x_2 \in \dot{x}$ we have $p_j(x_1 - x_2) = 0$ and from the inequality

$$|p_j(x_1) - p_j(x_2)| \leqslant p_j(x_1 - x_2)$$

we infer that $p_j(x_1) = p_j(x_2)$. By putting

$$q_j(\dot{x}) = p_j(x), \quad \dot{x} \in \mathscr{X}/G_j,$$

we get a norm on the quotient vector space $\mathscr{Y}_j = \mathscr{X}/G_j$.

Let l_j be the natural mapping of $\mathscr{X}$ onto $\mathscr{Y}_j$. By defining

$$S_j = \{x \in \mathscr{X};\ p_j(x) < 1\},$$

$$W_j = \{y \in \mathscr{Y};\ q_j(y) < 1\},$$

we get

$$S_j = \overset{-1}{l_j}(W_j).$$

Hence we infer that the topology τ of the space $\mathscr{X}$ is the weakest topology in $\mathscr{X}$, such that all linear mappings l_j are continuous. By putting

$$\mathscr{Y} = \prod_{j \in J} \mathscr{Y}_j,$$

$$l(x) = \{l_j(x)\}_{j \in J},$$

and by considering in $\mathscr{Y}$ the product topology, it follows that the topology τ is the weakest topology in $\mathscr{X}$, such that the mapping l of $\mathscr{X}$ onto $l(\mathscr{X})$ is continuous [9]).

Since the space $\mathscr{X}$ is Hausdorff, the mapping l is one-to-one and it is easy to check that l^{-1} is continuous. Consequently:

Any Hausdorff locally convex space is isomorphic to a subspace of a product of normed vector spaces.

Since any normed vector space may be embedded in a Banach space, it follows that any Hausdorff locally convex space is isomorphic to a subspace of a product of Banach spaces.

★

If $\{\mathscr{X}_j\}_{j \in J}$ is a family of locally convex spaces, $\mathscr{X}$ their algebraic direct sum and l_j the natural mapping of $\mathscr{X}_j$ into $\mathscr{X}$, then the vector space $\mathscr{X}$, endowed with the strongest locally convex topology such that all mappings l_j be continuous, is called the *topological direct sum* of the family $\{\mathscr{X}_j\}_{j \in J}$.

2.2.3 Barreled spaces and bornological spaces

Let $\mathscr{X}$ be a locally convex space.

A subset E of $\mathscr{X}$ is said to be a *barrel* if it is equilibrated, convex, absorbent and closed.

For any locally convex space $\mathscr{X}$, there exists a fundamental system of neighbourhoods of the origin, consisting of barrels.

The locally convex space $\mathscr{X}$ is said to be a *barreled space* if any barrel in $\mathscr{X}$ is a neighbourhood of the origin.

---

[9]) More generally, let $\{\mathscr{X}_j\}_{j \in J}$ be a family of arbitrary topological vector spaces, let $\mathscr{X}$ be any vector space and suppose that for any $j \in J$ a linear mapping f_j of $\mathscr{X}$ into $\mathscr{X}_j$ is given. Let $\mathscr{Y} = \prod_{j \in J} \mathscr{X}_j$ and $g(x) = \{f_j(x)\}_{j \in J}$. The weakest topology in $\mathscr{X}$, such that all mappings f_j are continuous, equals the weakest topology such that the mapping g of $\mathscr{X}$ onto $g(\mathscr{X})$ is continuous.

Let $\{\mathscr{X}_j\}_{j\in J}$ be a family of barreled spaces, $\mathscr{X}$ a topological vector space and, for each $j \in J$, let f_j be a continuous linear mapping of $\mathscr{X}_j$ into $\mathscr{X}$.

If E is a barrel in $\mathscr{X}$, then $\overset{-1}{f_j}(E)$ is a barrel in $\mathscr{X}_j$ and, therefore, a neighbourhood of the origin in this space. If τ is the strongest locally convex topology in $\mathscr{X}$ such that all mappings f_j be continuous, then E is a neighbourhood of the origin for the topology τ. Hence $\mathscr{X}$ is a barreled space for the topology τ.

From this remark we obtain, in particular, that: *if $\mathscr{X}$ is a barreled space and if $\mathscr{X}_0$ is a vector subspace of $\mathscr{X}$, then the quotient space $\mathscr{X}/\mathscr{X}_0$ is a barreled space; any direct sum of barreled spaces is a barreled space.*

Proposition 1. *Any locally convex space $\mathscr{X}$, which is not a meager space* [10]), *is barreled.*

Indeed, if E is a barrel in the space $\mathscr{X}$, then

$$\mathscr{X} = \bigcup_{n\in\mathbb{N}} (nE),$$

because E is absorbent and equilibrated. Since nE is closed (for any n), it follows that the set E has interior points. If $x_0 \in \operatorname{int} E$, then $-x_0 \in E$, because E is equilibrated. It follows that

$$\mathbf{0} = \frac{1}{2}(-x_0) + \frac{1}{2}x_0 \in \operatorname{int} E$$

(see the remark in Section 2.1.4), and this implies that E is a neighbourhood of the origin.

Consequently, $\mathscr{X}$ is a barreled space.

In particular, *any Banach space is a barreled space.*

Proposition 2. *If E is a barrel in a Hausdorff locally convex space, then E is absorbing any complete, bounded, convex, equilibrated subset of $\mathscr{X}$.*

Proof. Let τ be the topology of the space $\mathscr{X}$. Let A be a complete, bounded, convex, equilibrated subset of $\mathscr{X}$ and let us show that E absorbs A. Without restricting the generality, we can assume that the set A is also absorbent, because otherwise we can replace $\mathscr{X}$ by the linear hull of A.

If we denote by p the Minkowski semi-norm associated with A, then p is a norm on $\mathscr{X}$. Indeed, since the space $\mathscr{X}$ is Hausdorff, for any element $x \neq 0$ there exists a neighbourhood W of the origin, such that $x \notin W$. On the other hand, there exists a number $\lambda > 0$, such that $\lambda A \subset W$. It follows that there exists a $\lambda > 0$, such that $x \notin \lambda A$, and this implies that $p(x) \neq 0$.

For the topology τ_0, determined by the norm p, the system of all sets of the form αA, where $0 < \alpha \in \mathbb{R}$, is a fundamental system of neighbourhoods of the origin. Since A is bounded for the topology τ, it follows that $\tau_0 \geqslant \tau$.

---

[10]) We recall that a topological space $\mathscr{X}$ is said to be meager (of the first category) if it can be written as $\mathscr{X} = \bigcup_{n=1}^{\infty} A_n$, where $\operatorname{int} \overline{A_n} = \emptyset$, $n \in \mathbb{N}$.

If $\{x_\delta\}_{\delta \in \Delta}$ is a generalized sequence whose members belong to A, and if it is τ_0-Cauchy, then it is also τ-Cauchy and this implies that there exists an $x_0 \in A$, such that $x_\delta \xrightarrow{\tau} x_0$. From $x_\delta - x_{\delta'} \in \alpha A$ for $\delta, \delta' \geqslant \delta_\alpha$, it follows that $x_\delta \in \alpha A + x_0$, for any $\delta \geqslant \delta_\alpha$, and this shows that $x_\delta \xrightarrow{\tau_0} x_0$. Therefore, A is complete for the topology τ_0.

The sets αA ($\alpha > 0$) being complete for the topology τ_0, one infers that $\mathscr{X}$ is complete for the topology τ_0.

It follows that $\mathscr{X}$ is a Banach space with respect to the norm p and, in particular, $\mathscr{X}$ is a barreled space. In this space, the set E is a barrel (since E is closed for the topology τ, it is also closed for τ_0). It follows that E is a neighbourhood of the origin in the considered Banach space and therefore there exists a $\lambda > 0$ such that $\lambda A \subset E$. In other words, E absorbs the set A.

A locally convex space $\mathscr{X}$ is called a *bornological space* if any equilibrated convex subset, which absorbs [11]) any bounded subset of $\mathscr{X}$, is a neighbourhood of the origin.

In order that the locally convex space $\mathscr{X}$ be a bornological space, it is necessary and sufficient that $\mathscr{X}$ possess the following property: if φ is a semi-norm on $\mathscr{X}$, which is bounded on any bounded subset of $\mathscr{X}$, then φ is continuous on the space[12]) $\mathscr{X}$.

Any metrizable locally convex space is a bornological space.

Indeed, let $\mathscr{X}$ be a metrizable locally convex space; then the topology τ of $\mathscr{X}$ admits of a countable fundamental system of neighbourhoods of the origin, which are also equilibrated and convex. It follows that the topology τ is determined by a countable family $\{p_n\}_{n \in \mathbb{N}}$ of semi-norms. By defining

$$d(x, y) = \sum_{n=1}^{\infty} \frac{1}{2^n} \cdot \frac{p_n(x-y)}{1 + p_n(x-y)},$$

one gets an (invariant) distance on $\mathscr{X}$, whereas the topology determined by this distance coincides with the topology τ.

The space $\mathscr{X}$ is a quasi-normed [13]) vector space with respect to the quasi-norm $q(x) = d(x, 0)$.

If φ is a semi-norm on $\mathscr{X}$ and if it is bounded on any bounded subset of $\mathscr{X}$, then φ is continuous on $\mathscr{X}$ because $\mathscr{X}$ is a quasi-normed vector space [14]).

It follows that $\mathscr{X}$ is a bornological space.

2.2.4 Inductive limits of locally convex spaces [15])

Let $\mathscr{X}$ be a vector space and $\{\mathscr{X}_j\}_{j \in J}$ be a family of vector subspaces of $\mathscr{X}$, directed to the right for the inclusion. For any $j \in J$ let τ_j be a locally convex topology in

---

[11]) We recall that if A and B are two subsets of a vector space then one says that B absorbs A if there exists a number $\alpha > 0$, such that $\lambda A \subset B$ for any λ satisfying the condition $0 < |\lambda| \leqslant \alpha$.

[12]) For the proof the reader may consult R. Cristescu [31].

[13]) We recall that one calls *quasi-norm* on a vector space $\mathscr{X}$ any function $q: \mathscr{X} \to \mathbb{R}$, such that : (i) $q(x) = 0 \Leftrightarrow x = 0$; (ii) $q(x) = q(-x)$; (iii) $q(x+y) \leqslant q(x) + q(y)$. A quasi-normed (real) vector space is any (real) vector space on which a quasi-norm q has been given, such that: (iv) $\alpha_n \to 0$ ($\alpha_n \in \mathbb{R}$ and $x \in \mathscr{X}$) $\Rightarrow q(\alpha_n x) \to 0$, (v) $q(x_n) \to 0$ and $\alpha \in \mathbb{R} \Rightarrow q(\alpha x_n) \to 0$.

[14]) See R. Cristescu [31].

[15]) See R. Cristescu [31].

$\mathscr{X}_j$ such that if $\mathscr{X}_{j_1} \subset \mathscr{X}_{j_2}$, then the topology $\tau_{j_2 j_1}$ induced by τ_{j_2} on $\mathscr{X}_{j_1}$ be weaker than τ_{j_1}.

Let $\mathscr{W}$ be the set of all absorbent, convex, equilibrated subsets $W \subset \mathscr{X}$, such that for any $j \in J$, the set $W \cap \mathscr{X}_j$ be a neighbourhood of the origin in the space $\mathscr{X}_j$. The topology τ in $\mathscr{X}$, such that $\mathscr{W}$ is a fundamental system of neighbourhoods of the origin, is called the *inductive limit* of the topologies τ_j. The linear space $\mathscr{X}$, endowed with the locally convex topology τ, is called the *inductive limit* of the spaces $\mathscr{X}_j$, and it is denoted by $\mathscr{X} = \varinjlim_{j \in J} \mathscr{X}_j$.

If $\mathscr{X} = \varinjlim_{j \in J} \mathscr{X}_j$, if $\mathscr{X} = \bigcup_{j \in J} \mathscr{X}_j$ and if $\mathscr{X}_{j_1} \subset \mathscr{X}_{j_2}$ implies $\tau_{j_2 j_1} = \tau_{j_1}$, then $\mathscr{X}$ (with the topology τ of $\mathscr{X}$) is called the *strict inductive limit* of the family $\{\mathscr{X}_j\}_{j \in J}$ of the topologies τ_j, respectively.

If $\mathscr{X}$ is the strict inductive limit of an increasing sequence $\{\mathscr{X}_n\}_{n \in \mathbb{N}}$ of locally convex spaces, then the topology induced in $\mathscr{X}_n$ by the topology τ of $\mathscr{X}$ coincides with the original topology τ_n of $\mathscr{X}_n$.

Let $\mathscr{D}(\mathbb{R})$ be the set of all real functions defined on the real line, which are indefinitely differentiable and have compact supports. With the usual linear operations, $\mathscr{D}(\mathbb{R})$ becomes a vector space. For each compact subset $S \subset \mathbb{R}$, let $\mathscr{D}(\mathbb{R}; S)$ be the vector subspace of $\mathscr{D}(\mathbb{R})$ consisting of the functions whose support is contained in S. We shall consider in $\mathscr{D}(\mathbb{R}; S)$ the topology τ_S determined by the family of semi-norms $p_{S,n} (n = 0, 1, 2, \ldots)$, given by the formula

$$p_{S,n}(x) = \max_{\substack{t \in S \\ 0 \leqslant k \leqslant n}} |x^{(k)}(t)|.$$

Let τ be the (strict) inductive limit of the topologies τ_S. Endowed with the topology τ, the locally convex space $\mathscr{D}(\mathbb{R})$ is called the *Schwartz space*.

2.3 Uniform convergence topologies in spaces of linear continuous operators

2.3.1 Linear continuous operators

Let $\mathscr{X}$ and $\mathscr{Y}$ be (real) topological spaces.

A linear (or only additive) operator, which maps $\mathscr{X}$ into $\mathscr{Y}$ and which is continuous at a point x_0, is continuous at every point.

If U is a linear continuous operator (which maps $\mathscr{X}$ into $\mathscr{Y}$) then for any bounded set $A \subset \mathscr{X}$ the set $U(A)$ is bounded.

Indeed, if W' is a neighbourhood of the origin of the space $\mathscr{Y}$, then there exists a neighbourhood W of the origin of the space $\mathscr{X}$, such that $U(W) \subset W'$. Since the set A is bounded, there exists a number $\lambda > 0$ such that $\lambda A \subset W$, whence $\lambda U(A) \subset W'$.

Proposition 1. *A linear operator U, which maps a locally convex space $\mathscr{X}$ into a locally convex space $\mathscr{Y}$, is continuous iff the following condition is satisfied: if $\mathscr{P}$ and $\mathscr{P}'$ are directed sets of semi-norms, which generate the topologies of $\mathscr{X}$ and $\mathscr{Y}$, respectively, then for any $p' \in \mathscr{P}'$ there exists a $p \in \mathscr{P}$ and a number $\mu > 0$, such that*

$$p'(U(x)) \leqslant \mu p(x), \tag{1}$$

for any $x \in \mathscr{X}$.

Proof. If U is linear and continuous, then for any $p' \in \mathscr{P}'$ there exists a $p \in \mathscr{P}$ and a number $\eta > 0$, such that if $p(x) \leqslant \eta$, then

$$p'(U(x)) \leqslant 1.$$

Let now x be a non-zero element of $\mathscr{X}$. There then exists a number $\alpha > 0$, such that

$$p(\alpha x) = \alpha p(x) \leqslant \eta. \tag{2}$$

Hence we infer that

$$p'(U(x)) \leqslant \frac{1}{\alpha}. \tag{3}$$

If $p(x) = 0$, then (2) holds for any $\alpha > 0$ and from (3) we get that $p'(U(x)) = 0$. If $p(x) \neq 0$, then by putting

$$\alpha = \frac{\eta}{p(x)}$$

we get $p(\alpha x) = \eta$ and from (3) one obtains

$$p'(U(x)) \leqslant \frac{1}{\eta} p(x).$$

Consequently, inequality (1) holds, with $\mu = 1/\eta$.

Conversely, if inequality (1) holds, then from $p(x) < \dfrac{\varepsilon}{\mu}$ it follows that $p'(U(x)) < \varepsilon$. From the condition expressed in the statement of the proposition we infer that U is continuous at the origin. Since U is linear, it is continuous at any point.

Corollary. *A linear operator U, which maps a locally convex space $\mathscr{X}$ into a locally convex space $\mathscr{Y}$, is continuous iff for any continuous semi-norm q on $\mathscr{Y}$ there exists a continuous semi-norm p on $\mathscr{X}$ such that*

$$q(U(x)) \leqslant p(x),$$

for any $x \in \mathscr{X}$.

Remark. In the case of functionals, the condition given by proposition 1 may be reformulated in a simpler manner. If $\mathscr{X}$ is a locally convex space whose topology is defined by a directed family $\mathscr{P}$ of semi-norms, then a linear functional φ, defined on $\mathscr{X}$, is continuous iff there exists a $p \in \mathscr{P}$ and a number $\mu > 0$, such that

$$|\varphi(x)| \leqslant \mu p(x),$$

for any $x \in \mathscr{X}$.

Proposition 2. *Let $\mathscr{X}$ be a bornological space, $\mathscr{Y}$ a locally convex space and U a linear operator which maps $\mathscr{X}$ into $\mathscr{Y}$. In order that U be continuous it is necessary and sufficient that for any bounded set $A \subset \mathscr{X}$, the set $U(A)$ be bounded.*

Proof. We have to prove only the sufficiency. If W' is a convex, equilibrated and closed neighbourhood of the origin of $\mathscr{Y}$, and if $p_{W'}$ is the Minkowski semi-norm associated with W', then by putting

$$q(x) = p_{W'}(U(x)), \quad x \in \mathscr{X},$$

one gets a semi-norm on $\mathscr{X}$. If A is a bounded subset of $\mathscr{X}$, then $U(A)$ is bounded, by assumption. There then exists a number $\lambda > 0$ such that $U(A) \subset \lambda W'$; it follows that $q(A)$ is bounded. Since $\mathscr{X}$ is a bornological space, the semi-norm q is continuous on $\mathscr{X}$. By putting

$$W = \{x; q(x) \leqslant 1\},$$

one gets a neighbourhood of the origin in the space $\mathscr{X}$. Since $W = \overset{-1}{U}(W')$, the operator U is shown to be continuous.

2.3.2 Uniform convergence topologies

Let $\mathscr{X}$ and $\mathscr{Y}$ be (real) topological vector spaces and let $\mathscr{L}(\mathscr{X}, \mathscr{Y})$ be the set of all linear, continuous operators which map $\mathscr{X}$ into $\mathscr{Y}$. With the usual operations, the set $\mathscr{L}(\mathscr{X}, \mathscr{Y})$ becomes a vector space.

Let $\mathscr{M}$ be a set of bounded subsets of $\mathscr{X}$.

Let us consider a fundamental system $\mathscr{W}$ of equilibrated neighbourhoods of the origin of the space $\mathscr{Y}$. Let $A_0 \in \mathscr{M}$ and, for any $W \in \mathscr{W}$ let us define

$$E(A_0, W) = \{U \in \mathscr{L}(\mathscr{X}, \mathscr{Y}); \; U(A_0) \subset W\}.$$

It is easy to check that the set $E(A_0, W)$ is equilibrated and absorbent (because $U(A_0)$ is bounded). It is also easily verified that the system $\mathscr{E} = \{E(A_0, W)\}_{W \in \mathscr{W}}$ has the following properties

(i) for any $E_1, E_2 \in \mathscr{E}$ there exists a set $E_3 \in \mathscr{E}$, such that $E_3 \subset E_1 \cap E_2$;

(ii) for any $E \in \mathscr{E}$ there exists a set $E_1 \in \mathscr{E}$, such that $E_1 + E_2 \subset E$;

(iii) if $E \in \mathscr{E}$ and $0 \neq \lambda \in \mathbb{R}$, then $\lambda E \in \mathscr{E}$.

The system $\mathscr{E}$ then is a fundamental system of neighbourhoods of the origin for a topology τ_{A_0}, which is compatible with the structure of vector space of $\mathscr{L}(\mathscr{X}, \mathscr{Y})$ [16]. The topology

$$\tau_{\mathscr{M}} = \sup_{A \in \mathscr{M}} \tau_A$$

in the set $\mathscr{L}(\mathscr{X}, \mathscr{Y})$ may be obtained by taking as a fundamental system of neighbourhoods of the origin all finite intersections of the form

$$\bigcap_{i=1}^{n} E(A_i, W_i)$$

where $A_i \in \mathscr{M}$ and $W_i \in \mathscr{W}$.

If $\mathscr{Y}$ is a locally convex space, then $\tau_{\mathscr{M}}$ is a locally convex topology on $\mathscr{L}(\mathscr{X}, \mathscr{Y})$.

Let us observe that if $\mathscr{M}'$ is the system of all equilibrated hulls of all the sets which belong to $\mathscr{M}$, and if $\mathscr{M}''$ is the system of all the closures of the sets which belong to $\mathscr{M}$, then $\tau_{\mathscr{M}} = \tau_{\mathscr{M}'} = \tau_{\mathscr{M}''}$.

The space $\mathscr{L}(\mathscr{X}, \mathscr{Y})$, endowed with the topology $\tau_{\mathscr{M}}$ will be denoted by $\mathscr{L}_{\mathscr{M}}(\mathscr{X}, \mathscr{Y})$.

The topology $\tau_{\mathscr{M}}$ is called the *topology of uniform convergence on the sets of* $\mathscr{M}$.

According to the nature of the set $\mathscr{M}$, the topology $\tau_{\mathscr{M}}$ bears special names. If $\mathscr{M}$ is the set of all finite subsets of $\mathscr{X}$, then $\tau_{\mathscr{M}}$ is called the topology *of simple convergence* in $\mathscr{L}(\mathscr{X}, \mathscr{Y})$; if $\mathscr{M}$ is the set of all compact subsets of $\mathscr{X}$, then $\tau_{\mathscr{M}}$ is called the *topology of compact convergence;* if $\mathscr{M}$ is the set of all bounded subsets of $\mathscr{X}$, then τ is called the *topology of bounded convergence.*

Proposition 1. *If $\mathscr{Y}$ is a Hausdorff space and if* [17] $\overline{\mathrm{Sp}}(\cup \mathscr{M}) = \mathscr{X}$, *then the space $\mathscr{L}_{\mathscr{M}}(\mathscr{X}, \mathscr{Y})$ is also a Hausdorff space.*

Proof. Let $M = \cup \mathscr{M}$ and let $\mathbf{0} \neq U \in \mathscr{L}_{\mathscr{M}}(\mathscr{X}, \mathscr{Y})$. Then there exists an $x_0 \in M$, such that $U(x_0) \neq \mathbf{0}$. There also exists a neighbourhood W of the origin of $\mathscr{Y}$, such that $U(x_0) \notin W$. It therefore follows that if $A \in \mathscr{M}$ and $x_0 \in A$, then $U \notin E(A, W)$. The proposition is proved.

Proposition 2. *If $\mathscr{Y}$ is a locally convex space and if $\mathscr{P}$ is a directed set of semi-norms which determines the topology of $\mathscr{Y}$, then the topology $\tau_{\mathscr{M}}$ is determined by the family $\{p_A; p \in \mathscr{P}, A \in \mathscr{M}\}$ of the semi-norms given by the formulae*

$$p_A(U) = \sup_{x \in A} p(U(x)).$$

---

[16]) See, for example, R. Cristescu [31].

[17]) By $\overline{\mathrm{Sp}}(A)$ we have denoted the closed vector subspace generated by A.

Proof. It is obvious that p_A is a semi-norm on $\mathscr{L}(\mathscr{X}, \mathscr{Y})$. Let $p \in \mathscr{P}$, $0 < \varepsilon \in \mathbb{R}$ and

$$W = \{y; p(y) \leqslant \varepsilon\}.$$

Then the inclusion $U(A) \subset W$ is obviously equivalent to the condition $p_A(U) \leqslant \varepsilon$, i.e.,

$$E(A, W) = \{U; p_A(U) \leqslant \varepsilon\}.$$

The proposition is proved.

Remarks. (1) In the foregoing proposition, the family of seminorms $\{p_A; p \in \mathscr{P}, A \in \mathscr{M}\}$ may be replaced by any subfamily $\{p_A: p \in \mathscr{P}, A \in \mathscr{M}_0\}$, where $\mathscr{M}_0$ is a subset of $\mathscr{M}$, such that any set $A \in \mathscr{M}$ be included in a set of the form λA_0, where $A_0 \in \mathscr{M}_0$ and $\lambda \in \mathbb{R}$. Indeed, we have only to observe that

$$p_{\lambda A_0}(U) = |\lambda| p_{A_0}(U).$$

(2) If $\mathscr{X}$ and $\mathscr{Y}$ are normed vector spaces, then the topology of bounded convergence in $\mathscr{L}(\mathscr{X}, \mathscr{Y})$ is determined by the norm given by the formula

$$||U|| = \sup_{||x|| \leqslant 1} ||U(x)||.$$

Proposition 3. *Let $\mathscr{X}$ and $\mathscr{Y}$ be topological vector spaces, $\mathscr{M}$ a set of bounded subsets of $\mathscr{X}$ and $\mathscr{G} \subset \mathscr{L}_{\mathscr{M}}(\mathscr{X}, \mathscr{Y})$. Then $\mathscr{G}$ is bounded iff for any neighbourhood W of the origin of $\mathscr{Y}$, the set $\bigcap_{\mathscr{U} \in \mathscr{G}} \overset{-1}{U}(W)$ absorbs any set from $\mathscr{M}$.*

Proof. The condition that $\mathscr{G}$ be bounded for the topology $\tau_{\mathscr{M}}$ means that for any $A \in \mathscr{M}$ and any neighbourhood W of the origin of $\mathscr{Y}$, there should exist a number $\lambda > 0$ such that $\mathscr{G} \subset \lambda E(A, W)$, i.e., $U(A) \subset \lambda W$, for any $U \in \mathscr{G}$.

We have only to state that the condition $U(A) \subset \lambda W$, for any $U \in \mathscr{G}$, is equivalent to the condition

$$A \subset \lambda \bigcap_{\mathscr{U} \in \mathscr{G}} \overset{-1}{U}(W).$$

Theorem. *Let $\mathscr{X}$ and $\mathscr{Y}$ be Hausdorff locally convex spaces and $\mathscr{G}$ a subset of $\mathscr{L}(\mathscr{X}, \mathscr{Y})$. Let $\mathscr{M}$ be the family of all complete, bounded, convex, equilibrated subsets of $\mathscr{X}$. If the set $\mathscr{G}$ is bounded for the topology of simple convergence, then $\mathscr{G}$ is bounded for the topology $\tau_{\mathscr{M}}$.*

Proof. By taking into account proposition 3 we must show that if W is an equilibrated, convex and closed neighbourhood of the origin in the space $\mathscr{Y}$, then the set

$$B = \bigcap_{U \in \mathscr{G}} \overset{-1}{U}(W)$$

absorbs any set from $\mathscr{M}$.

The set B is a barrel in $\mathscr{X}$. Indeed, it is obvious that B is equilibrated, convex and closed. Since $\mathscr{G}$ is bounded in the topology of simple convergence, for any element $x \in \mathscr{X}$ there exists a number $\lambda \neq 0$ such that

$$\lambda \mathscr{G} \subset E(x; W).$$

From $U \in \mathscr{G}$ it follows that $\lambda U(x) \in W$ and therefore $\lambda x \in \overset{-1}{U}(W)$; this implies the fact that $\lambda x \in B$. Since B is equilibrated and since for any x there exists a $\lambda \neq 0$ such that $\lambda x \in B$, the set B is absorbent. It follows that B is a barrel in $\mathscr{X}$.

Since $\mathscr{X}$ is a Hausdorff locally convex space and B is a barrel in $\mathscr{X}$, it follows that the set B absorbs any set from $\mathscr{M}$ (see Section 2.2.3, proposition 2).

2.4 Duality between vector spaces[18])

2.4.1 Separation of convex sets in topological vector spaces

Proposition 1. *Let $\mathscr{X}$ be a topological vector space, A a (non-empty) convex, open subset of $\mathscr{X}$ and E an affine subspace such that $E \cap A = \varnothing$. Then there exists a closed hyperplane H, such that $H \supset E$ and $H \cap A = \varnothing$.*

Proof. One can asume that $\mathbf{0} \in A$, because otherwise we can choose an element $x_0 \in A$ and we can replace the sets A and E by $A - x_0$ and $E - x_0$. Let[19]) $\mathscr{X}_0 = \mathrm{Sp}(E)$. The set E is a hyperplane in $\mathscr{X}_0$ hence there exists a linear functional $f_0 : \mathscr{X}_0 \to \mathbb{R}$, such that

$$E = \{x \in \mathscr{X}_0; f_0(x) = 1\}.$$

Let p_A be the Minkowski functional associated to A, i.e.,

$$p_A(x) = \inf \{\mu; x \in \mu A, \mu > 0\}.$$

This functional is sub-linear. Since

$$A = \{x; p_A(x) < 1\},$$

from $E \cap A = \varnothing$ we infer that

$$f_0(x) = 1 \leqslant p_A(x), \quad x \in E.$$

It follows that if $\lambda > 0$, then

$$f_0(\lambda x) \leqslant p_A(\lambda x), \quad x \in E,$$

---

[18]) As in the preceding spaces, by a topological vector space we mean a real topological vector space.

[19]) We denote by Sp(E) the vector subspace generated by E.

whereas for $\lambda \leqslant 0$ we get that

$$f_0(\lambda x) \leqslant 0 \leqslant p_A(\lambda x), \quad x \in E.$$

It follows that

$$f_0(x) \leqslant p_A(x), \quad x \in \mathscr{X}_0.$$

With the well-known Hahn-Banach theorem we infer that f_0 may be extended to the entire space $\mathscr{X}$, to a linear functional f such that

$$f(x) \leqslant p_A(x), \quad x \in \mathscr{X}.$$

If we define

$$H = \{x; f(x) = 1\},$$

then H is a hyperplane. Since

$$f(x) \leqslant p_A(x) < 1, \quad x \in A$$

it follows that f is bounded on $A \cap (-A)$ and from the fact that A is an open set, we deduce that f is continuous on $\mathscr{X}$ (Section 2.1.5, proposition 1). It follows that H is a closed hyperplane and, obviously, $H \supset E$ and $H \cap A = \varnothing$.

Remarks. (1) The proposition can be also stated in the following way :

If A is a (non-empty) convex and open set and if E is an affine subspace such that $E \cap A = \varnothing$, then there exists a linear continuous functional $f: \mathscr{X} \to \mathbb{R}$, and a number μ, such that $f(x) = \mu$ for any $x \in E$ and $f(x) < \mu$ for $x \in A$.

(2) From proposition 1 we infer

If A is a (non-empty) convex, open subset and if E is a vector subspace such that $E \cap A = \varnothing$, then there exists a linear continuous functional f such that $f(x) = 0$ for any $x \in E$ and $f(x) > 0$ for $x \in A$.

Proposition 2. *Let $\mathscr{X}$ be a topological vector space and A_1 and A_2 two (non-empty) convex, disjoint subsets. If at least one of the sets A_1, A_2 is open, then there exists a closed hyperplane which separates the given sets.*

Proof. If A_1 is open, then the set $A = A_1 - A_2$ is open. Since A_1 and A_2 are convex, the set A is convex, too. From $A_1 \cap A_2 = \varnothing$ we infer that $\mathbf{0} \notin A$. By taking into account remark (2), it follows that there exists a linear continuous functional $f: \mathscr{X} \to \mathbb{R}$ such that $f(z) > 0$ for any $z \in A$. It follows that $f(x) > f(y)$ for any $x \in A_1$ and $y \in A_2$. By defining

$$\lambda = \inf_{x \in A_1} f(x),$$

it is easy to check that A_1 and A_2 are separated by the closed hyperplane

$$H = \{z; f(z) = \lambda\}.$$

Proposition 3. *Let $\mathscr{X}$ be a locally convex space and A_1 and A_2 (non-empty) convex, disjoint subsets. If one of the sets A_1, A_3 is closed and the other is compact, then there exists a closed hyperplane H which strictly separates the given sets.*

Proof. Let us assume that A_1 is compact and A_2 is closed. There then exists a neighbourhood W of the origin, which is an open convex set [20]), such that

$$(W + A_1) \cap (W + A_2) = \emptyset.$$

Since the sets $W + A_1$, $W + A_2$ are open and convex, there exists a closed hyperplane H which separates $W + A_1$ from $W + A_2$ (proposition 2). It follows that H strictly separates A_1 from A_2.

2.4.2 Duality between vector spaces and weak topologies

Let $\mathscr{X}$ and $\mathscr{Y}$ be two (real) vector spaces and F a bilinear functional defined on $\mathscr{X} \times \mathscr{Y}$: One says that the spaces $\mathscr{X}$ and $\mathscr{Y}$ are *in duality* with respect to F, if

(i) for any $x \neq \mathbf{0}$ $(x \in \mathscr{X})$ there exists an $y \in \mathscr{Y}$, such that $F(x, y) \neq 0$;

(ii) for any $y \neq \mathbf{0}$, $(y \in \mathscr{Y})$, there exists an $x \in \mathscr{X}$, such that $F(x, y) \neq 0$.

One also says that $\mathscr{X}$ and $\mathscr{Y}$ are a *dual system*.

We shall denote by $\mathscr{X}^l$ the set of all linear functionals defined on the vector space $\mathscr{X}$.

We shall suppose that the vector spaces $\mathscr{X}$ and $\mathscr{Y}$ are in duality with respect to the bilinear functional F.

If for any element $y \in \mathscr{Y}$ we put

$$F^y(x) = F(x, y), \quad x \in \mathscr{X},$$

then $F^y \in \mathscr{X}^l$ and the mapping $y \mapsto F^y$ of $\mathscr{Y}$ into $\mathscr{X}^l$ is an isomorphism of $\mathscr{Y}$ into $\mathscr{X}^l$.

Analogously, if for any element $x \in \mathscr{X}$ we put

$$F_x(y) = F(x, y), \quad y \in \mathscr{Y},$$

then $F_x \in \mathscr{Y}^l$ and the mapping $x \mapsto F_x$ of $\mathscr{X}$ into $\mathscr{Y}^l$ is an isomorphism of $\mathscr{X}$ into $\mathscr{Y}^l$.

The weakest topology in $\mathscr{X}$ for which all functionals F^y are continuous is called the *weak topology* with respect to F in $\mathscr{X}$ and it is denoted by $\sigma_F(\mathscr{X}; \mathscr{Y})$.

Analogously, one can define the weak topology $\sigma_F(\mathscr{Y}; \mathscr{X})$ in the space $\mathscr{Y}$.

The topology $\sigma_F(\mathscr{X}; \mathscr{Y})$ is defined by the family of semi-norms $\mathscr{P} = \{p_y\}_{y \in \mathscr{Y}}$, where

$$p_y(x) = | F(x, y) |. \tag{1}$$

---

[20]) Let $\mathscr{V}$ be the set of all neighbourhoods of the origin. If $(V + A_1) \cap (V + A_2) \neq \emptyset$, for any V, then $A_1 \cap (V + A_2) \neq \emptyset$, for any $V \in \mathscr{V}$. If for any $V \in \mathscr{V}$ we consider an element $x_V \in A_1 \cap (V + A_2)$, then the generalized sequence $\{x_V\}_{V \in \mathscr{V}}$ has a limit point x and one can see that $x \in A_1 \cap A_2$, which contradicts the disjointness of the sets A_1, A_2.

The proof can be easily made.

It follows that the weak topology (with respect to F) is a locally convex topology. If A is a subset of $\mathscr{X}$, which is closed for the topology $\sigma_F(\mathscr{X}, \mathscr{Y})$, then we shall say that A is weakly closed with respect to F or, shortly, that A is weakly closed. Analogously, we shall speak of weakly open sets (with respect to F) or of weakly continuous functionals (with respect to F).

If φ is a linear functional on $\mathscr{X}$ and if it is weakly continuous (with respect to F), then there exists a unique element $y \in \mathscr{Y}$, such that

$$\varphi(x) = F(x, y), \tag{2}$$

for any $x \in \mathscr{X}$.

Indeed, with the help of the semi-norms (1), from the fact that φ is linear and continuous for the topology $\sigma_F(\mathscr{X}, \mathscr{Y})$, it will follow that there exist elements $y_i \in \mathscr{Y}$, $i = 1, 2, \ldots, n$, such that

$$|\varphi(x)| \leqslant \max_{1 \leqslant i \leqslant n} |F(x, y_i)|. \tag{3}$$

If we put

$$\varphi_i(x) = F(x, y_i), \qquad i = 1, 2, \ldots, n,$$

then from (2) it follows that if $\varphi_i(x) = 0$, $i = 1, 2, \ldots, n$, then $\varphi(x) = 0$. But then we infer that φ may be written as [21])

$$\varphi(x) = \sum_{i=1}^{n} \lambda_i \varphi_i(x),$$

i.e.,

$$\varphi(x) = F\left(x, \sum_{i=1}^{n} \lambda_i y_i\right).$$

If we define

$$y = \sum_{i=1}^{n} \lambda_i y_i,$$

we get equality (2).

The uniqueness of the representation follows from condition (ii) in the definition of the duality.

If $\mathscr{X}$ is a topological vector space, with the topology τ, we shall denote by $\mathscr{X}_\tau^*$ the set of all linear, continuous functionals on $\mathscr{X}$ (that is, the "algebraic and topological conjugate space" of $\mathscr{X}$).

---

[21]) See R. Cristescu, *Matematici generale*, vol. 2, Edit. did. și pedag., Bucharest, 1969 (in Romanian) or M. M. Day, *Normed linear spaces*, Springer-Verlag, Berlin-Göttingen-Heidelberg, 1958.

From the representation of the linear and weakly continuous functionals on $\mathscr{X}$ it follows that $\mathscr{X}^*_{\sigma_F(x,y)} = \mathscr{Y}$ (if we identify the two spaces by an isomorphism).

If $\mathscr{X}$ is a Hausdorff locally convex space [22]) and if τ is its topology, by defining

$$F(x, \varphi) = \varphi(x), \quad x \in \mathscr{X}, \quad \varphi \in \mathscr{X}^*_\tau,$$

then F is a bilinear functional on $\mathscr{X} \times \mathscr{X}^*_\tau$ and the spaces $\mathscr{X}, \mathscr{X}^*_\tau$ are in duality with respect to F.

The weak topology with respect to this functional will be denoted by $\sigma(\mathscr{X}, \mathscr{X}^*_\tau)$ and it will be called the *weakened topology* in $\mathscr{X}$.

It is easy to remark that the weak topology (with respect to F) in the space $\mathscr{X}^*_\tau$ is the topology of simple convergence in the space $\mathscr{L}(\mathscr{X}, \mathbb{R})$.

2.4.3 The notion of polar of a set

Let $\mathscr{X}$ and $\mathscr{Y}$ be two (real) vector spaces in duality with respect to a bilinear functional F.

If A is any subset of $\mathscr{X}$, the following subset of $\mathscr{Y}$:

$$A^0 = \{y \in \mathscr{Y};\ F(x, y) \leqslant 1,\ x \in A\}$$

is called the *polar* of A.

It is easy to prove the following properties of the polar

(1) $(\lambda A)^0 = \lambda^{-1} A^0$, for any λ, $\lambda \neq 0$;
(2) if $A \subset B$, then $B^0 \subset A^0$;
(3) if B absorbs A, then A^0 absorbs B^0,
(4) $(\bigcup_{j \in J} A_j)^0 = \bigcap_{j \in J} A_j^0$;
(5) if A is equilibrated, then A^0 is equilibrated and the following equality holds

$$A^0 = \{y \in \mathscr{Y};\ |F(x, y)| \leqslant 1,\ x \in A\};$$

(6) for any subset A of $\mathscr{X}$, the polar of A is a convex, weakly closed subset of $\mathscr{Y}$.

We note that the fact that A^0 is closed in the topology $\sigma_F(\mathscr{X}, \mathscr{Y})$, for any A, is a consequence of the equality

$$A^0 = \bigcap_{x \in A} \{y \in \mathscr{Y};\ F_x(y) \leqslant 1\},$$

if we take into account the fact that all functionals F_x are continuous for the topology $\sigma_F(\mathscr{Y}, \mathscr{X})$.

---

[22]) We recall that for any non-zero element x_0 of a Hausdorff locally convex space $\mathscr{X}$, there exists a continuous linear functional φ on $\mathscr{X}$, such that $\varphi(x_0) \neq 0$ (see R. Cristescu [31]).

If A is any subset of $\mathscr{X}$, then one can also consider the *bipolar* A^{00} of A, that is, the set $(A^0)^0$, which is a subset of $\mathscr{X}$.

The bipolar of a set A equals the convex, weakly closed hull of the set $A \cup \{\mathbf{0}\}$.

Indeed, if

$$B = \mathrm{co}(A \cup \{\mathbf{0}\}),$$

then $B^0 = A^0$, as one can easily prove. We can therefore assume from the beginning that A is convex and $\mathbf{0} \in A$. On the other hand, we obviously have $A \subset A^{00}$. It follows that if $\bar{A}$ denotes the closure of the set A for the topology $\sigma_F(\mathscr{X}, \mathscr{Y})$, then $\bar{A} \subset A^{00}$, because the polar of any set is a weakly closed set.

If an element $x_0 \notin \bar{A}$, then there exists a weakly closed hyperplane H in $\mathscr{X}$, which strictly separates x_0 from $\bar{A}$ (Section 2.4.1, proposition 3). Since $\mathbf{0} \notin H$, it follows that there exists a $y_0 \in \mathscr{Y}$, such that

$$H = \{x;\ F(x, y_0) = 1\}.$$

Hence from $x \in \bar{A}$ we get that $F(x, y_0) < 1$ (because $\mathbf{0} \in A$ and this shows that $F(\mathbf{0}, y_0) < 1$). Consequently, $y_0 \in A^0$ and from $F(x_0, y_0) > 1$ we get that $x_0 \notin A^{00}$. We have thus shown that $\complement\bar{A} \subset \complement A^{00}$, i.e. $\bar{A} \supset A^{00}$. It follows that $A^{00} = \bar{A}$.

If E is a vector subspace of $\mathscr{X}$, then

$$E^0 = \{y \in \mathscr{Y};\ F(x, y) = 0,\ \forall x \in E\}.$$

Indeed, if $y \in E^0$, then $|F(x, y)| \leqslant 1$, for any $x \in E$. But from $x \in E$ it follows that $\lambda x \in E$, and this implies that $|F(\lambda x, y)| \leqslant 1$, for any $\lambda \in \mathbb{R}$ and $x \in E$. It immediately follows that $F(x, y) = 0$.

In particular, *if E is a vector subspace of* $\mathscr{X}$, *then* E^0 *is a weakly closed vector subspace of* $\mathscr{Y}$ *and* $E^{00} = \bar{E}$.

2.4.4 Locally convex topologies compatible with a given duality

Let $\mathscr{X}$ and $\mathscr{Y}$ be two vector spaces, in duality with respect to a bilinear functional F.

As we have seen in Section 2.4.2, the space $\mathscr{Y}$ is isomorphic (as a vector space) to a subspace of $\mathscr{X}^l$, by the mapping $y \mapsto F^y$. By identifying $\mathscr{Y}$ to its image in $\mathscr{X}^l$ by this isomorphism, we shall assume that $\mathscr{Y} \subset \mathscr{X}^l$.

A locally convex topology τ in $\mathscr{X}$ is said to be *compatible* with the given duality if $\mathscr{X}^*_\tau = \mathscr{Y}$.

Proposition 1. *If* τ_1 *and* τ_2 *are two Hausdorff locally convex topologies in* $\mathscr{X}$, *which are compatible with the given duality, then any convex set* $A \subset \mathscr{X}$, *which is closed for one of the topologies, is also closed for the other.*

Proof. By hypothesis we have

$$\mathscr{X}^*_{\tau_1} = \mathscr{X}^*_{\tau_2} = \mathscr{Y}. \tag{1}$$

If the convex set A is closed for the topology τ_1, then A is the intersection of all (τ_1)-closed semi-spaces which contain it. Indeed, with proposition 3 from section 2.4.1, we conclude that for any element x_0, which does not belong to A, there exists a closed (for the topology τ_1) hyperplane, which strictly separates x_0 from A. It follows that if $\{S_j\}_{j\in J}$ is the family of all closed semi-spaces which contain A, then there exists an index j_0, such that $x_0 \in \complement S_{j_0}$ and, therefore, $x_0 \in \complement(\bigcap_{j\in J} S_j)$. It follows that $A \supset \bigcap_{j\in J} S_j$, whereas the opposite inclusion is obvious.

We have now only to observe that since any closed (for the topology τ_1) semi-space is of the form

$$S = \{x; f(x) \leqslant \lambda\},$$

where $f \in \mathscr{X}^*_{\tau_1}$, it is also closed for the topology τ_2, by virtue of relation (1).

Remarks. (1) Since for any locally convex topology, the closure of any convex set is convex, it follows that the closure of a convex set A is the intersection of all closed, convex sets which contain A.

By taking into account proposition 1, the closure of a convex set A is the same for the topologies τ_1 and τ_2.

(2) If $\mathscr{X}$ is a Hausdorff locally convex space, with the topology τ, then the weakened topology of $\mathscr{X}$ is compatible with the duality of the system $(\mathscr{X}, \mathscr{X}^*_\tau)$ given by the bilinear functional $F(x, \varphi) = \varphi(x)$. With proposition 1 we infer that the convex sets of $\mathscr{X}$ have the same closure in the weakened topology as in the topology τ.

In particular, if A is a convex set which contains $\mathbf{0}$, then the closure of A for the topology τ is equal to A^{00}.

Lemma. *If $\mathscr{X}$ is a Hausdorff locally convex space (with the topology τ) and if Φ is a set of equi-continuous* [23]) *linear functionals, defined on $\mathscr{X}$, then Φ is a relatively compact set for the weak topology of $\mathscr{X}^*_\tau$.*

Proof. There exists a neighbourhood W of the origin of $\mathscr{X}$, such that $|f(x)| \leqslant 1$ for any $x \in W$ and $f \in \Phi$. If x is any element of $\mathscr{X}$, then there exists a $\lambda > 0$, such that $\lambda x \in W$. It follows that $|f(x)| \leqslant 1/\lambda$, for any $f \in \Phi$. Consequently, for any element $x \in \mathscr{X}$, the set

$$\Phi(x) = \{f(x); f \in \Phi\}$$

is bounded and, therefore, relatively compact in $\mathbb{R}$. By taking into consideration the product space $\mathbb{R}^{\mathscr{X}}$, the set Φ is relatively compact in this space [24]). On the other hand, the closure $\overline{\Phi}$ of Φ in the space $\mathbb{R}^{\mathscr{X}}$ is contained in $\mathscr{X}^*_\tau$, as one can easily show.

It follows that Φ is relatively compact for the weak topology of $\mathscr{X}^*_\tau$.

In the following theorem we shall consider two vector spaces, which are in duality with respect to a bilinear functional F, and we shall identify the space $\mathscr{X}$ to the space $\mathscr{L}(\mathscr{Y}, \mathbb{R})$, where $\mathscr{Y}$ is taken with the weak topology.

---

[23]) Equi-continuity of a set Φ of linear functionals is equivalent to their equi-continuity at the origin: for any $\varepsilon > 0$ there exists a neighbourhood W of the origin of $\mathscr{X}$, such that $|f(x)| < \varepsilon$ for any $x \in W$ and $f \in \Phi$.

[24]) If a subset A of a product Hausdorff topological space has the property that any of its projections is a relatively compact set, then A is a relatively compact set. This proposition is a consequence of Tikhonov's theorem.

Theorem [25]). *Let $\mathscr{X}$ and $\mathscr{Y}$ be two vector spaces, in duality (with respect to F), and let τ be a Hausdorff locally convex topology in $\mathscr{X}$.*

In order that the topology τ be compatible with the duality, it is necessary and sufficient that τ be the topology of uniform convergence on a set $\mathscr{M}$ of equilibrated, convex subsets of $\mathscr{Y}$, which are compact for the topology $\sigma_F(\mathscr{Y}, \mathscr{X})$, and which should cover the space $\mathscr{Y}$.

Proof. The necessity. Let us assume that $\mathscr{X}^*_\tau = \mathscr{Y}$. Let us consider a fundamental system $\mathscr{W}$ of neighbourhoods of the origin in the space $\mathscr{X}$, consisting of equilibrated, convex (τ)-closed sets, and such that $W \in \mathscr{W}$ and $\lambda \in \mathbb{R}$, $\lambda \neq 0$, implies $\lambda W \in \mathscr{W}$.

If $W \in \mathscr{W}$, then, by taking into account the preceding remark (2) we infer that $W = W^{00}$. Let us define

$$\mathscr{M} = \{W^0; W \in \mathscr{W}\}.$$

Since W^0 is a set of equi-continuous linear functionals, with the help of the lemma we infer that W^0 is a relatively compact set (and, consequently, a compact one), for the topology $\sigma_F(\mathscr{Y}, \mathscr{X})$.

One can therefore consider in $\mathscr{X}$ the topology of uniform convergence on the sets of $\mathscr{M}$.

From the definition of the topology $\tau_{\mathscr{M}}$ (and by taking into account the properties of the sets of $\mathscr{W}$) it follows that $\mathscr{W}$ is a fundamental system of neighbourhoods of the origin for the topology $\tau_{\mathscr{M}}$.

The set $\mathscr{M}$ covers $\mathscr{Y}$. Indeed, let $y \in \mathscr{Y}$. Since the functional F is continuous for the topology $\sigma_F(\mathscr{X}, \mathscr{Y})$ and, therefore, for the topology τ, too, there exists a $W \in \mathscr{W}$, such that $|F^y(x)| \leqslant 1$, for any $x \in W$; it follows that $y \in W^0$.

By taking into account the properties of the polars, the set $\mathscr{M}$ consists of equilibrated, convex sets.

Sufficiency. Let $\mathscr{M}$ have the properties from the statement of the theorem. Without loss of generality, one can assume that if $M \in \mathscr{M}$, and $\lambda \neq 0$, $\lambda \in \mathbb{R}$, then $\lambda M \in \mathscr{M}$; and if $M_i \in \mathscr{M}$ $(i = 1, 2, \ldots, n)$, then the convex, equilibrated hull of the set $\bigcup_{i=1}^{n} M_i$ belongs to $\mathscr{M}$. Indeed if this be not true, we shall consider the set $\mathscr{M}'$ obtained from $\mathscr{M}$ by forming all subsets of the indicated [26]) form and we have only to state that $\tau_{\mathscr{M}'} = \tau_{\mathscr{M}}$.

Due to the properties imposed on $\mathscr{M}$ and to the definition of the topology $\tau_{\mathscr{M}}$, the system $\mathscr{W} = \{M^0; M \in \mathscr{M}\}$ is a fundamental system of neighbourhoods of the origin for the topology $\tau_{\mathscr{M}}$. We shall denote $\tau = \tau_{\mathscr{M}}$.

From the fact that any finite subset of $\mathscr{Y}$ is contained in a set M from $\mathscr{M}$, it follows that $\sigma_F(\mathscr{X}, \mathscr{Y}) \leqslant \tau$. Consequently, $\mathscr{Y}$ may be considered to be a linear subspace of $\mathscr{X}^*_\tau$ (because $\mathscr{Y} = \mathscr{X}^*_{\sigma_F(\mathscr{X}, \mathscr{Y})}$).

We have only to show that $\mathscr{X}^*_\tau \subset \mathscr{Y}$.

---

[25]) Mackey's theorem.

[26]) We recall that in a real Hausdorff topological vector space the convex hull of the union of a finite number of convex, compact sets is a compact set (see Section 2.1.4, proposition 2).

Let $f \in \mathscr{X}^*_\tau$. Then there exists an $M \in \mathscr{M}$, such that $|f(x)| \leqslant 1$, for any $x \in M^0$. By denoting by M^1 the polar of M^0 in the space $\mathscr{X}^*_\tau$(with respect to the dual system $(\mathscr{X}, \mathscr{X}^*_\tau)$), we have $f \in M^1$. On the other hand, the set M^1 is the weak closure of M in the space $\mathscr{X}^*_\tau$. Since the weak topology of $\mathscr{X}^*_\tau$ induces on $\mathscr{Y}$ the topology $\sigma_F(\mathscr{Y}, \mathscr{X})$, and since M is a compact set for this topology, it follows that M is weakly closed in $\mathscr{X}^*_\tau$. Consequently, $M = M^1$ and this implies that $f \in \mathscr{Y}$.

We have thus shown that $\mathscr{X}^*_\tau = \mathscr{Y}$, and, therefore, the topology τ is compatible with the duality. The theorem is proved.

If $\mathscr{M}$ is the set of all convex, equilibrated and weakly compact subsets of the space $\mathscr{Y}$, then the topology $\tau_{\mathscr{M}}$ of the space $\mathscr{X}$ is called the *Mackey topology* (for the given duality) and it will be denoted by $\mu_F(\mathscr{X}, \mathscr{Y})$.

From the preceding theorem it follows that a Hausdorff locally convex topology τ in $\mathscr{X}$ is compatible with a given duality iff

$$\sigma_F(\mathscr{X}, \mathscr{Y}) \leqslant \tau \leqslant \mu_F(\mathscr{X}, \mathscr{Y}).$$

If $\mathscr{X}$ is a Hausdorff locally convex space, with the topology τ, then the Mackey topology, which corresponds to the natural duality between $\mathscr{X}$ and $\mathscr{X}^*_\tau$ (see Section 2.4.2) will be denoted by $\mu(\mathscr{X}, \mathscr{X}^*_\tau)$.

Proposition 2. *If $\mathscr{X}$ and $\mathscr{Y}$ are vector spaces in duality, and if τ_1 and τ_2 are two Hausdorff locally convex topologies in $\mathscr{X}$, compatible with the duality, then any set, which is bounded for one of the topology, is bounded, for the other, too.*

Proof. Let $\mathscr{M}$ be the set of all equilibrated, convex and weakly compact subsets of the space $\mathscr{Y}$. If $M \in \mathscr{M}$, then M is weakly complete in $\mathscr{Y}$. By virtue of the theorem from Section 2.3.2, any set $A \subset \mathscr{X} = \mathscr{L}(\mathscr{Y}, \mathbb{R})$, which is bounded for the topology $\sigma_F(\mathscr{X}, \mathscr{Y})$, is also bounded for the topology $\tau_{\mathscr{M}} = \mu_F(\mathscr{X}, \mathscr{Y})$.

If τ_1 and τ_2 are two Hausdorff locally convex topologies, which are compatible with the duality, then they are finer than $\sigma_F(\mathscr{X}, \mathscr{Y})$, and are weaker than $\mu_F(\mathscr{X}, \mathscr{Y})$. The proposition now immediately follows.

In particular, if A is a subset of a Hausdorff locally convex space $\mathscr{X}$, and if A is bounded for the weakened topology of $\mathscr{X}$, then A is bounded for the topology τ of the locally convex space $\mathscr{X}$.

2.4.5 Mackey spaces

A Hausdorff locally convex space $\mathscr{X}$ (with the topology τ) is said to be a *Mackey space* if

$$\tau = \mu(\mathscr{X}, \mathscr{X}^*_\tau).$$

Proposition 1. *Any Hausdorff barreled space is a Mackey space.*

Proof. Let $\mathscr{X}$ be a Hausdorff barreled space and τ be its topology. If B is an equilibrated, weakly bounded subset of $\mathscr{X}^*_\tau$, then B^0 is absorbent in $\mathscr{X}$, as one can immediately show. On the other hand, the polar of a set is always a convex, weakly closed set, whereas the polar of an equilibrated set is equilibrated. It follows that B^0 is a barrel in $\mathscr{X}$. Since $\mathscr{X}$ is a barreled space, the set B^0 is a neighbourhood of the origin for the topology τ.

Conversely, if E is a barrel in $\mathscr{X}$, then $E = E^{00}$ (see Section 2.4.3, as well as Section 2.4.4, proposition 1) and this implies that E^0 is weakly bounded (if the polar A^0 of a set A is absorbent, then A is weakly bounded).

It follows that the topology τ of $\mathscr{X}$ equals the finest topology of uniform convergence, when $\mathscr{X} = \mathscr{L}(\mathscr{X}_\tau^*, \mathbb{R})$ and $\mathscr{X}_\tau^*$ is endowed with the topology $\sigma(\mathscr{X}_\tau^*, \mathscr{X})$. Because τ is compatible with the duality of the system $(\mathscr{X}, \mathscr{X}_\tau^*)$, it follows that $\tau = \mu(\mathscr{X}, \mathscr{X}_\tau^*)$.

Proposition 2. *Any Hausdorff bornological space is a Mackey space.*

Proof. Let $\mathscr{X}$ be a Hausdorff bornological space, with the topology τ. Since τ is compatible with the duality of the system $(\mathscr{X}, \mathscr{X}_\tau^*)$, we have $\tau \leqslant \mu(\mathscr{X}, \mathscr{X}_\tau^*)$. We have only to prove that $\mu(\mathscr{X}, \mathscr{X}_\tau^*) \leqslant \tau$.

Let W be an equilibrated and convex neighbourhood of the origin for the topology $\mu(\mathscr{X}, \mathscr{X}_\tau^*)$. It absorbs any set which is bounded for this topology. Since the bounded sets for the topologies τ and $\mu(\mathscr{X}, \mathscr{X}_\tau^*)$ are the same (Section 2.4.4, proposition 2), it follows that W absorbs any set which is bounded for the topology τ. Since the space $\mathscr{X}$ is bornological, the set W is a neighbourhood of the origin for the topology τ.

The proposition is proved.

Bibliographical comments

The topological vector spaces have been introduced by J. von Neumann and A. N. Kolmogorov.

An exposition of the theory of the topological vector spaces can be found in the books by N. Bourbaki [3], A. Grothendieck [1], G. Köthe [1], H. H. Schaefer [8], A. P. Robertson—W. Robertson [1], G. Marinescu [4], etc.

Particular cases of locally convex spaces have been studied by G. Köthe.

The general theory of the locally convex spaces has developed especially as a consequence of research by J. Dieudonné and G. W. Mackey.

3

Ordered vector spaces

An ordered vector space is a set in which a vector space structure has been given, as well as an order relation, which is bound to satisfy some compatibility relations to the vector space structure. The most interesting ordered vector spaces are those which, at the same time, are lattices.

3.1 Ordered vector spaces. Directed spaces and Archimedean spaces

3.1.1 The notion of ordered vector space: General properties

Definition 1. A real vector space $\mathscr{X}$ is said to be an *ordered vector space* if an order relation has been given in it, such that the following conditions are satisfied [1])

(i) If $x_1, x_2 \in \mathscr{X}$ and $x_1 \leqslant x_2$, then $x_1 + x \leqslant x_1 + x$ for any $x \in \mathscr{X}$.

(ii) if $x_1, x_2 \in \mathscr{X}$ and $x_1 \leqslant x_2$, then $\alpha x_1 \leqslant \alpha x_2$, for any $\alpha \in \mathbb{R}$, $\alpha \geqslant 0$.

From condition (i) it follows that if $x_1 \leqslant x_2$ and $x_3 \leqslant x_4$, then $x_1 + x_3 \leqslant x_2 + x_4$.

Indeed, from $x_1 \leqslant x_2$ it follows that $x_1 + x_3 \leqslant x_2 + x_3$, whereas from $x_3 \leqslant x_4$ we infer that $x_2 + x_3 \leqslant x_2 + x_4$.

One has also the following proposition.

Proposition 1. *Let $\mathscr{X}$ be an ordered vector space, x, x_1, x_2 elements of $\mathscr{X}$ and α, β real numbers.*

(1) *if* $x_1 \geqslant \mathbf{0}$ *and* $x_2 \geqslant \mathbf{0}$, *then* $x_1 + x_2 \geqslant \mathbf{0}$;

(2) *if* $x \geqslant \mathbf{0}$ *and* $\alpha \geqslant 0$, *then* $\alpha x \geqslant \mathbf{0}$;

(3) *if* $x \geqslant \mathbf{0}$ *and* $-x \geqslant \mathbf{0}$, *then* $x = \mathbf{0}$;

---

[1]) $\mathbb{R}$ stands for the set of real numbers.

(4) *if* $x_1 \leqslant x_2$ *and* $\alpha \leqslant 0$, *then* $\alpha x_1 \geqslant \alpha x_2$,

(5) *if* $x \geqslant \mathbf{0}$ *and* $\alpha \leqslant \beta$, *then* $\alpha x \leqslant \beta x$.

The proof is easily made.

Definition 2. In an ordered vector space $\mathscr{X}$ an element x is said to be *positive* if $x \geqslant \mathbf{0}$; it is called a *negative* element if $x \leqslant \mathbf{0}$.

Remark. If $\mathscr{X}$ is an ordered vector space, then the relation $x_1 \leqslant x_2$, in this space is equivalent to the relation $x_2 - x_1 \geqslant \mathbf{0}$. Conversely, let $\mathscr{X}$ be a real vector space and P a non-empty subset of $\mathscr{X}$, such that

(1) $x_1, x_2 \in P \Rightarrow x_1 + x_2 \in P$;

(2) $x \in P$ and $0 \leqslant \alpha \in \mathbb{R} \Rightarrow \alpha x \in P$;

(3) $x \in P$ and $-x \in P \Rightarrow x = \mathbf{0}$.

Then $\mathscr{X}$ becomes an ordered vector space, in which P is the set of the positive elements if we define $x_1 \leqslant x_2$ to mean the same thing as $x_2 - x_1 \in P$.

Proposition 2. *Let* $\{x_j\}_{j \in J}$ *be a system of elements of an ordered vector space. If the element* $\bigvee_{j \in J} x_j$ *exists, then* $\bigwedge_{j \in J} (-x_j)$ *exists too, and the following equality*

$$\bigvee_{j \in J} x_j = -\bigwedge_{j \in J} (-x_j) \tag{1}$$

holds.

Proof. Let us denote

$$a = \bigvee_{j \in J} x_j.$$

From $x_j \leqslant a$ $(j \in J)$ it follows that $-a \leqslant -x_j$ $(\forall j \in J)$. If x is any element such that $x \leqslant -x_j$ $(j \in J)$, then $x_j \leqslant -x$ $(\forall j \in J)$, whence we infer that $a \leqslant -x$, i.e., $x \leqslant -a$. This shows that the equality

$$-a = \bigwedge_{j \in J} (-x_j),$$

which is equivalent to (1), is proved.

Remark. An analogous proposition can be obtained by interchanging the symbols $\bigvee$ and $\bigwedge$.

Proposition 3. *Let* $\{x_j\}_{j \in J}$ *and* $\{y_k\}_{k \in K}$ *be two systems of elements of an ordered vector space.*

If $\bigvee_{j\in J} x_j$ *and* $\bigvee_{k\in K} y_k$ *exist, then* $\bigvee_{j,k}(x_j + y_k)$ *exists too, and the equality*

$$\bigvee_{\substack{j\in J\\ k\in K}} (x_j + y_k) = (\bigvee_{j\in J} x_j) + (\bigvee_{k\in K} y_k) \tag{2}$$

holds.

Proof. By putting

$$a = \bigvee_{j\in J} x_j, \quad b = \bigvee_{k\in K} y_k,$$

we obviously have

$$x_j + y_k \leqslant a + b, \ (\forall j \in J, \forall k \in K).$$

Now let z be an element, such that

$$x_j + y_k \leqslant z, \quad \forall j \in J, \forall k \in K.$$

By fixing the index k, from

$$x_j \leqslant z - y_k, \ (\forall j \in J),$$

we infer that $a \leqslant z - y_k$ and, therefore,

$$y_k \leqslant z - a, \ (\forall k \in K);$$

this implies that $b \leqslant z - a$, i.e., $a + b \leqslant z$. It follows that

$$a + b = \bigvee_{\substack{j\in J\\ k\in K}} (x_j + y_k)$$

and equality (2) is proved.

Remark. An analogous proposition holds for the greatest lower bound.

Proposition 4. *Let* $\{x_j\}_{j\in J}$ *be a system of elements of an ordered vector space. If the element* $\bigvee_{j\in J} x_j$ *exists and if* $0 < \alpha \in \mathbb{R}$, *then the element* $\bigvee_{j\in J}(\alpha x_j)$ *exists too, and the following equality holds*:

$$\bigvee_{j\in J} (\alpha x_j) = \alpha (\bigvee_{j\in J} x_j). \tag{3}$$

Proof. By putting

$$a = \bigvee_{j\in J} x_j,$$

we obviously have

$$\alpha x_j \leqslant \alpha a, \ (\forall j \in J).$$

If x is an element such that

$$\alpha x_j \leqslant x, \quad (\forall j \in J),$$

then, from

$$x_j \leqslant \frac{1}{\alpha} x, \quad (\forall j \in J),$$

we infer that $a \leqslant \frac{1}{\alpha} x$, i.e., $\alpha a \leqslant x$.

It follows that

$$\alpha a = \bigvee_{j \in J} (\alpha x_j),$$

that is, equality (3) holds.

Corollary. *If* $\bigvee_{j \in J} x_j$ *exists and if* $\alpha < 0$, *then* $\bigwedge_{j \in J} (\alpha x_j)$ *exists too, and the following equality holds*

$$\bigwedge_{j \in J} (\alpha x_j) = \alpha \left(\bigvee_{j \in J} x_j\right).$$

This easily follows by using proposition 2.

Remark. A proposition analogous to proposition 4 holds for the greatest lower bound.

Proposition 5. *In an ordered vector space the (o)-convergence has the following properties*

(1) $x_n \xrightarrow{o} x$ *and* $y_n \xrightarrow{o} y \Rightarrow x_n + y_n \xrightarrow{o} x + y$;

(2) $x_n \xrightarrow{o} x$ *and* $\alpha \in \mathbb{R} \Rightarrow \alpha x_n \xrightarrow{o} \alpha x$.

Proof. (1) In virtue of the definition given in Section 1.3.1, there exist sequences

$$\{a_n\}_{n \in \mathbb{N}}, \ \{b_n\}_{n \in \mathbb{N}}, \ \{a'_n\}_{n \in \mathbb{N}}, \ \{b'_n\}_{n \in \mathbb{N}},$$

such that

$$a_n \leqslant x_n \leqslant b_n, \quad a'_n \leqslant y_n \leqslant b'_n;$$

$$a_n \uparrow x; \ b_n \downarrow x; \ a'_n \uparrow y; \ b'_n \downarrow y.$$

We therefore have

$$a_n + a'_n \leqslant x_n + y_n \leqslant b_n + b'_n,$$

whereas the sequences $\{a_n + a'_n\}_{n \in \mathbb{N}}$, $\{b_n + b'_n\}_{n \in \mathbb{N}}$ are increasing and decreasing, respectively.

Let us show that

$$x + y = \bigvee_{n \in \mathbb{N}} (a_n + a'_n). \tag{4}$$

First of all, it is obvious that

$$a_n + a'_n \leqslant x + y, \ (\forall n \in \mathbb{N}).$$

Let z be any element such that

$$a_n + a'_n \leqslant z, \quad (\forall n \in \mathbb{N}).$$

Since the sequences $\{a_n\}_{n \in \mathbb{N}}$, $\{a'_n\}_{n \in \mathbb{N}}$ are increasing, we have

$$a_m + a'_n \leqslant z,$$

for any m and n. By taking into account proposition 3, we get

$$x + y \leqslant z,$$

and relation (4) is therefore proved.

Similarly, one can show that

$$x + y = \bigwedge_{n \in \mathbb{N}} (b_n + b'_n).$$

It follows that

$$x + y = (o)\text{-}\lim_n (x_n + y_n).$$

(2) Let $a_n \leqslant x_n \leqslant b_n$, such that $a_n \uparrow x$ and $b_n \downarrow x$. If $\alpha > 0$, then $\alpha a_n \leqslant \alpha x_n \leqslant \leqslant \alpha b_n$, and proposition 4, together with the remark following it, implies $\alpha a_n \uparrow \alpha x$ and $\alpha b_n \downarrow \alpha x$.

If $\alpha < 0$, then $\alpha b_n \leqslant \alpha x_n \leqslant \alpha a_n$ and $\alpha b_n \uparrow \alpha x$, $\alpha a_n \downarrow \alpha x$. In both cases, $\alpha x_n \xrightarrow{o} \alpha x$.

Definition 3. Two ordered vector spaces $\mathscr{X}$ and $\mathscr{Y}$ are said to be *isomorphic* if there exists a one-to-one mapping f of $\mathscr{X}$ onto $\mathscr{Y}$, such that the following conditions be satisfied

(i) for any elements $x_1, x_2 \in \mathscr{X}$ and any numbers $\alpha, \beta \in \mathbb{R}$ we have

$$f(\alpha x_1 + \beta x_2) = \alpha f(x_1) + \beta f(x_2);$$

(ii) if $x_1, x_2 \in \mathscr{X}$, then

$$x_1 \leqslant x_2 \Leftrightarrow f(x_1) \leqslant f(x_2).$$

3.1.2 The notion of directed vector space

If an ordered vector space is a set directed to the right, then it is obviously directed to the left, and conversely.

A *directed vector space* is any ordered vector space which is a directed set.

In order that an ordered vector space be a directed vector space it is necessary and sufficient that for any element x there should exist a positive element x', such that $x' \geqslant x$.

Indeed, if x_1 and x_2 are any elements and if a is a positive element such that $x_1 - x_2 \leqslant a$, then $x_1 \leqslant a + x_2$ and $x_2 \leqslant a + x_2$.

If $\mathscr{X}$ is a directed vector space and if $x \in \mathscr{X}$, then x may be represented as $x = x_1 - x_2$, where x_1 and x_2 are positive.

Indeed, there exists a positive element $x_1 \geqslant x$ and one can take $x_2 = x_1 - x$.

Conversely, if in an ordered vector space $\mathscr{X}$ any element may be represented as the difference of two positive elements, then $\mathscr{X}$ is directed.

Indeed, if $x \in \mathscr{X}$, and $x = x_1 - x_2$, where x_1 and x_2 are positive elements, then $x_1 \geqslant x$, and this implies that $\mathscr{X}$ is directed.

It is also easy to verify that an ordered vector space $\mathscr{X}$ is directed iff the linear hull of the set of all positive elements of $\mathscr{X}$ coincides with $\mathscr{X}$.

3.1.3 Archimedean spaces

Definition. A directed vector space $\mathscr{X}$ is called an *Archimedean space*[2]) if for any element x, which is not negative, the set $\{\alpha x;\ 0<\alpha \in \mathbb{R}\}$ is not bounded from above.

Proposition 1. *In order that a directed vector space be Archimedean it is necessary and sufficient that from the conditions* $\mathbf{0} < x \in \mathscr{X}$ $0 < \alpha_j \in \mathbb{R} (j \in J)$ *and* $\bigwedge_{j \in J} \alpha_j = 0$ *one could infer that*

$$\mathbf{0} = \bigwedge_{j \in J} (\alpha_j x). \tag{1}$$

Proof. Let us assume that the space $\mathscr{X}$ is Archimedean. If $\mathbf{0} < x \in \mathscr{X}$ and $0 < \alpha \in \mathbb{R}$, then $\alpha x > 0$. If $y \leqslant \alpha x$, for any $\alpha > 0$, then $\lambda y \leqslant x$ for any $\lambda > 0$. By taking into account the definition of the Archimedean space, it follows that $y \leqslant 0$, hence

$$\mathbf{0} = \bigwedge_{0<\alpha \in \mathbb{R}} \alpha x.$$

If x and α_j are as in the statement of the proposition and if y is an element such that

$$y \leqslant \alpha_j x, \ (\forall j \in J),$$

---

[2]) The notion of Archimedean ordered vector space may be introduced without assuming the ordered vector space to be directed.

then $y \leqslant \alpha x$ for any $\alpha > 0$, hence $y \leqslant 0$. Consequently, equality (1) holds.

Conversely, let $\mathscr{X}$ be a directed vector space and let us assume that for any two elements $x, y \in \mathscr{X}$ we have $\alpha x \leqslant y$, for any $\alpha > 0$. Since the space is directed, one can assume that $y \geqslant \mathbf{0}$ and we have $x \leqslant \frac{1}{\alpha} y$ for any $\alpha > 0$. If the condition in the statement of the proposition is satisfied, it follows that $x \leqslant \mathbf{0}$, and so the space $\mathscr{X}$ is Archimedean.

Corollary. *In order that a directed vector space be Archimedean, it is necessary and sufficient that for any element* $x > \mathbf{0}$ *one should have* $\frac{1}{n} x \underset{n \in \mathbb{N}}{\downarrow} \mathbf{0}$.

Proposition 2. *In order that a directed vector space* $\mathscr{X}$ *be Archimedean, it is necessary and sufficient that it should satisfy the following condition*

$$\alpha_n x \leqslant y \textit{ and } \alpha_n \to \alpha \Rightarrow \alpha x \leqslant y.$$

Proof. Let $\mathscr{X}$ be Archimedean and assume that $\alpha_n \uparrow \alpha$ $(\alpha_n, \alpha \in \mathbb{R})$. If we denote $\varepsilon_n = \alpha - \alpha_n$, then $\varepsilon_n \downarrow 0$ and, with proposition 1, we infer that $\mathbf{0} = \bigwedge_{n \in \mathbb{N}} \varepsilon_n z$, for any $z \geqslant \mathbf{0}$. If $\alpha_n x \leqslant y$, then by considering a positive element $z \geqslant x$, we have

$$\alpha x - y \leqslant \varepsilon_n z$$

and, consequently, $\alpha x - y \leqslant \mathbf{0}$, i.e., $\alpha x \leqslant y$.

If $\alpha_n \downarrow \alpha$, then $-\alpha_n \uparrow -\alpha$ and from $\alpha_n x = (-\alpha_n)(-x) \leqslant y$ it follows that $\alpha x \leqslant y$.

If $\{\alpha_n\}_{n \in \mathbb{N}}$ is not monotonous, then there exists a monotonous subsequence $\alpha_{j_n} \to \alpha$.

Conversely, let us assume that the condition from the statement of the proposition is satisfied. From $\alpha x \leqslant y$, $(\alpha > 0)$, it follows that $\frac{1}{\alpha}(-y) \leqslant -x$. From

$$\frac{1}{n}(-y) \leqslant -x, \quad (\forall n \in N),$$

we infer that $\mathbf{0} \leqslant -x$, i.e., $x \leqslant \mathbf{0}$. This implies that $\mathscr{X}$ is Archimedean.

Corollary. *In order that a directed vector space* $\mathscr{X}$ *be Archimedean, it is necessary and sufficient that for any element* $x \nleqslant \mathbf{0}$, *the sequence* $\{nx\}_{n \in \mathbb{N}}$ *be not bounded from above.*

Indeed, if the space is Archimedean, then from $nx \leqslant y$, i.e., from $\frac{1}{n}(-y) \leqslant -x$, for any $n \in \mathbb{N}$, one infers that $\mathbf{0} \leqslant -x$, that is, $x \leqslant \mathbf{0}$, if one also takes into consideration proposition 2.

The sufficiency of the condition is obvious.

Proposition 3. *If $\mathscr{X}$ is an Archimedean directed vector space, then for any element $x > \mathbf{0}$ of $\mathscr{X}$ and any bounded from above family $\{\alpha_j\}_{j \in J}$ of real numbers, there exists the element $\bigvee_{j \in J} (\alpha_j x)$ and the following equality holds*

$$\bigvee_{j \in J} (\alpha_j x) = (\bigvee_{j \in J} \alpha_j)\, x. \tag{2}$$

Proof. Let $x > \mathbf{0}$ and

$$\alpha = \bigvee_{j \in J} \alpha_j.$$

Obviously, $\alpha_j x \leqslant \alpha x$, for any $j \in J$. Now let y be an element such that

$$\alpha_j x \leqslant y, \ (\forall j \in J).$$

Let $\alpha_{j_n} \to \alpha$. From

$$\alpha_{j_n} x \leqslant y, \ (\forall n \in N),$$

by taking into account proposition 2, we infer that $\alpha x \leqslant y$. Consequently, equality (2) holds.

Remark. An analogous proposition holds, by replacing the phrase "bounded from above" by "bounded from below" and the sign $\bigvee$ by the sign $\bigwedge$.

3.1.4 The topology of (o)-boundedness

Let $\mathscr{X}$ be an ordered vector space.

One calls the *topology of* (o)*-boundedness* in the space $\mathscr{X}$, and one denotes it by θ, the finest locally convex topology in $\mathscr{X}$ such that any (o)-bounded set be bounded for the topology.

We shall denote by $\mathscr{M}_o$ the set of all (o)-bounded subsets of $\mathscr{X}$. If τ is any topology in $\mathscr{X}$, we shall denote by $\mathscr{M}_\tau$ the set of all (τ)-bounded subsets of $\mathscr{X}$.

A fundamental system of neighbourhoods of the origin for the topology θ is the set $\mathscr{W}$ of all equilibrated and convex subsets of $\mathscr{X}$ which absorb any (o)-bounded set.

Indeed, any set from $\mathscr{W}$ is absorbent. If W_1 and $W_2 \in \mathscr{W}$, then $W_1 \cap W_2 \in \mathscr{W}$, as one can immediately check. If $W \in \mathscr{W}$ and $0 \neq \alpha \in \mathbb{R}$, then $\alpha W \in \mathscr{W}$. In particular, if $W \in \mathscr{W}$ and if we put $W_1 = \frac{1}{2} W$, then $W_1 + W_1 = W$. It follows that $\mathscr{W}$ is a fundamental system of neighbourhoods of the origin for a locally convex topology τ_0. It is obvious that $\mathscr{M}_o \subset \mathscr{M}_{\tau_0}$.

Let now τ be a locally convex topology such that $\mathscr{M}_o \subset \mathscr{M}_\tau$. If V is an equilibrated and convex neighbourhood of the origin for the topology τ, and if $A \in \mathscr{M}_o$, then $A \in \mathscr{M}_\tau$, and this shows that V absorbs A. It follows that V is a neighbourhood of the origin for the topology τ_0 and this shows that $\tau \leqslant \tau_0$. By taking into account the definition of θ, we infer that $\tau_0 = \theta$.

For the topology θ, it is obvious that any ordered vector space is a bornological space.

3.2 Vector lattices

3.2.1 The notion of vector lattice. General properties

Definition. One calls a *vector lattice* any ordered vector space, which is a lattice.

Proposition 1. *If $\mathscr{X}$ is a vector lattice, then for any elements $x_1, x_2 \in \mathscr{X}$ the following equality holds*

$$x_1 + x_2 = (x_1 \vee x_2) + (x_1 \wedge x_2). \tag{1}$$

Indeed, by taking into account the propositions from Section 3.1.1, we have

$$x_1 - (x_1 \vee x_2) = x_1 + ((-x_1) \wedge (-x_2)) = \mathbf{0} \wedge (x_1 - x_2)$$

$$= (x_1 \wedge x_2) - x_2,$$

and relation (1) is thus proved.

Proposition 2. *In any vector lattice the infinite distributivity laws hold.*

Proof. Let $\{x_j\}_{j \in J}$ be an arbitrary family of elements and assume that its least upper bound exists. Let

$$a = \bigvee_{j \in J} x_j.$$

If x is any element of the space, then

$$x \wedge x_j \leqslant x \wedge a, \qquad (\forall j \in J).$$

Let y be an element such that

$$x \wedge x_j \leqslant y, \qquad (\forall j \in J).$$

By taking into account proposition 1, one can write

$$x + x_j - (x \vee x_j) \leqslant y, \qquad (\forall j \in J),$$

i.e.,

$$x_j \leqslant y - x + (x \vee x_j), \qquad (j \in J).$$

Hence

$$a \leqslant y - x + x \vee a,$$

that is,

$$x + a - (x \vee a) \leqslant y$$

and, by taking into account proposition 1 again, we get $x \wedge a \leqslant y$. Consequently,

$$x \wedge a = \bigvee_{j \in J} (x \wedge x_j),$$

or, with another notation,

$$x \wedge (\bigvee_{j \in J} x_j) = \bigvee_{j \in J} (x \wedge x_j).$$

By taking into account proposition 2 from Section 3.1.1, it follows that if $\bigwedge_{j \in J} x_j$ exists, then $\bigwedge_{j \in J} (x \vee x_j)$ exists too and the following equality holds

$$x \vee (\bigwedge_{j \in J} x_j) = \bigwedge_{j \in J} (x \vee x_j).$$

Proposition 3. *Let $\mathscr{X}$ be a vector lattice. If*

$$\sum_{i=1}^{n} x_i = \sum_{j=1}^{m} y_j,$$

where x_i, y_j are positive elements of $\mathscr{X}$, then there exist positive elements z_{ij} ($i = 1, 2, \ldots$ $\ldots, n; j = 1, 2, \ldots, m$), such that

$$\sum_{j=1}^{m} z_{ij} = x_i, \qquad \sum_{i=1}^{n} z_{ij} = y_j.$$

Proof. We shall prove the proposition for $m = 2$. For an arbitrary m, the proposition can be proved by induction. Let then

$$\sum_{i=1}^{n} x_i = y_1 + y_2$$

and let us show that there exist z_{i1}, z_{i2}, ($i = 1, 2, \ldots, n$), such that

$$z_{i1} + z_{i2} = x_i, \qquad (i = 1, 2, \ldots, n),$$

$$\sum_{i=1}^{n} z_{i1} = y_1, \qquad \sum_{i=1}^{n} z_{i2} = y_2.$$

For $n = 1$, that is, for

$$x = y_1 + y_2,$$

the proposition is obviously true, because one can take $z_{11} = y_1$, $z_{12} = y_2$.

Let us assume that the proposition is true for a natural number n and let

$$x = \sum_{i=1}^{n+1} x_i = y_1 + y_2, \tag{2}$$

where $x_i \geqslant \mathbf{0}$ and $y_1, y_2 \geqslant \mathbf{0}$.

Let us put

$$z_{n+1,1} = x_{n+1} \wedge y_1.$$

From (2) one infers that

$$\sum_{i=1}^{n} x_i = x - x_{n+1} \geqslant \mathbf{0} \vee (y_1 - x_{n+1}), \tag{3}$$

because $x - x_{n+1} \geqslant \mathbf{0}$ and $x \geqslant y_1$ (consequently $x - x_{n+1} \geqslant y_1 - x_{n+1}$). By taking into account the propositions from Section 3.1.1, we get

$$\mathbf{0} \vee (y_1 - x_{n+1}) = y_1 + ((-y_1) \vee (-x_{n+1})) = y_1 - (y_1 \wedge x_{n+1}) = y_1 - z_{n+1,1}.$$

From (3) we then have

$$\sum_{i=1}^{n} x_i \geqslant y_1 - z_{n+1,1} \geqslant \mathbf{0} \tag{4}$$

and, from the definition of $z_{n+1,1}$, we get

$$\mathbf{0} \leqslant z_{n+1,1} \leqslant x_{n+1}. \tag{5}$$

Let us now put

$$z_{n+1,2} = x_{n+1} - z_{n+1,1}.$$

From (5) we have $z_{n+1,2} \geqslant \mathbf{0}$. On the other hand, from

$$x_{n+1} = z_{n+1,1} + z_{n+1,2} \tag{6}$$

and from (2) we get

$$\sum_{i=1}^{n} x_i = (y_1 - z_{n+1,1}) + (y_2 - z_{n+1,2});$$

whereas from (4) we get $y_2 - z_{n+1,2} \geqslant \mathbf{0}$.

By virtue of the hypothesis, there exist positive elements z_{i1}, z_{i2}, $(i=1, 2, \ldots, n)$, such that

$$x_i = z_{i1} + z_{i2}, \qquad (i = 1, 2, \ldots, n), \tag{7}$$

and such that

$$\sum_{i=1}^{n} z_{i1} = y_1 - z_{n+1,1}, \qquad \sum_{i=1}^{n} z_{i2} = y_2 - z_{n+1,2}.$$

It follows that

$$\sum_{i=1}^{n+1} z_{ij} = y_j, \qquad (j = 1, 2),$$

and from (6) and (7) we get that

$$z_{i1} + z_{i2} = x_i, \qquad (i = 1, 2, \ldots, n+1).$$

The proposition is therefore proved for any n.

Corollary. *If x_1, x_2 are positive elements and if $\mathbf{0} \leqslant y \leqslant x_1 + x_2$, then there exist elements y_1, y_2, such that $\mathbf{0} \leqslant y_1 \leqslant x_1$, $\mathbf{0} \leqslant y_2 \leqslant x_2$ and $y = y_1 + y_2$.*

Proposition 4. *If $x_1, x_2, \ldots, x_n$ are elements of a vector lattice, such that $x_i \wedge x_j = \mathbf{0}$ for $i \neq j$, then*

$$\sum_{i=1}^{n} x_i = \bigvee_{i=1}^{n} x_i. \tag{8}$$

Proof. For $n = 2$, equality (8) immediately follows from equality (1).

Let now (8) be true for a given natural number n and let us consider the elements $x_1, x_2, \ldots, x_{n+1}$, such that $x_i \wedge x_j = \mathbf{0}$ if $i \neq j$. By putting

$$x = \sum_{i=1}^{n} x_i = \bigvee_{i=1}^{n} x_i,$$

we have $x \wedge x_{n+1} = \mathbf{0}$, whence

$$\bigvee_{i=1}^{n+1} x_i = \left(\bigvee_{i=1}^{n} x_i \right) \vee x_{n+1} = x \vee x_{n+1} = x + x_{n+1} = \sum_{i=1}^{n+1} x_i.$$

The proposition is proved.

3.2.2 The positive part, the negative part and the absolute value of an element

Let $\mathscr{X}$ be a vector lattice.

Definition. Let x be any element of $\mathscr{X}$. Then the *positive part of* x is, by definition, the element $x_+ = x \vee \mathbf{0}$; *the negative part* of x is the element $x_- = (-x) \vee \mathbf{0}$; the *absolute value* of x is the element $|x| = x \vee (-x)$.

From the definition one immediately infers that the elements x_+, x_- and $|x|$ are positive. Also, the following equalities hold

$$x = x_+ - x_-, \tag{1}$$

$$|x| = x_+ + x_-. \tag{2}$$

Indeed, by taking into account proposition 1 from Section 3.2.1, we have

$$x_+ - x_- = (x \vee \mathbf{0}) - (-(x \wedge \mathbf{0})) = (x \vee \mathbf{0}) + (x \wedge \mathbf{0}) = x,$$

i.e., equality (1) holds. Equality (2) can be proved in the following way

$$|x| = x \vee (-x) = (2x \vee \mathbf{0}) - x = 2x_+ - (x_+ - x_-) = x_+ + x_-.$$

Proposition 1. *For any element x of the vector lattice $\mathscr{X}$, from $x = x_1 - x_2$ and $x_1 \wedge x_2 = \mathbf{0}$ one infers that $x_1 = x_+$, $x_2 = x_-$, and conversely.*

Proof. Let us first show that $x_+ \wedge x_- = \mathbf{0}$. Let $a = x_+ \wedge x_-$ and let us define

$$b = x_+ - a, \qquad c = x_- - a.$$

The elements b, c are obviously positive and since $x = b - c$, one infers that $b \geqslant x_+$. But then $-a \geqslant \mathbf{0}$ and, from $a \geqslant \mathbf{0}$, one gets $a = \mathbf{0}$. Let us now assume that $x = x_1 - x_2$, where $x_1 \wedge x_2 = \mathbf{0}$. Since the elements x_1, x_2 are positive, it follows that $x_1 \geqslant x_+$. By defining $y = x_1 - x_+$, we get $x_1 = y + x_+$ and, therefore, $x_2 = y + x_-$. But in this case

$$x_1 \wedge x_2 \geqslant y \geqslant \mathbf{0},$$

whence $y = \mathbf{0}$. It follows that $x_1 = x_+$ and $x_2 = x_-$.

Remark. From $x_+ \wedge x_- = \mathbf{0}$, from equality (2) and from proposition 1, Section 3.2.1, we deduce

$$|x| = x_+ \vee x_-.$$

The following two propositions are easy to prove.

Proposition 2. *For any elements $x, y \in \mathscr{X}$, and any number $\alpha > 0$, the following relations hold:*

$$(x + y)_+ \leqslant x_+ + y_+; \qquad (x + y)_- \leqslant x_- + y_-;$$

$$(\alpha x)_+ = \alpha x_+; \qquad (\alpha x)_- = \alpha x_-.$$

Proposition 3. *If* $\bigvee_{j \in J} x_j$ *exists, then* $\bigvee_{j \in J} (x_j)_+$ *and* $\bigwedge_{j \in J} (x_j)_-$ *exist too, and the following equalities hold*

$$(\bigvee_{j \in J} x_j)_+ = \bigvee_{j \in J} (x_j)_+$$

$$(\bigvee_{j \in J} x_j)_- = \bigwedge_{j \in J} (x_j)_- .$$

Proposition 4. *In order that an ordered vector space $\mathscr{X}$ be a vector lattice it is necessary and sufficient that for any element $x \in \mathscr{X}$ the element $x_+ = x \vee \mathbf{0}$ should exist.*

Indeed, if $x, y \in \mathscr{X}$, then

$$(x - y)_+ + y = ((x - y) \vee \mathbf{0}) + y = x \vee y$$

$$-(x - y)_+ + x = x \wedge y.$$

Proposition 5. *In a vector lattice, the absolute value has the following properties:*

(i) $|x + y| \leqslant |x| + |y|$;

(ii) $|\alpha x| = |\alpha|\,|x|$;

(iii) $||x| - |y|| \leqslant |x - y|$;

(iv) $|x - y| = (x \vee y) - (x \wedge y)$;

(v) $|x - y| = |(x \vee z) - (y \vee z)| + |(x \wedge z) - (y \wedge z)|$,

for any elements x, y, z.

Proof. (i) It is an immediate consequence of proposition 2.

(ii) For $\alpha > 0$, the property is obvious; for $\alpha = -1$, it is an immediate consequence of the definition of the absolute value; for $\alpha < 0$, it is an immediate consequence of the equality $|\alpha x| = |-\alpha x|$.

(iii) From the equality

$$x = (x - y) + y$$

it follows that

$$|x| \leqslant |x - y| + |y|,$$

hence

$$|x| - |y| \leqslant |x - y|.$$

By interchanging the rôles of the elements x and y, we get

$$|y| - |x| \leqslant |x - y|.$$

One has now to apply the definition of the absolute value.

(iv) We have first

$$| x - y | = (x - y)_+ + (x - y)_-. \tag{3}$$

On the other hand

$$(x - y)_+ + y = x \vee y \tag{4}$$

$$(x - y)_- - y = -((x - y) \wedge \mathbf{0}) - y = -(x \wedge y). \tag{5}$$

We must now take into account equalities (4) and (5) in (3).

(v) With the preceding property, we have

$$| (x \vee z) - (y \vee z) | = (x \vee y \vee z) - (x \wedge y) \vee z$$

$$| (x \wedge z) - (y \wedge z) | = (x \vee y) \wedge z - (x \wedge y \wedge z).$$

We have now only to observe that, by using proposition 1 from Section 3.2.1, we have

$$(x \vee y \vee z) - (x \wedge y) \vee z + (x \vee y) \wedge z - (x \wedge y \wedge z)$$

$$= ((x \vee y) + z) - ((x \wedge y) + z) = x \vee y - (x \wedge y) = | x - y |.$$

Corollary. *For any elements x, y, z the following inequalities hold*

$$| (x \vee z) - (y \vee z) | \leqslant | x - y |,$$

$$| (x \wedge z) - (y \wedge z) | \leqslant | x - y |.$$

Proposition 6. *In a vector lattice $\mathscr{X}$ a set E is bounded iff the set $E' = \{| x |; x \in E\}$ is bounded.*

Proof. If E is bounded, then there exist $a, b \in \mathscr{X}$, such that $a \leqslant x \leqslant b$, for any $x \in E$. From $a \leqslant x \leqslant b$ we infer that $x \leqslant | b |$ and $-x \leqslant | a |$ and, therefore, $| x | \leqslant | a | \vee | b |$. It follows that the set E' is bounded.

Conversely, if E' is bounded, there exists a $c \in \mathscr{X}$, such that $| x | \leqslant c$, for any $x \in E'$. From $| x | \leqslant c$ one gets $x \leqslant c$ and $-x \leqslant c$, hence $-c \leqslant x \leqslant c$. The set E is therefore bounded.

3.2.3 Orthogonality. Discrete elements

Let $\mathscr{X}$ be a vector lattice.

Definition 1. Two elements $x_1, x_2 \in \mathscr{X}$ are said to be *orthogonal* if $| x_1 | \wedge | x_2 | = \mathbf{0}$. One writes then $x_1 \perp x_2$.

The definition can be extended to the subsets of $\mathscr{X}$. Two such subsets A_1, A_2 are said to be orthogonal if $x_1 \perp x_2$ for any $x_1 \in A_1$, $x_2 \in A_2$.

Lemma. *For any positive elements x, x_1, x_2 of $\mathscr{X}$ the inequality*

$$x \wedge (x_1 + x_2) \leqslant (x \wedge x_1) + (x \wedge x_2)$$

holds.

Proof. Let us define

$$y = x \wedge (x_1 + x_2).$$

Since

$$0 \leqslant y \leqslant x_1 + x_2,$$

in virtue of the corollary of proposition 3 from Section 3.2.1, there exist elements y_1, y_2, such that $y = y_1 + y_2$ and

$$0 \leqslant y_i \leqslant x_i, \quad (i = 1,2).$$

From the inequality $y \leqslant x$ one gets $y_1 \leqslant x$ and $y_2 \leqslant x$, which imply

$$y_i \leqslant x \wedge x_i, \quad i = 1,2,$$

whence

$$y = y_1 + y_2 \leqslant (x \wedge x_1) + (x \wedge x_2).$$

Proposition 1. *If $x \perp x_1$ and $x \perp x_2$, then $x \perp \alpha x_1 + \beta x_2$, for any $\alpha, \beta \in \mathbb{R}$.*

Proof. Indeed, from $|x| \wedge |x_1| = 0$, one gets

$$|x| \wedge |\alpha x_1| \leqslant (\max(1, |\alpha|))\,(|x| \wedge |x_1|) = 0$$

and, therefore, $|x| \wedge |\alpha x_1| = 0$. Analogously, one has $|x| \wedge |\beta x_2| = 0$. From the preceding lemma, one gets

$$|x| \wedge (|\alpha x_1| + |\beta x_2|) = 0$$

and, consequently

$$|x| \wedge |\alpha x_1 + \beta x_2| = 0.$$

Proposition 2. *If $x \perp y$, then*

$$(x + y)_+ = x_+ + y_+ \tag{1}$$

$$(x + y)_- = x_- + y_- \tag{2}$$

$$|x + y| = |x| + |y|. \tag{3}$$

Proof. If $|x| \wedge |y| = \mathbf{0}$, then $x_+ \wedge y_- = \mathbf{0}$. On the other hand, $x_+ \wedge x_- = \mathbf{0}$, With the lemma, we get

$$x_+ \wedge (x_- + y_-) = \mathbf{0}.$$

In the same manner, we get the equality

$$y_+ \wedge (x_- + y_-) = \mathbf{0}.$$

From the same lemma, we infer that

$$(x_+ + y_+) \wedge (x_- + y_-) = \mathbf{0}.$$

From the equality

$$x + y = (x_+ + y_+) - (x_- + y_-)$$

and from proposition 1, Section 3.2.2, we get equalities (1) and (2). Equality (3) now immediately follows from (1) and (2).

Proposition 3. *If* $x = \bigvee_{j \in J} x_j$ *and* $a \perp x_j$, $(\forall j \in J)$, *then* $a \perp x$.

Proof. If $x_j \geqslant \mathbf{0}$ for any $j \in J$, then, by taking into account proposition 2 from Section 3.2.1, we get

$$|a| \wedge x = \bigvee_{j \in J} (|a| \wedge x_j) = \mathbf{0}.$$

Let us now assume that the elements x_j are arbitrary. From the existence of $\bigvee_{j \in J} x_j$ we infer the existence of $\bigvee_{j \in J} (x_j)_+$ and of $\bigwedge_{j \in J} (x_j)_-$ (proposition 3, Section 3.2.2) and the following equalities hold

$$x_+ = \bigvee_{j \in J} (x_j)_+, \quad x_- = \bigwedge_{j \in J} (x_j)_-.$$

Since the elements $(x_j)_+$ are positive, from $a \perp x_j$, hence $a \perp (x_j)_+$, one infers that $a \perp x_+$. On the other hand, from $a \perp (x_j)_-$ we obviously infer that $a \perp x_-$. From proposition 1 we conclude that $a \perp x$.

Remark. If $x \perp y$, then $x \vee y \geqslant 0$. Indeed, we have

$$(x \vee y)_- = x_- \wedge y_- = \mathbf{0}.$$

Definition 2. An element $e \in \mathscr{X}$ is said to be *discrete* if there exist no orthogonal elements a, b, such that $\mathbf{0} < a \leqslant |e|$ and $\mathbf{0} < b \leqslant |e|$.

Zero is obviously a discrete element.

A discrete element is either positive, or negative, since

$$\mathbf{0} \leqslant e_+ \leqslant |e| \text{ and } \mathbf{0} \leqslant e_- \leqslant |e|, \text{ whereas } e_+ \perp e_-.$$

Proposition 4. *Let $\mathscr{X}$ be an Archimedean vector lattice and e a discrete element of $\mathscr{X}$. If $0 < x \leqslant |e|$, then there exists a number α, such that $e = \alpha x$.*

Proof. Without loss of generality we can assume that $e > \mathbf{0}$. Since the space $\mathscr{X}$ is Archimedean, the set of all numbers μ, such that $\mu x \leqslant e$ is bounded from above and, by putting

$$\alpha = \sup \{\mu; \mu x \leqslant e\},$$

we have $\alpha x \leqslant e$.

Let us now define $v = e - \alpha x$ and let $\lambda \in \mathbb{R}$ be such that $0 < \lambda \leqslant 1$. By denoting $z = v \wedge \lambda x$, we have

$$(v - z) \wedge (\lambda x - z) = (v \wedge \lambda x) - z = \mathbf{0}.$$

On the other hand, the following inequalities hold

$$\mathbf{0} \leqslant v - z \leqslant e,$$

$$\mathbf{0} \leqslant \lambda x - z \leqslant e,$$

whereas $\lambda x \neq z$ because otherwise $(\lambda + \alpha) x \leqslant e$, which leads to a contradiction. Since the element e is discrete, we then infer that $v = z$ and, therefore, $v \leqslant \lambda x$. Since λ was arbitrary in the interval (0,1] and since $\mathscr{X}$ is Archimedean, it follows that $v \leqslant \mathbf{0}$. Since v is also positive, it follows that $v = \mathbf{0}$ and this shows that $e = \alpha x$.

Corollary. *If e_1, e_2 are two discrete elements of an Archimedean vector lattice, then they are either orthogonal, or there exists a number α such that $e_1 = \alpha e_2$.*

Indeed, by putting $a = |e_1| \wedge |e_2|$ and, assuming that $a \neq \mathbf{0}$, from $\mathbf{0} < a \leqslant |e_1|$ and $\mathbf{0} < a \leqslant |e_2|$, we get $a = \alpha_1 e_1$ and $a = \alpha_2 e_2$.

3.2.4 Convergence with respect to the order relation in vector lattices

Since in a vector lattice the infinite distributivity laws hold, from what we have proved in section 1.3.1 (proposition 2) it follows that:

If in a vector lattice $x_n \xrightarrow{o} x$ and $y_n \xrightarrow{o} y$, then $x_n \vee y_n \xrightarrow{o} x \vee y$ and $x_n \wedge y_n \xrightarrow{o} x \wedge y$.

From this we immediately infer that if

$$x_n \xrightarrow{o} x, \text{ then } (x_n)_+ \xrightarrow{o} x_+, (x_n)_- \xrightarrow{o} x_-, |x_n| \xrightarrow{o} |x|.$$

Proposition 1. *If $\{x_n\}_{n \in \mathbb{N}}$ is a sequence of elements in a vector lattice, then $x_n \xrightarrow{o} x$ iff there exists a sequence $\{r_n\}_{n \in \mathbb{N}}$ of elements of $\mathscr{X}$, such that $r_n \downarrow 0$ and*

$$|x_n - x| \leqslant r_n, \ (\forall\, n \in \mathbb{N}). \tag{1}$$

Proof. If $x_n \xrightarrow{o} x$, then $x_n - x \xrightarrow{o} \mathbf{0}$ and, therefore, $|x_n - x| \xrightarrow{o} \mathbf{0}$. There then exists a sequence $\{r_n\}_{n\in\mathbb{N}}$ with the desired properties. Conversely, if (1) holds, where $r_n \downarrow 0$, then $|x_n - x| \xrightarrow{o} 0$. From the inequalities

$$-r_n \leqslant -|x_n - x| \leqslant x_n - x \leqslant |x_n - x| \leqslant r_n$$

and from $-r_n \uparrow 0$ we infer that $x_n - x \xrightarrow{o} 0$ and hence $x_n \xrightarrow{o} x$.

Proposition 2. *If $\{x_n\}_{n\in\mathbb{N}}$ is a sequence of elements belonging to an Archimedean vector lattice, and if $\{\alpha_n\}_{n\in\mathbb{N}}$ is a sequence of real numbers, then from $x_n \xrightarrow{o} x$ and $\alpha_n \to \alpha$ it follows that $\alpha_n x_n \xrightarrow{o} \alpha x$.*

Proof. Let β be a number such that $\beta \geqslant |\alpha_n|$, for any n. Let $\{r_n\}_{n\in\mathbb{N}}$ be a sequence as in proposition 1, and let $\{\gamma_n\}_{n\in\mathbb{N}}$ be a sequence of numbers, such that $\gamma_n \downarrow 0$ and $|\alpha_n - \alpha| \leqslant \gamma_n$. The following relations then hold

$$|\alpha_n x_n - \alpha x| \leqslant |\alpha_n x_n - \alpha_n x| + |\alpha_n x - \alpha x| = |\alpha_n|\,|x_n - x| + |\alpha_n - \alpha|\,|x| \leqslant \beta r_n + \gamma_n|x|.$$

By putting

$$v_n = \beta r_n + \gamma_n|x|,$$

we have

$$|\alpha_n x_n - \alpha x| \leqslant v_n, \quad (\forall\, n \in \mathbb{N}).$$

Since the space is Archimedean, we have $\gamma_n|x| \downarrow \mathbf{0}$. On the other hand, $\beta r_n \downarrow \mathbf{0}$ and this implies that $v_n \downarrow \mathbf{0}$. From proposition 1 one gets that $\alpha_n x_n \xrightarrow{o} \alpha x$.

Definition 1. A generalized sequence $\{x_\delta\}_{\delta\in\Delta}$ of elements of a vector lattice $\mathscr{X}$ is said to (ω)-converge to an element x, if there exists a generalized sequence $\{r_\lambda\}_{\lambda\in\Lambda}$ of elements of $\mathscr{X}$, such that $r_\lambda \downarrow \mathbf{0}$ and such that for any $\lambda \in \Lambda$ there exists an index $\delta' \in \Delta$, such that

$$|x_\delta - x| \leqslant r_\lambda,$$

for any $\delta \geqslant \delta'$.

In this case one writes $x_\delta \xrightarrow[\delta\in\Delta]{\omega} x$ or $x = (\omega)\text{-}\lim\limits_{\delta\in\Delta} x_\delta$.

Definition 2. Let $\mathscr{X}$ be an Archimedean vector lattice and $\{x_\delta\}_{\delta\in\Delta}$ a generalized sequence of elements of $\mathscr{X}$. Then the generalized sequence is said to *(ρ)-converge* to x if there exists an element $r \in \mathscr{X}$, such that for any number $\varepsilon > 0$ there exists a $\delta' \in \Delta$, having the property that

$$|x_\delta - x| \leqslant \varepsilon r,$$

for any $\delta \geqslant \delta'$.

In this case one writes $x_\delta \xrightarrow[\delta\in\Delta]{\rho} x$ or $x = (\rho)\text{-}\lim\limits_{\delta\in\Delta} x_\delta$. The element r which appears in definition 2 is called a *convergence regulator* of the generalized sequence $\{x_\delta\}_{\delta\in\Delta}$, and the (ρ)-convergence is also called *convergence with regulator.*

In an Archimedean vector lattice, from $x_\delta \xrightarrow[\delta\in\Delta]{\rho} x$ one infers that $x_\delta \xrightarrow[\delta\in\Delta]{\omega} x$.

Indeed, the role of the generalized sequence $\{r_\lambda\}_{\lambda\in\Lambda}$ from definition 1 may be taken by the sequence $\left\{\frac{1}{n} r\right\}_{n\in\mathbb{N}}$.

The convergence with regulator can be considered, in particular, for usual sequences. If $\{x_n\}_{n\in\mathbb{N}}$ is a sequence of elements of an Archimedean vector lattice, then from $x_n \xrightarrow{\rho} x$ one can infer[3]) that $x_n \xrightarrow{o} x$. The converse is generally not true.

3.2.5 The notion of σ-complete vector lattice

Definition. One calls a *σ-complete vector lattice* any ordered vector space which is a relatively σ-complete lattice.

Proposition 1. *In order that a vector lattice $\mathscr{X}$ be a σ-complete vector lattice it is necessary and sufficient that any of the following two equivalent conditions be satisfied:*

(i) *If $\{x_n\}_{n\in\mathbb{N}}$ is an increasing, bounded from above, sequence of positive elements, then the element $\bigvee\limits_{n\in\mathbb{N}} x_n$ exists;*

(ii) *If $\{x_n\}_{n\in\mathbb{N}}$ is a decreasing sequence of positive elements, then $\bigwedge\limits_{n\in\mathbb{N}} x_n$ exists.*

Proof. Let $\mathscr{X}$ be a vector lattice and $\{x_n\}_{n\in\mathbb{N}}$ an arbitrary, bounded from above, sequence of elements of $\mathscr{X}$. If $x_n \leqslant a$ and if we put

$$x'_n = \bigvee_{i=1}^{n} x_i, \quad n \in \mathbb{N},$$

then

$$0 \leqslant x'_n - x_1 \leqslant a - x_1,$$

and, therefore, the sequence $\{x'_n - x_1\}_{n\in\mathbb{N}}$ is bounded from above and it consists of positive elements. This sequence is also increasing. If condition (i) is satisfied, then there exists the element

$$x' = \bigvee_{n\in\mathbb{N}} (x'_n - x_1).$$

---

[3]) If we denote by ε_n the smallest number ε such that $|x_n - x| \leqslant \varepsilon r$, then $\varepsilon_n \to 0$ and, therefore, $\varepsilon_n r \xrightarrow{o} 0$.

It follows that

$$x' + x_1 = \bigvee_{n\in\mathbb{N}} x'_n = \bigvee_{n\in\mathbb{N}} x_n,$$

if we take into account proposition 3 from Section 3.1.1.

If $\{x_n\}_{n\in\mathbb{N}}$ is a bounded from below sequence of elements of $\mathscr{X}$, then the sequence $\{-x_n\}_{n\in\mathbb{N}}$ is bounded from above. From what we have shown it follows that there exists the element $\bigwedge_{n\in\mathbb{N}} x_n$ (Section 3.1.1, proposition 2).

It follows that if (i) is satisfied, then $\mathscr{X}$ is a σ-complete vector lattice.

Condition (ii) implies condition (i).

Indeed, let $\{x_n\}_{n\in\mathbb{N}}$ be an increasing and bounded from above sequence of positive elements. If $x_n \leqslant a$ for any $n \in \mathbb{N}$, then the sequence $\{a - x_n\}_{n\in\mathbb{N}}$ is a decreasing sequence of positive elements. By putting

$$b = \bigwedge_{n\in\mathbb{N}} (a - x_n)$$

it follows that

$$a - b = \bigvee_{n\in\mathbb{N}} x_n$$

(Section 3.1.1, propositions 2 and 3).

Consequently, (ii) $\Rightarrow$ (i). Since conditions (i) and (ii) are obviously necessary, the proposition is proven.

Proposition 2. *Any σ-complete vector lattice $\mathscr{X}$ is an Archimedean space.*

Proof. From the corollary of proposition 1 from Section 3.1.3, it is sufficient to prove that if x is any positive element of $\mathscr{X}$, then $\frac{1}{n} x \downarrow \mathbf{0}$.

Since the space $\mathscr{X}$ is σ-complete (as a vector lattice), the sequence $\left\{\frac{1}{n} x\right\}_{n\in\mathbb{N}}$ has a greatest lower bound. Then let

$$x_0 = \bigwedge_{n\in\mathbb{N}} \frac{1}{n} x.$$

Since $n\, x_0 \leqslant x$, $\forall\, n \in \mathbb{N}$, there also exists the element

$$x_1 = \bigvee_{n\in\mathbb{N}} n\, x_0.$$

From the relations

$$x_1 + x_0 = (\bigvee_{n\in\mathbb{N}} n\, x_0) + x_0 = \bigvee_{2\leqslant n\in\mathbb{N}} (n\, x_0) \leqslant \bigvee_{n\in\mathbb{N}} n\, x_0 = x_1,$$

it follows that $x_0 \leqslant \mathbf{0}$. But it is obvious that $x_0 \geqslant \mathbf{0}$ and this shows that $x_0 = \mathbf{0}$. The proposition is proved.

3.2.6 (o)-Convergent sequences and series in σ-complete vector lattices

Let $\mathscr{X}$ be a σ-complete vector lattice.

Proposition 1. *In order that a sequence $\{x_n\}_{n\in\mathbb{N}}$ of elements of the space $\mathscr{X}$ be (o)-convergent, it is necessary*[4]*) and sufficient that a sequence $\{r_n\}_{n\in\mathbb{N}}$ of elements should exist, such that $r_n \downarrow 0$ and*

$$|x_i - x_j| \leqslant r_n, \quad i, j \geqslant n. \tag{1}$$

Proof. If $x_n \xrightarrow{o} x$, then there exists a sequence $\{s_n\}_{n\in\mathbb{N}}$ of elements of $\mathscr{X}$, such that $s_n \downarrow 0$ and

$$|x_n - x| \leqslant s_n, \quad n \in \mathbb{N}.$$

If $i \geqslant n$, then $s_i \leqslant s_n$ and, therefore, by denoting $r_n = 2s_n$, for any $i, j \geqslant n$ we have

$$|x_i - x_j| \leqslant |x_i - x| + |x_j - x| \leqslant r_n.$$

Conversely, let us assume, that (1) holds, with $r_n \downarrow \mathbf{0}$.

If $i, j \geqslant n$, then $x_i \leqslant x_j + r_n$, and this implies that the sequence $\{x_j\}_{j\in N}$ is bounded and

$$\bigvee_{i\geqslant n} x_i \leqslant \bigwedge_{j\geqslant n} x_j + r_n.$$

From

$$\mathbf{0} \leqslant \bigvee_{i\geqslant n} x_i - \bigwedge_{j\geqslant n} x_j \leqslant r_n,$$

one gets

$$\overline{\lim}\, x_n - \underline{\lim}\, x_n = (o)\text{-}\lim_n \Big(\bigvee_{i\geqslant n} x_i - \bigwedge_{j\geqslant n} x_j\Big) = \mathbf{0}.$$

Consequently, $\overline{\lim}\, x_n = \underline{\lim}\, x_n$, and this shows that the sequence $\{x_n\}_{n\in\mathbb{N}}$ is (o)-convergent.

Remarks. (1) The condition from proposition 1 can be stated in the following form

$$(o)\text{-}\lim_n \Big(\bigvee_{i,j\geqslant n} |x_i - x_j|\Big) = \mathbf{0}.$$

This can be checked easily.

---

[4]) The necessity also holds in arbitrary vector lattices.

(2) From proposition 1 one can easily deduce an analogous proposition for the case of the convergence with regulator:

If $\{x_n\}_{n\in\mathbb{N}}$ *is a sequence of elements of the space* $\mathcal{X}$ *and if there exists an element* r, *such that for any number* $\varepsilon > 0$ *one has*

$$|x_i - x_j| \leqslant \varepsilon r, \text{ if } i, j \geqslant n_\varepsilon,$$

then the sequence $\{x_n\}_{n\in\mathbb{N}}$ *if* (ρ)*-convergent to an element* x, *with the regulator* r. The converse is also true.

The notion of (o)*-convergent series* can be introduced in a natural manner:

$$(o)\text{-}\sum_{n=1}^{\infty} x_n = (o)\text{-}\lim_m \sum_{n=1}^{m} x_n,$$

if the right-hand member limit exists.

Proposition 2. *Let* $\{x_n\}_{n\in\mathbb{N}}$, $\{y_n\}_{n\in\mathbb{N}}$ *be two sequences of elements of the space* $\mathcal{X}$, *such that* $|x_n| \leqslant |y_n|$, *for any* $n \in \mathbb{N}$. *If* $y = (o)\text{-}\sum_{n=1}^{\infty} |y_n|$ *exists, then* $x = (o)\text{-}\sum_{n=1}^{\infty} x_n$ *exist too and one has the inequality* $|x| \leqslant y$.

Proof. By putting

$$s_n = \sum_{i=1}^{n} x_i,$$

we have

$$|s_{n+m} - s_n| = \left|\sum_{i=n+1}^{n+m} x_i\right| \leqslant \sum_{i=n+1}^{n+m} |x_i| \leqslant \sum_{i=n+1}^{n+m} |y_i|$$

and from proposition (1) it follows that the element $x = (o)\text{-}\lim s_n$ exists.

The inequality $|x| \leqslant y$ can be easily checked.

From proposition 2 it follows that if $(o)\text{-}\sum_{n=1}^{\infty} |x_n|$ exists, then $(o)\text{-}\sum_{n=1}^{\infty} x_n$ exists too. The last series is then said to be *absolutely* (o)*-convergent*.

Proposition 3. *If* $\{x_n\}_{n\in\mathbb{N}}$ *is a bounded sequence of elements of* $\mathcal{X}$, *which are mutually orthogonal, then it determines an absolutely* (o)*-convergent series.*

Proof. From $|x_n| \leqslant a$ we infer that

$$\sum_{i=1}^{n} |x_i| = \bigvee_{i=1}^{n} |x_i| \leqslant a,$$

for any n (by taking into account proposition 4 from Section 3.2.1). Consequently, the element

$$(o)\text{-}\lim \sum_{i=1}^{n} |x_i| = \bigvee_{n \in \mathbb{N}} \sum_{i=1}^{n} |x_i|$$

exists, and the proposition is proved.

3.2.7 Complete vector lattices

Definition 1. One calls a *complete vector lattice* any ordered vector space which is a relatively complete lattice.

Proposition 1. *In order that a vector lattice $\mathscr{X}$ be a complete vector lattice it is necessary and sufficient that one of the following two equivalent conditions be satisfied:*

(i) *If E is a non-empty bounded from above set of positive elements of $\mathscr{X}$, and if it is directed to the right, then $\sup E$ exists.*

(ii) *If E is a non-empty set of positive elements of $\mathscr{X}$, which is directed to the left, then $\inf E$ exists.*

Proof. Let $\mathscr{X}$ be a vector lattice and A a non-empty, bounded from above, subset of $\mathscr{X}$. Let B be the set of all lowest upper bounds of the finite subsets of A. Let us consider an element $x_0 \in B$ and define

$$B_0 = \{x;\ x_0 \leqslant x \in B\},$$

$$E = B_0 - x_0 = \{x - x_0;\ x \in B_0\}.$$

Since the set B is obviously directed to the right, the set E is directed to the right. Moreover, the set E consists of positive elements and it is bounded from above. If we assume condition (i) to be satisfied, then the element $a = \sup E$ exists and we have

$$a + x_0 = (\sup E) + x_0 = \sup B_0 = \sup B = \sup A.$$

It follows that if (i) is satisfied, then $\mathscr{X}$ is a complete vector lattice.

Condition (ii) implies condition (i).

Indeed, let A be a non-empty, bounded from above, set of positive elements of $\mathscr{X}$, and assume that A is directed to the right. Let x_0 be an upper bound for A. The set $E = x_0 - A$ is formed of positive elements and it is directed to the left. By defining $a = \inf E$, it follows that $x_0 - a = \sup A$.

It follows that (ii) $\Rightarrow$ (i). Since conditions (i), (ii) are obviously necessary, the proposition is proved.

Definition 2. Let $\mathscr{X}$ and $\mathscr{Y}$ be two ordered vector spaces. The space $\mathscr{Y}$ is said to be an *extension* of $\mathscr{X}$, and $\mathscr{X}$ is said to be *embeddable* in $\mathscr{Y}$, if $\mathscr{X}$ is isomorphic (as an ordered vector space) to a subspace of $\mathscr{Y}$.

Theorem. *Any Archimedean vector lattice is embeddable in a complete vector lattice.*

Proof. Let $\mathscr{X}$ be an Archimedean vector lattice and let $\tilde{\mathscr{X}}$ be the extension of the ordered set $\mathscr{X}$, obtained by application of the Dedekind extension (Section 1.2.2). We shall define in $\tilde{\mathscr{X}}$ the operations of addition and multiplication by scalars. If $A, B \in \mathscr{X}$ and if $\alpha \in \mathbb{R}$, we shall put (see the notations from the proof of the theorem in the cited section)

$$A + B = i(s(\{x + y;\ x \in A, y \in B\})) \tag{1}$$

and

$$\alpha A = \begin{cases} \{\alpha x;\ x \in A\}, & \text{if } \alpha > 0 \\ i(\mathbf{0}) & \text{if } \alpha = 0 \\ \{\alpha y;\ y \in s(A), & \text{if } \alpha < 0. \end{cases}$$

It is obvious that $A + B \in \tilde{\mathscr{X}}$ and one can show[5]) that $\alpha A \in \tilde{\mathscr{X}}$, for any $\alpha \in \mathbb{R}$.

The addition, as defined above, is obviously commutative. Let us show that it is also associative.

Let A, B, C be elements of $\tilde{\mathscr{X}}$ and let us define

$$D = \{x + y + z;\ x \in A,\ y \in B,\ z \in C\}.$$

It is easy to prove that

$$i(s(D)) \subset (A + B) + C. \tag{2}$$

Now let $v \in (A + B) + C$. By putting $E = A + B$, we consequently have

$$v \in i(s(\{w + z;\ w \in E,\ z \in C\})).$$

It follows that $v \leqslant p$, for any p such that

$$p \geqslant w + z,\ \forall w \in E,\ \forall z \in C.$$

On the other hand

$$w \in i(s(\{x + y;\ x \in A,\ y \in B\}))$$

and therefore $w \leqslant q$, for any q such that

$$q \geqslant y + x,\ \forall x \in A,\ \forall y \in B.$$

---

[5]) We leave this for the reader.

Let $f \in s(D)$; then

$$f \geqslant x + y + z, \ \forall x \in A, \ \forall y \in B, \ \forall z \in C.$$

By considering an arbitrary element $z \in C$, we have

$$f - z \geqslant x + y, \quad \forall x \in A, \ \forall y \in B,$$

hence $w \leqslant f - z$, i.e., $w + z \leqslant f$, for any $z \in C$.

It follows that $v \leqslant f$ and, since f is arbitrary in the set $s(D)$, we have $v \in i(s(D))$. We have thus obtained the opposite inclusion to (2) and, therefore, the equality

$$i(s(D)) = (A + B) + C$$

is proved.

In a similar manner one can prove the equality

$$i(s(D)) = A + (B + D),$$

and this shows that the addition, as defined by formula (1), is associative.

It is not difficult to check that $i(\mathbf{0})$ plays the rôle of the zero-element.

Let us now show that for any $A \in \tilde{\mathscr{X}}$, by defining

$$B = \{-y; y \in s(A)\},$$

one has

$$A + B = i(\mathbf{0}). \tag{3}$$

To this effect, we shall first show that for any $A \in \tilde{\mathscr{X}}$ the equality

$$\mathbf{0} = \inf\{y - x; \ x \in A, \ y \in s(A)\} \tag{4}$$

holds.

We obviously have $y - x \geqslant 0$, for any $x \in A$, $y \in s(A)$. Now let z be an element of $\mathscr{X}$, such that

$$z \leqslant y - x, \ \forall x \in A, \ \forall y \in s(A).$$

We consequently have $x + z \leqslant y$, for any $y \in s(A)$, and any $x \in A$. Since

$$A = i(s(A))$$

it follows that $x + z \in A$, for any $x \in A$.

By induction one can infer that $x + nz \in A$, for any $n \in \mathbb{N}$ and this shows that the sequence $\{nz\}_{n \in \mathbb{N}}$ is bounded from above. Since the space is Archimedean, we infer that $z \leqslant \mathbf{0}$. Consequently, equality (4) holds.

In order to establish equality (3), let us take into account the fact that, by defining

$$C = \{x - y;\ x \in A,\ y \in s(A)\},$$

we have

$$A + B = i(s(C)).$$

We have, therefore, to prove the equality

$$i(s(C)) = i(\mathbf{0}). \tag{5}$$

It is obvious that $C \subset i(\mathbf{0})$ and, therefore,

$$i(s(C)) \subset i(\mathbf{0}).$$

On the other hand, from (4) we have

$$\mathbf{0} = \sup C$$

and this implies that

$$s(C) = s(\mathbf{0}),$$

whence

$$i(s(C)) \supset i(s(\mathbf{0})) = i(\mathbf{0}).$$

Thus equality (5) is proved, and this implies (3).

The proofs for the properties of the multiplication by scalars of the elements from $\tilde{\mathscr{X}}$ are not difficult [6]).

It is also easy to prove that the two operations are "compatible with the order relation", that is, $\tilde{\mathscr{X}}$ is an ordered vector space. Since $\tilde{\mathscr{X}}$ is a relatively complete lattice, it follows that $\tilde{\mathscr{X}}$ is a complete vector lattice.

Let us now state that if $x, y \in \mathscr{X}$ and $\alpha \in \mathbb{R}$, then

$$i(x) + i(y) = i(x + y),$$

$$i(\alpha x) = \alpha i(x),$$

in the space $\tilde{\mathscr{X}}$.

---

[6]) The properties $1.A = A$ and $\alpha(\beta A) = (\alpha\beta)A$, where $A \in \tilde{\mathscr{X}}$ and $\alpha, \beta \in \mathbb{R}$, can be immediately proved. The properties $\gamma(A + B) = \gamma A + \gamma B$ and $(\alpha + \beta)A = \alpha A + \beta A$ will be first proved for the case when γ, α and β are positive numbers.

If we then define (see Section 1.2.2)

$$\tilde{\mathscr{X}}_0 = \{i(x);\, x \in \mathscr{X}\},$$

the mapping $x \mapsto i(x)$ is an isomorphism of the ordered vector space $\mathscr{X}$ onto $\tilde{\mathscr{X}}_0$.

Remark. Any Archimedean vector lattice may be considered as an ordered vector subspace of a complete vector lattice $\mathscr{Y}$, such that:

(1) the bounds of the subsets of $\mathscr{X}$, if they exist in the space $\mathscr{X}$, coincide with the bounds of the same subsets when considered as subsets of $\mathscr{Y}$;

(2) any element $y \in \mathscr{Y}$ may be represented in the following form

$$y = \bigwedge_{h \in H} a_h = \bigvee_{j \in J} b_j,$$

where $\{a_h\}_{h \in H}$ and $\{b_j\}_{j \in J}$ are suitable families of elements of $\mathscr{X}$.

Up to an isomorphism, the space $\mathscr{Y}$ having the preceding properties is given by the Dedekind extension $\tilde{\mathscr{X}}$ of $\mathscr{X}$, with the ordered vector space structure described in the theorem just proved.

3.2.8 Quasi-σ-complete spaces

Definition 1. Any vector subspace $\mathscr{X}_0$ of a vector lattice $\mathscr{X}$, which is also a sublattice, is called a *vector sublattice.*

Proposition 1. *In order that a vector subspace $\mathscr{X}_0$ of the vector lattice $\mathscr{X}$ be a vector sublattice, it is necessary and sufficient that $x \in \mathscr{X}_0$ should imply $x_+ \in \mathscr{X}_0$.*

The proof is immediate.

It is obvious that if $\mathscr{X}_0$ is a vector sublattice of $\mathscr{X}$, then $x \in \mathscr{X}_0$ implies not only that $x_+ \in \mathscr{X}_0$, but also that $x_-, |x| \in \mathscr{X}_0$.

Definition 2. A vector lattice $\mathscr{X}$ is said to be *quasi-σ-complete* if $\mathscr{X}$ is isomorphic to a vector sublattice $\mathscr{X}'$ of a σ-complete vector lattice $\mathscr{Y}$, such that for any positive element $y \in \mathscr{Y}$ there exist sequences $\{a_n\}_{n \in \mathbb{N}}, \{b_n\}_{n \in \mathbb{N}}$ of positive elements of $\mathscr{X}'$, having the property that $a_n \uparrow y$ and $b_n \downarrow y$.

If $\mathscr{Y}$ is an arbitrary vector lattice, for any subset $E \subset \mathscr{Y}$ we shall write

$$U(E) = \{y;\, \exists\, y_n \in E,\ (n \in \mathbb{N}),\ y_n \downarrow y\},$$

$$V(E) = \{y;\, \exists\, y_n \in E,\ (n \in \mathbb{N}),\ y_n \uparrow y\}.$$

A sequence $\{x_n\}_{n \in \mathbb{N}}$ of elements of a vector lattice $\mathscr{X}$ is said to be an (*o*)-*Cauchy sequence* if it satisfies the condition from the statement of proposition 1, Section 3.2.6, i.e., if there exists a sequence $\{r_n\}_{n \in \mathbb{N}}$ of elements, such that $r_n \downarrow \mathbf{0}$ and

$$|x_i - x_j| \leqslant r_n,\ (i, j \geqslant n).$$

Proposition 2. *In order that a vector lattice $\mathscr{X}$ be quasi-σ-complete, it is necessary and sufficient that any monotonous, bounded sequence of elements of $\mathscr{X}$ be an (o)-Cauchy sequence.*

Proof. Let $\mathscr{X}$ be quasi-σ-complete and let $\mathscr{X}'$ and $\mathscr{Y}$ be as in the preceding definition. If $0 \leqslant x_n \downarrow$ in the space $\mathscr{X}$, then, by identifying $\mathscr{X}$ with $\mathscr{X}'$, there exists a positive element $y \in \mathscr{Y}$, such that $x_n \downarrow y$. On the other hand, there exists a sequence $\{a_n\}_{n \in \mathbb{N}}$ of elements of $\mathscr{X}$, such that $\mathbf{0} \leqslant a_n \uparrow y$. By putting $r_n = x_n - a_n$, we have $r_n \downarrow 0$ and $\mathbf{0} \leqslant x_i - x_j \leqslant r_n$, for $j \geqslant i \geqslant n$. The case of the arbitrary monotonous sequences reduces to the preceding one.

Conversely, let us assume that, in the space $\mathscr{X}$, any bounded, monotonous sequence is an (o)-Cauchy sequence. Then $\mathscr{X}$ is Archimedean. Indeed, if $\mathbf{0} < x \in \mathscr{X}$ and

$$\left| \frac{1}{i} x - \frac{1}{j} x \right| \leqslant r_n, \quad (i, j \geqslant n),$$

where $r_n \downarrow \mathbf{0}$, then by making $i = n$ and $j = n + 1$, we get

$$\mathbf{0} \leqslant \frac{1}{n(n+1)} x \leqslant r_n,$$

whence $\frac{1}{m} x \downarrow \mathbf{0}$.

Let us now consider the Dedekind extension $\tilde{\mathscr{X}}$ of $\mathscr{X}$. We shall first prove that $U(\mathscr{X}_+) = V(\mathscr{X}_+)$ in $\tilde{\mathscr{X}}$.

Let $\mathbf{0} \leqslant y \in U(\mathscr{X}_+)$ and $x_n \downarrow y$ in $\tilde{\mathscr{X}}$, where $\mathbf{0} \leqslant x_n \in \mathscr{X}$ (we identify $\mathscr{X}$ with its image in $\tilde{\mathscr{X}}$). Let $r_n \downarrow \mathbf{0}$ in $\mathscr{X}$ be such that

$$\mathbf{0} \leqslant x_n - x_{n+k} \leqslant r_n, \ n \in \mathbb{N}.$$

It follows that $x_n - r_n \leqslant y (\forall n \in \mathbb{N})$ and therefore, by putting

$$a_n = \bigvee_{j=1}^{n} (x_j - r_j)_+$$

we have $\mathbf{0} \leqslant a_n \in \mathscr{X}$ and $a_n \uparrow y$ in $\tilde{\mathscr{X}}$. In other words, $y \in V(\mathscr{X}_+)$. We thus showed that $U(\mathscr{X}_+) \subset V(\mathscr{X}_+)$. Analogously, the opposite inclusion can be proved. If we now define

$$\mathscr{Y} = V(\mathscr{X}_+) - V(\mathscr{X}_+),$$

then we can see that $\mathscr{Y}$ is a σ-complete vector lattice, whereas $\mathscr{Y}_+ = V(\mathscr{X}_+) = U(\mathscr{X}_+)$.

The proposition is proved.

3.2.9 Product spaces

Let $\{\mathscr{X}_j\}_{j\in J}$ be a family of ordered vector spaces and let $\mathscr{X} = \prod_{j\in J} \mathscr{X}_j$.

The set $\mathscr{X}$, endowed with the structure of product vector space, becomes an ordered vector space if one defines

$$\{x_j\}_{j\in J} \leqslant \{y_j\}_{j\in J} \Leftrightarrow x_j \leqslant y_j, (\forall j \in J).$$

This order relation in $\mathscr{X}$ is called the *product order relation* and the space $\mathscr{X}$, with the product vector space and product order relation structures is called a *product ordered vector space.*

Proposition. *The product ordered vector space $\mathscr{X}$ is a vector lattice (σ-complete, complete vector lattice) iff each factor space $\mathscr{X}_j$ is a vector lattice (a σ-complete, complete vector lattice, respectively).*

Proof. Let us assume that $\mathscr{X}$ is a vector lattice. Let $a_j, b_j \in \mathscr{X}_j$ and let us define [7]

$$a = \{\delta_{ij}\, a_j\}_{j\in J},\ b = \{\delta_{ij} b_j\}_{j\in J},$$

and

$$c = a \vee b = \{c_i\}_{i\in J}.$$

It follows that

$$c_i = \delta_{ij} a_j \vee \delta_{ij} b_j,\ (i \in J)$$

and this shows that

$$c_j = a_j \vee b_j,$$

which shows that $\mathscr{X}_j$ is a vector lattice.

Let us now assume that $\{\mathscr{X}_j\}_{j\in J}$ is a family of vector lattices and let a, b be two arbitrary elements of the product ordered vector space $\mathscr{X}$. If $a = \{a_j\}_{j\in J}$, $b = \{b_j\}_{j\in J}$ and if we define $c = \{a_j \vee b_j\}_{j\in J}$, then we obviously have $c = a \vee b$.

The case of the σ-complete or that of the complete vector lattices may be treated analogously.

3.3 Decompositions of a vector lattice

3.3.1 Normal subspaces

Let $\mathscr{X}$ be a vector lattice.

---

[7]) δ_{ij} is the Kronecker symbol.

Definition 1. A subset E of $\mathscr{X}$ is said to be *solid* if $|y| \leqslant |x|$ and $x \in E$ implies $y \in E$.

If we denote

$$Z(x) = \{y; |y| \leqslant |x|\},$$

then the condition that E be solid may be stated in the following way: if $x \in E$, then $Z(x) \subset E$.

Definition 2. One calls a *normal subspace* of $\mathscr{X}$ any vector subspace $\mathscr{X}_0$ which, at the same time, is a solid subset of $\mathscr{X}$.

Proposition 1. *Any normal subspace is a vector sublattice.*

The proof is immediate.

In the following proposition we shall denote by $\sup_{\mathscr{X}_0} A$ the least upper bound in $\mathscr{X}_0$ (of a subset A of $\mathscr{X}_0$) and by $\sup A$ its least upper bound in $\mathscr{X}$. Similar notations will hold for the greatest lower bounds.

Proposition 2. *Let $\mathscr{X}_0$ be a normal subspace of $\mathscr{X}$ and $A \subset \mathscr{X}_0$.*

In order that $a = \sup_{\mathscr{X}_0} A$ (or $a = \inf_{\mathscr{X}_0} A$) it is necessary and sufficient that $a = \sup A$ ($a = \inf A$, respectively) and that the set A be bounded from above (bounded from below, respectively) in $\mathscr{X}_0$.

Proof. If $a = \sup_{\mathscr{X}_0} A$, then $x \leqslant a$, for any $x \in A$. If $y \in \mathscr{X}$ and $x \leqslant y$ for any $x \in A$, then the element $b = a \wedge y$ is an upper bound for A. It follows that:

$$x \leqslant b \leqslant a, \quad (\forall x \in A).$$

From the inequality

$$|b| \leqslant |x| \vee |a|, \; x \in A,$$

it follows that $b \in \mathscr{X}_0$. From the definition of a it follows that $a \leqslant b$, hence $a = b$. But this implies that $a \leqslant y$, and this shows that $a = \sup A$.

Conversely, let $y \in \mathscr{X}_0$, such that y be an upper bound for A. Let $a = \sup A$. Then we have

$$x \leqslant a \leqslant y, \; (\forall x \in A),$$

and this implies that $a \in \mathscr{X}_0$ and $a = \sup_{\mathscr{X}_0} A$.

The proposition is proved.

Corollary. *If $\mathscr{X}$ is a complete vector lattice (or a σ-complete vector lattice) and if $\mathscr{X}_0$ is a normal subspace of $\mathscr{X}$, then $\mathscr{X}_0$ is a complete vector lattice (a σ-complete vector lattice, respectively).*

Indeed, if $\mathscr{X}$ is, for instance, a complete vector lattice, if $A \subset \mathscr{X}_0$, and if A is bounded from above in $\mathscr{X}_0$, then A is bounded from above in $\mathscr{X}$ too and, consequently, $\sup A$ exists in $\mathscr{X}$. From proposition 2 we infer that $\sup A = \sup_{\mathscr{X}_0} A$.

Proposition 3. *Let $\mathscr{X}_0$ be a normal subspace of the vector lattice $\mathscr{X}$, $\{x_n\}_{n \in \mathbb{N}}$ a sequence of elements of $\mathscr{X}_0$ and $x \in \mathscr{X}_0$. In order that $x_n \overset{o}{\to} x$ in $\mathscr{X}_0$, it is necessary and sufficient that $x_n \overset{o}{\to} x$ in $\mathscr{X}$ and that the sequence $\{x_n\}_{n \in \mathbb{N}}$ be bounded in $\mathscr{X}_0$.*

Proof. Let $x_n \overset{o}{\to} x$ in $\mathscr{X}_0$. From what we have shown in Section 3.2.4 (proposition 1), there exists a sequence $\{r_n\}_{n \in \mathbb{N}}$ of elements of $\mathscr{X}_0$, such that $r_n \downarrow \mathbf{0}$ in $\mathscr{X}_0$ and, moreover,

$$|x_n - x| \leqslant r_n, \ (\forall n \in \mathbb{N}). \tag{1}$$

From the fact $r_n \downarrow \mathbf{0}$ in $\mathscr{X}$, which is a consequence of proposition 2, and from (1) it follows that $x_n \overset{o}{\to} x$ in $\mathscr{X}$. The sequence $\{x_n\}_{n \in \mathbb{N}}$ is bounded in $\mathscr{X}_0$, since it is (o)-convergent in this space.

Conversely, let $y \in \mathscr{X}_0$ be such that $|x_n| \leqslant y$, for any $n \in \mathbb{N}$, and let us assume that $x_n \overset{o}{\to} x$ in the space $\mathscr{X}$. Then we get that $|x| \leqslant y$. On the other hand, there exists a sequence $\{r_n\}_{n \in \mathbb{N}}$ of elements of $\mathscr{X}$, such that $r_n \downarrow \mathbf{0}$ in $\mathscr{X}$ and such that (1) holds. Let us denote

$$s_n = r_n \wedge 2y$$

and let us now observe that from (1) and from the inequalities

$$|x_n - x| \leqslant 2y, \ \forall n \in \mathbb{N}$$

one gets

$$|x_n - x| \leqslant s_n, \ (\forall n \in \mathbb{N}). \tag{2}$$

From the inequalities

$$\mathbf{0} \leqslant s_n \leqslant 2y,$$

and from $2y \in \mathscr{X}_0$ it follows that $s_n \in \mathscr{X}_0$ and, obviously, $s_n \downarrow \mathbf{0}$ in $\mathscr{X}_0$. But, in this case, relation (2) shows that $x_n \overset{o}{\to} x$ in $\mathscr{X}_0$. The proposition is proved.

3.3.2 The orthogonal complement. Semi-components

Let $\mathscr{X}$ be a vector lattice.

If $A \subset \mathscr{X}$ and if x is an element orthogonal to A (i.e., is orthogonal to any element of A) we shall write $x \perp A$.

Definition 1. *If $A \subset \mathscr{X}$, then the orthogonal complement of A is, by definition, the set* $A^{\perp} = \{x; x \perp A\}$.

We shall denote by $A^{\perp\perp}$ the orthogonal complement of the set $A^{\perp}$. It is easily verified that $A \subset B$ implies $B^{\perp} \subset A^{\perp}$ and that $A \subset A^{\perp\perp}$, for any A.

Proposition 1. *The orthogonal complement* $A^{\perp}$ *of any set* A *is a normal subspace of* $\mathscr{X}$, *with the property: if* $B \subset A^{\perp}$ *and if* $b = \sup B$ *exists, then* $b \in A^{\perp}$.

This proposition immediately follows from the properties of orthogonal elements (Section 3.2.3).

Definition 2. One calls a *semi-component* of $\mathscr{X}$ any subset $E \subset \mathscr{X}$, such that $E = E^{\perp\perp}$.

Proposition 2. A *subset* E *of* $\mathscr{X}$ *is a semi-component of* $\mathscr{X}$ *if, and only if, there exists a subset* $A \subset \mathscr{X}$, *such that* $E = A^{\perp}$.

Proof. The necessity is obvious. The converse follows from the fact that for any subset A of $\mathscr{X}$ the following equality

$$A^{\perp\perp\perp} = A^{\perp}$$

holds; its proofs is easy.

It is obvious that for any subset A of $\mathscr{X}$, the set $A^{\perp\perp}$ is the smallest semi-component which contains A.

Definition 3. A set A of elements of $\mathscr{X}$ is said to be *total* (in the space $\mathscr{X}$) if $A^{\perp} = \{0\}$. A family $\{A_j\}_{j\in J}$ of subsets of $\mathscr{X}$ is total in $\mathscr{X}$ if the union of the elements of the family is total.

Lemma. *Let* E *be a semi-component of* $\mathscr{X}$ *and* $\{E_j\}_{j\in J}$ *any family of semi-components of* $\mathscr{X}$, *contained in* E. *In order that the family* $\{E_j\}_{j\in J}$ *be total in* E *it is necessary and sufficient that*

$$E = (\bigcup_{j\in J} E_j)^{\perp\perp}. \tag{1}$$

Proof. The condition is obviously sufficient. Let us prove its necessity.

Let us assume that the family $\{E_j\}_{j\in J}$ is total in E and let us denote

$$F = (\bigcup_{j\in J} E_j)^{\perp\perp}.$$

It is obvious that $F \subset E$. If $F \neq E$, then there exist an $x \in E$ and a $y \in F^{\perp}$, such that the relation $x \perp y$ be not satisfied. (if this is not true, one would have $E \subset F^{\perp\perp} = F$). By putting $z = |x| \wedge |y|$, we have $z \neq 0$ and, from proposition 1, we would infer that $z \in E \cap F^{\perp}$. In particular,

$$z \perp \bigcup_{j\in J} E_j,$$

and this contradicts the fact that the system $\{E_j\}_{j\in J}$ is total in E. It follows that $F = E$, and the lemma is proved.

Proposition 3. *The set* $\mathscr{C}$ *of all semi-components of the vector lattice* $\mathscr{X}$ *is a Boolean lattice with respect to the inclusion, as an order relation.*

Proof. In the set $\mathscr{C}$, in which the inclusion has been taken to be the order relation $\{\mathbf{0}\}$ is the smallest element and $\mathscr{X}$ is the greatest element.

Let us now observe that for any family, $\{A_j\}_{j\in J}$ of subsets of $\mathscr{X}$ the following equality

$$(\bigcup_{j\in J} A_j)^{\perp} = \bigcap_{j\in J} A_j^{\perp} \tag{2}$$

holds; its proof is an easy matter.

From equality (2) we infer that if $\{E_j\}_{j\in J}$ is a family of elements of $\mathscr{C}$, then its greatest lower bound exists and it is given by the equality

$$\bigwedge_{j\in J} E_j = \bigcap_{j\in J} E_j.$$

The least upper bound exists too, and it is given by

$$\bigvee_{j\in J} E_j = (\bigcup_{j\in J} E_j)^{\perp\perp}. \tag{3}$$

Let us now prove the distributivity law for the lattice $\mathscr{C}$. Let E, E_1, E_2 be elements of $\mathscr{C}$. We obviously have the following inclusions:

$$E \wedge E_1 \subset E \wedge (E_1 \vee E_2),$$

$$E \wedge E_2 \subset E \wedge (E_1 \vee E_2).$$

Let $x \neq \mathbf{0}$, be such that

$$x \in E \wedge (E_1 \vee E_2).$$

Since the system $\{E_1, E_2\}$ is total in $E_1 \vee E_2$, at least in one of the semi-components E_1, E_2 (for instance, in E_1), there exists a y, such that the relation $x \perp y$ should not be satisfied. By defining $z = |x| \wedge |y|$, we have $z \neq \mathbf{0}$, $z \in E_1$, and

$$z \in E \wedge (E_1 \vee E_2).$$

From $z \in E$ and $z \in E_1$ it follows that $z \in E_1 \wedge E$ and from $|x| \wedge z \neq 0$ we get

$$x \notin ((E \wedge E_1) \cup (E \wedge E_2))^{\perp}.$$

It follows that the system $\{E \wedge E_1, E \wedge E_2\}$ is total in the space $E \wedge (E_1 \vee E_2)$. From the preceding lemma and from equality (3) we get

$$E \wedge (E_1 \vee E_2) = (E \wedge E_1) \vee (E \wedge E_2).$$

It follows that $\mathscr{C}$ is a distributive lattice. For any $E \in \mathscr{C}$ we have

$$E \wedge E^{\perp} = \{\mathbf{0}\}, \; E \vee E^{\perp} = \mathscr{X}.$$

It follows that $\mathscr{C}$ is a Boolean lattice.
Corollary. *Let E be a semi-component of $\mathscr{X}$ and $\{E_j\}_{j \in J}$ a family of semi-components of $\mathscr{X}$, which are assumed to be contained in E. The system $\{E_j\}_{j \in J}$ is total in E iff*

$$E = \bigvee_{j \in J} E_j$$

in the Boolean lattice $\mathscr{C}$ of all semi-components.
Proposition 4. *If E is a normal subspace of an Archimedean vector lattice $\mathscr{X}$, then for any positive element $x \in \mathscr{X}$ the equality*

$$x = \sup \{y + z; \; \mathbf{0} \leqslant y \leqslant x, \; y \in E, \; \mathbf{0} \leqslant z \leqslant x, \; z \in E^{\perp}\}$$

holds.

Proof. The set

$$G = \{y + z; y \in E, z \in E^{\perp}\},$$

is obviously a vector subspace of $\mathscr{X}$. Let $x \in \mathscr{X}$, $y \in E$ and $z \in E^{\perp}$ be such that

$$|x| \leqslant |y + z| = |y| + |z|.$$

Then there exist[8]) $x_1, x_2 \in \mathscr{X}$, such that $x = x_1 + x_2$, $|x_1| \leqslant |y|$, $|x_2| \leqslant |z|$. Since E and $E^{\perp}$ are normal subspaces, it follows that $x_1 \in E$ and $x_2 \in E^{\perp}$, hence $x \in G$. The set G is consequently a normal subspace of $\mathscr{X}$.

Let now x be a positive element of $\mathscr{X}$. If $x \notin G$, let us consider the set

$$H = \{x'; \mathbf{0} \leqslant x' \leqslant x, \; x' \in G\}.$$

Let us show that

$$x = \sup H. \tag{4}$$

If (4) does not hold, then there exists an element y, such that

$$x' \leqslant y < x, (\; \forall x' \in H).$$

Since the set G is obviously total in $\mathscr{X}$, there exists an element $x_0 \in G$, such that $x - y$ and x_0 be not orthogonal. Let us put

$$x_1 = (x - y) \wedge |x_0|.$$

---

[8]) The corollary of proposition 3 from Section 3.2.1 is applied to the elements x_+, x_-.

From $0 < x_1 \leqslant |x_0|$ it follows that $x_1 \in G$, because G is a normal subspace. On the other hand, $x_1 + y \leqslant x$ and, therefore, if $x' \in H$ it follows that $x' + x_1 \in H$. By induction, it follows that $x' + nx_1 \in H$, for any $n \in \mathbb{N}$, and this implies that

$$nx_1 \leqslant x - x', \ \forall n \in \mathbb{N}.$$

Since the space is Archimedean, it follows that $x_1 \leqslant 0$, and this is a contradiction.
Equality (4) therefore holds and the proposition is proved.

Corollary. *If $\mathscr{X}$ is an Archimedean vector lattice and if E is a normal and total subspace of $\mathscr{X}$, then any positive element $x \in \mathscr{X}$ can be expressed in the form*

$$x = \sup \{x'; 0 \leqslant x' \leqslant x, \ x' \in E\}.$$

3.3.3 Components and projections

Definition 1. A subset Q of a vector lattice $\mathscr{X}$ is called a *component* of $\mathscr{X}$ if any element $x \in \mathscr{X}$ may be written as $x = x' + x''$, where $x' \in Q$ and $x'' \in Q^{\perp}$. The element x' is called the *projection* of x onto Q and it is denoted by $x' = [Q]x$.

It is easy to verify that the projection of any element on a component is unique.

It is also easy to verify that

$$[Q](\alpha x + \beta y) = \alpha [Q]x + \beta [Q]y, \tag{1}$$

for any $x, y \in \mathscr{X}$ and any $\alpha, \beta \in \mathbb{R}$; we also have

$$[Q] \ [Q]x = [Q]x \tag{2}$$

for any $x \in \mathscr{X}$, and if $x \geqslant 0$, then [9])

$$0 \leqslant [Q]x \leqslant x. \tag{3}$$

From (1) and (3) it immediately follows that if $x \leqslant y$, then $[Q]x \leqslant [Q]y$, and also that $|[Q]x| \leqslant |x|$, for any x.

Proposition 1. *Any component is a semi-component.*

Proof. Since $Q \subset Q^{\perp\perp}$ for any Q, we have only to show that if Q is a component, then $Q^{\perp\perp} \subset Q$.

Let $x \in Q^{\perp\perp}$. From $x = x' + x''$, where $x' \in Q$ and $x'' \in Q^{\perp}$, it follows that $x \perp x''$. On the other hand, $x' \perp x''$ and, therefore, $x - x' \perp x''$. From the fact that $x - x' = x''$ it follows that $x'' = 0$ and, therefore, $x \in Q$.

Corollary. If *Q is a component, then $Q^{\perp}$ is a component too.*

---

[9]) If $a + b \geqslant 0$ and $a \perp b$, then $|a| + |b| = a + b$, whence it immediately follows that $a_- = b_- = 0$.

Proposition 2. *If Q_1 and Q_2 are components, then $Q_1 \cap Q_2$ is a component too.*

Proof. For any element $x \in \mathscr{X}$ let us put

$$x' = [Q_1]x, \ x'' = [Q_1^{\perp}]\, x,$$

$$y = [Q_2]x', \ z = [Q_2^{\perp}]\, x'.$$

We have

$$x = x' + x'' = (y + z) + x'' = y + (z + x'').$$

From $|y| \leqslant |x'|$ it follows that $y \in Q_1$ and, therefore, $y \in Q_1 \cap Q_2$. On the other hand, from $x'' \perp Q_1$ it follows that $x'' \perp Q_1 \cap Q_2$, i.e. $x'' \in (Q_1 \cap Q_2)^{\perp}$. We also have $z \in (Q_1 \cap Q_2)^{\perp}$ and, therefore,

$$z + x'' \in (Q_1 \cap Q_2)^{\perp}.$$

By taking into account definition 1, we infer that the set $Q_1 \cap Q_2$ is a component.

Proposition 3. *If Q is a component of a vector lattice $\mathscr{X}$, then for any positive element $x \in \mathscr{X}$ one has the equality*

$$[Q]x = \sup\{y; 0 \leqslant y \leqslant x, \ y \in Q\}. \tag{4}$$

Proof. Let us denote

$$E = \{y; \mathbf{0} \leqslant y \leqslant x, \ y \in Q\}.$$

If $y \in E$, then $y = [Q]y$ and, therefore,

$$y \leqslant [Q]x \leqslant x.$$

Moreover, $[Q]x \in Q$ and hence $[Q]x \in E$. It follows that

$$[Q]x = \sup E,$$

i.e. (4).

Proposition 4. *If $\mathscr{X}$ is a σ-complete vector lattice, then for any element $v \in \mathscr{X}$ the set $v^{\perp\perp}$ is a component of $\mathscr{X}$ and for any positive element $x \geqslant \mathbf{0}$ one has the equality*

$$[v^{\perp\perp}]x = (o)\text{-}\lim_n\,(x \wedge n|v|). \tag{5}$$

Proof. Let x be any positive element of $\mathscr{X}$. The sequence $\{x \wedge n|v|\}_{n\in\mathbb{N}}$ is increasing and bounded. By putting

$$x' = \bigvee_{n\in\mathbb{N}} (x \wedge n|v|) \tag{6}$$

we have $x' \in v^{\perp\perp}$. On the other hand, by putting $x'' = x - x'$, we have

$$x'' \wedge |v| = (x - \bigvee_{n\in\mathbb{N}} (x \wedge n|v|)) \wedge |v| = x \wedge (|v| + \bigvee_{n\in\mathbb{N}} (x \wedge n|v|)) - \bigvee_{n\in\mathbb{N}} (x \wedge n|v|) =$$

$$= \bigvee_{n\in\mathbb{N}} (x \wedge (x + |v|) \wedge (n+1)|v|) - \bigvee_{n\in\mathbb{N}} (x \wedge n|v|) = \bigvee_{n\in\mathbb{N}} (x \wedge (n+1)|v|) - \bigvee_{n\in\mathbb{N}} (x \wedge n|v|) = \mathbf{0}.$$

Hence $x'' \perp v$, i.e., $x'' \in v^{\perp}$. But from $v^{\perp} = v^{\perp\perp\perp}$ we infer that $v^{\perp\perp}$ is a component and $[v^{\perp\perp}]x = x'$. Equality (6) is equivalent to (5) because the sequence is increasing.

Remark. In a σ-complete vector lattice the set $v^{\perp\perp}$ is the smallest component which contains the element v. It is called the *component generated by v.*

Proposition 5. *In order that a subset Q of a complete vector lattice $\mathscr{X}$ be a component it is necessary and sufficient that Q be a normal subspace having the property: if $A \subset Q$ and if $a = \sup A$ exists, then $a \in Q$.*

Proof. The necessity is a consequence of proposition 1. Let us prove the sufficiency. Let Q be a normal subspace having the property from the statement of the proposition. Let x be any positive element of $\mathscr{X}$ and let us put

$$x' = \sup\{y; \mathbf{0} \leqslant y \leqslant x, \; y \in Q\}.$$

By assumption, $x' \in Q$. Let y be an arbitrary element of Q and define $z = (x - x') \wedge \wedge |y|$.

From $0 \leqslant z \leqslant |y|$ and $y \in Q$ we infer that $z \in Q$. From $z \leqslant x - x'$ it follows that

$$\mathbf{0} \leqslant z + x' \leqslant x,$$

and, since $z + x' \in Q$, we get the inequality $z + x' \leqslant x'$. It follows that $z \leqslant \mathbf{0}$ and, since $z \geqslant \mathbf{0}$, this implies that $z = \mathbf{0}$. It follows that $x - x' \perp y$ and, since y has been chosen arbitrarily in Q, we have $x - x' \in Q^{\perp}$. The proposition is proved.

Corollary. *In a complete vector lattice, the notions of component and semi-component are equivalent.*

Remark. In a complete vector lattice, for any set A of elements, the set $A^{\perp\perp}$ is the smallest component which contains A. It is called the *component generated by A.*

Definition 2. If Q is a component of a vector lattice $\mathscr{X}$, the mapping $x \mapsto [Q]x$, of $\mathscr{X}$ into $\mathscr{X}$, is called the *projector* (defined by Q) and it is denoted by $[Q]$.

Proposition 6. If $\mathscr{X}$ is a complete vector lattice, then a mapping U of $\mathscr{X}$ into $\mathscr{X}$ is a projector iff the following conditions are satisfied:

(i) $U(x+y) = U(x) + U(y)$, *for any* $x, y \in \mathscr{X}$;

(ii) if $x \geqslant \mathbf{0}$, *then* $\mathbf{0} \leqslant U(x) \leqslant x$;

(iii) $U(U(x)) = U(x)$, *for any* $x \in \mathscr{X}$.

Proof. Let U be a mapping of $\mathscr{X}$ into $\mathscr{X}$ satisfying conditions (i), (ii), (iii). We shall first show that the set

$$Q = \{x;\, U(x) = x\}$$

is a component of $\mathscr{X}$. In order to do this, let us show that Q is a normal subspace having the property from the statement of proposition 5.

Let us state that from condition (i) it follows that $U(\mathbf{0}) = \mathbf{0}$ and $U(-x) = -U(x)$ consequently, from $x \in Q$ it follows that $-x \in Q$. By virtue of (i), from $x, y \in Q$ it follows that $x + y \in Q$, whereas from $x \in Q$ it follows that $nx \in Q$, for any $n \in \mathbb{N}$.

Let

$$\mathbf{0} \leqslant y \leqslant x \in Q.$$

We have $U(y) \leqslant y$ and

$$U(x-y) \leqslant x - y;$$

consequently,

$$y = x - (x-y) \leqslant x - U(x-y) = U(y) \leqslant y,$$

whence $U(y) = y$. It follows that $y \in Q$.

If x is any element of Q, we have

$$x = U(x_+) - U(x_-). \tag{7}$$

On the other hand, from the inequalities

$$\mathbf{0} \leqslant U(x_+) \leqslant x_+,$$

$$\mathbf{0} \leqslant U(x_-) \leqslant x_-,$$

we infer that

$$U(x_+) \perp U(x_-),$$

whereas from (7) we get

$$U(x_+) = x_+,\ U(x_-) = x_-.$$

It follows that $x_+, x_- \in Q$ and, therefore, $|x| \in Q$. If $x \in Q$ and $|y| \leqslant |x|$, we get

$$\mathbf{0} \leqslant y_+,\ y_- \leqslant |x|,$$

and, from what we have already shown, it follows that $y_+, y_- \in Q$; this implies that $y \in Q$. If $x \in Q$ and $\alpha \in \mathbb{R}$, then there exists an $n \in \mathbb{N}$, such that $|\alpha| \leqslant n$. From $|\alpha x| \leqslant n |x|$ it follows that $\alpha x \in Q$.

We have thus shown that the set Q is a normal subspace. Now let A be a set of positive elements of Q and $a = \sup A$. If $x \in A$, then $\mathbf{0} \leqslant x \leqslant a$, whence

$$x = U(x) \leqslant U(a),$$

and this implies that $a \leqslant U(a)$. But $U(a) \leqslant a$, if we take (ii) into account, and consequently, $U(a) = a$. If the elements of A are not positive, we shall take into account the fact that $a = \sup A$ implies

$$a_+ = \sup_{x \in A} x_+, \quad a_- = \inf_{x \in A} x_-.$$

From the inclusion

$$\{x_+;\ x \in A\} \subset Q$$

and from what we have already shown, we deduce that $a_+ \in Q$. On the other hand, $a_- \in Q$, because Q is a normal subspace, and, therefore, $a \in Q$.

We have thus shown that Q is a component of $\mathscr{X}$. Let us now show that $[Q] = U$. Let x be any element of $\mathscr{X}$ and $x = x' + x''$, where $x' \in Q$, $x'' \in Q^\perp$, be the decomposition of x corresponding to the component Q. From condition (i) we get

$$U(x') = U(x') + U(x''),$$

whereas from condition (iii)

$$U(x'') = U(U(x'')),$$

and this implies that $U(x'') \in Q$. On the other hand, from [10]

$$|U(x'')| \leqslant |x''|$$

and from $x'' \in Q^\perp$ we infer that $U(x'') \in Q^\perp$. Consequently, $U(x'') = 0$. It follows that

$$U(x) = U(x') = x' = [Q]\, x$$

and, since x was arbitrary, we infer that $U = [Q]$.

Conversely, any projector has the properties (i), (ii), (iii), as we have already remarked.

3.3.4 The properties of the projectors

Let $\mathscr{X}$ be a vector lattice and Q a component of $\mathscr{X}$. As we have already noted in Section 3.3.3, we have

$$|[Q]\, x| \leqslant |x|, \tag{1}$$

for any element $x \in \mathscr{X}$.

[10]) We have $|U(a)| = |U(a_+) - U(a_-)| \leqslant a_+ + a_- = |a|$.

From (1) one immediately infers that if $x \perp y$, then $[Q]\,x \perp [Q]\,y$.

Proposition 1. *If* $x = \bigvee_{j\in J} x_j$ *(or* $x = \bigwedge_{j\in J} x_j$*), then* $[Q]\,x = \bigvee_{j\in J} [Q]\,x_j$, $([Q]\,x = \bigwedge_{j\in J} [Q]\,x_j$, respectively).

Proof. We shall consider the case of the least upper bound. Let $x = \bigvee_{j\in J} x_j$. From $x_j \leqslant x$, $j \in J$, we get

$$[Q]\,x_j \leqslant [Q]\,x, \quad \forall j \in J.$$

Let y be an element of $\mathscr{X}$, such that

$$[Q]\,x_j \leqslant y, \quad \forall j \in J.$$

For any $j \in J$ the following relations hold

$$x_j = [Q]\,x_j + [Q^\perp]\,x_j \leqslant y + [Q^\perp]\,x,$$

whence

$$x \leqslant y + [Q^\perp]\,x,$$

which implies that

$$[Q]\,x = x - [Q^\perp]\,x \leqslant y.$$

It follows that

$$[Q]\,x = \bigvee_{j\in J} [Q]\,x_j.$$

Corollary. *For any element* $x \in \mathscr{X}$ *the following equalities hold:*

$$([Q]\,x)_+ = [Q]\,x_+, \tag{2}$$

$$([Q]\,x)_- = [Q]\,x_- \tag{3}$$

$$|\,[Q]\,x\,| = [Q]\,|\,x\,|. \tag{4}$$

Remark. Equalities (2), (3) (and consequently (4)) are also a consequence of the fact that

$$[Q]\,x = [Q]\,x_+ - [Q]\,x_-,$$

$$([Q]\,x_+) \wedge ([Q]\,x_-) = 0.$$

Proposition 2. *If* $x_n \xrightarrow{o} x$, *then* $[Q]\,x_n \xrightarrow{o} [Q]\,x$.

Proof. Let $r_n \downarrow 0$ be such that

$$| x_n - x | \leqslant r_n, \quad \forall n \in \mathbb{N},$$

We then have

$$| [Q] x_n - [Q] x | = | [Q] (x_n - x) |$$

$$= [Q] | x_n - x | \leqslant [Q] r_n \leqslant r_n,$$

for any $n \in \mathbb{N}$, and this implies that $[Q] x_n \xrightarrow{o} [Q] x$.

Lemma. *If Q_1, Q_2 are two components of the vector lattice $\mathscr{X}$, then* [11])

$$Q_1 \cap Q_2 = [Q_1] Q_2. \tag{5}$$

Proof. From proposition 2 (Section 3.3.3) we infer that the set $Q_1 \cap Q_2$ is a component and we obviously have

$$[Q_1] Q_2 \subset Q_2.$$

It follows that

$$Q_1 \cap Q_2 = [Q_1] (Q_1 \cap Q_2) \subset [Q_1] Q_2 \subset Q_1 \cap Q_2,$$

and this shows that (5) holds.

The following proposition concerns the composition of two projectors $[Q_1]$, $[Q_2]$, i.e., the operator given by the formula

$$([Q_1] \; [Q_2]) (x) = [Q_1] \; ([Q_2] x).$$

Proposition 3. *If Q_1 and Q_2 are two components of the vector lattice $\mathscr{X}$, then $[Q_1] [Q_2]$ is a projector, namely*

$$[Q_1] \; [Q_2] = [Q_1 \cap Q_2]. \tag{6}$$

Proof. If x is an arbitrary element of $\mathscr{X}$, then

$$x = [Q_1] x + [Q_1^\perp] x. \tag{7}$$

On the other hand,

$$x = [Q_2] x + [Q_2^\perp] x$$

and from (7) we get

$$x = [Q_1] \; [Q_2] x + ([Q_1] \; [Q_2^\perp] x + [Q_1^\perp] x). \tag{8}$$

---

[11]) We denote $[Q]A = \{ [Q]x; \; x \in A\}$.

Since $[Q_2]\,x \in Q_2$, we have

$$[Q_1]\;([Q_2]\,x) \in [Q_1]\,Q_2$$

and, from the preceding lemma, we get

$$[Q_1]\;[Q_2]\,x \in Q_1 \cap Q_2. \tag{9}$$

On the other hand, from $[Q_2^{\perp}]\,x \in Q_2^{\perp}$ it follows that

$$[Q_1]\;([Q_2^{\perp}]\,x) \in Q_1 \cap Q_2^{\perp};$$

since

$$Q_1 \cap Q_2^{\perp} \subset Q_2^{\perp} \subset (Q_1 \cap Q_2)^{\perp},$$

we get

$$[Q_1]\;[Q_2^{\perp}]\,x \in (Q_1 \cap Q_2)^{\perp}. \tag{10}$$

We also have

$$[Q_1^{\perp}]\,x \in (Q_1 \cap Q_2)^{\perp},$$

and, therefore,

$$[Q_1]\;[Q_2^{\perp}]\,x + [Q_2^{\perp}]\,x \in (Q_1 \cap Q_2)^{\perp}.$$

From (8) we infer that

$$[Q_1]\;[Q_2]\,x = [Q_1 \cap Q_2]\,x;$$

since x was arbitrarily chosen in $\mathscr{X}$, formula (6) holds.

Corollary. $Q_1 \perp Q_2$ *iff* $[Q_1]\;[Q_2] = \mathbf{0}$.

Definition 1. If Q_1 and Q_2 are two components of the vector lattice $\mathscr{X}$, one writes $[Q_1] \leqslant [Q_2]$ iff $[Q_1]\,x \leqslant [Q_2]\,x$, for any $x \geqslant \mathbf{0}$.

Proposition 4. *For any components Q_1, Q_2 of the vector lattice $\mathscr{X}$, the following conditions are equivalent.*

(i) $[Q_1] \leqslant [Q_2]$; (ii) $Q_1 \subset Q_2$; (iii) $[Q_1]\;[Q_2] = [Q_1]$.

Proof. (i) implies (ii), since from $x \in Q_1$ we infer that

$$|\,x\,| = [Q_1]\,|\,x\,| \leqslant [Q_2]\,|\,x\,|,$$

and $[Q_2]\,x \in Q_2$ then implies $x \in Q_2$.

(ii) implies (iii), if we take into consideration proposition 3.

(iii) implies (i), since for any $x \geqslant \mathbf{0}$ we have

$$\mathbf{0} \leqslant [Q_1]\,x \leqslant x$$

and therefore,

$$[Q_1]\,x = [Q_1]\,[Q_2]\,x = [Q_2]\,[Q_1]\,x \leqslant [Q_2]\,x.$$

The proposition is proved.

Remark. The set of all projectors (in a vector lattice $\mathscr{X}$) is a Boolean lattice in which

$$[Q_1] \wedge [Q_2] = [Q_1 \cap Q_2],$$

$$[Q_1] \vee [Q_2] = [(Q_1 \cup Q_2)^{\perp\perp}].$$

In what follows we shall assume that $\mathscr{X}$ is a σ-complete vector lattice. From proposition 4 from Section 3.3.3, for any element $v \in \mathscr{X}$ the set $v^{\perp\perp}$ is a component of $\mathscr{X}$.

Definition 2. For any element v of the σ-complete vector lattice $\mathscr{X}$, the projector determined by $v^{\perp\perp}$ is called a *principal projector* and it is denoted by $[v]$. One says that $[v]$ is the projector determined by v.

It is obvious that $[v]\,v = v$, for any v, and $[v] = \mathbf{0}$ iff $v = \mathbf{0}$; one also has $[v] = [|v|] = [\alpha v]$, for any $\alpha \neq 0$. Also, one immediately checks that the equality $[v]\,x = 0$ holds iff $v \perp x$.

In particular, we infer that $[v_+]\,v = v_+$ and $[v_-]\,v = v_-$.

Proposition 5. *If Q is a component of the σ-complete vector lattice $\mathscr{X}$, then for any element $v \in \mathscr{X}$ the equality*

$$[\,[Q]\,v] = [Q]\,[v] \tag{11}$$

holds.

Proof. Let A be a subset of $\mathscr{X}$. As one knows, from $x \perp y$ we can infer that $[Q]\,x \perp [Q]\,y$; consequently, $x \perp A$ implies $[Q]\,x \perp [Q]\,A$.

It follows that the inclusion

$$[Q]\,A^{\perp} \subset ([Q]\,A)^{\perp}$$

holds. Since $[Q]\,[Q] = [Q]$, one has the inclusion

$$[Q]\,A^{\perp} \subset [Q]\,([Q]\,A)^{\perp}. \tag{12}$$

On the other hand, from

$$x \perp [Q]\,A$$

it follows that

$$[Q]\,x \perp [Q]\,A,$$

whence one immediately infers that

$$[Q]\,x \perp A.$$

Hence the inclusion

$$[Q]([Q]A)^{\perp} \subset A^{\perp}$$

holds, whence

$$[Q]([Q]A)^{\perp} \subset [Q]A^{\perp}.$$

From (12) we get the equality

$$[Q]([Q]A)^{\perp} = [Q]A^{\perp}. \tag{13}$$

By replacing A by $v^{\perp}$ in this quality, we get

$$[Q]([Q]v^{\perp})^{\perp} = [Q]v^{\perp\perp}. \tag{14}$$

If we now replace A by $([Q]v)^{\perp}$ in (13), we get

$$[Q]([Q]([Q]v)^{\perp})^{\perp} = [Q]([Q]v)^{\perp\perp} \tag{15}$$

Since

$$([Q]v)^{\perp\perp} \subset Q^{\perp\perp} = Q,$$

equality (15) becomes

$$[Q]([Q]([Q]v)^{\perp})^{\perp} = ([Q]v)^{\perp\perp}. \tag{16}$$

Since from (13) we get that

$$[Q]([Q]v)^{\perp} = [Q]v^{\perp},$$

(14) and (16) imply that

$$[Q]v^{\perp\perp} = ([Q]v)^{\perp\perp},$$

whence, (from (5) and (6)) we get that

$$[[Q]v] = [([Q]v)^{\perp\perp}] = [[Q]v^{\perp\perp}] = [Q][v^{\perp\perp}] = [Q][v].$$

The proposition is proved.

Corollary. *If* $v_1 \perp v_2$, *then* $[v_1][v_2] = 0$, *and conversely.*

Proposition 6. *For any elements* v_1, v_2 *of the* σ*-complete vector lattice* $\mathscr{X}$, *the following equalities hold:*

$$[|v_1| \vee |v_2|] = [|v_1| + |v_2|] = [v_1] \vee [v_2], \tag{17}$$

$$[|v_1| \wedge |v_2|] = [v_1][v_2] = [v_1] \wedge [v_2]. \tag{18}$$

Proof. If $x \perp v_1$ and $x \perp v_2$, then

$$x \perp |v_1| \vee |v_2|,$$

whence

$$\{v_1, v_2\}^{\perp} \subset (|v_1| \vee |v_2|)^{\perp}. \tag{19}$$

On the other hand, if $x \in (|v_1| \vee |v_2|)^{\perp}$, then $x \perp v_1$, and $x \perp v_2$; hence the inclusion opposite to (19) holds, too. Consequently, the equality

$$\{v_1, v_2\}^{\perp\perp} = (|v_1| \vee |v_2|)^{\perp\perp}.$$

One can analogously show that

$$\{v_1, v_2\}^{\perp\perp} = (|v_1| + |v_2|)^{\perp\perp},$$

whence equalities (17) immediately follow. In order to prove equalities (18), it is sufficient to consider the case in which v_1 and v_2 are positive.

Let us first show that, on the basis of proposition 3 and of the remark made immediately after proposition 4, the following relations hold

$$[v_1 \wedge v_2] \leqslant [v_1] \wedge [v_2] = [v_1]\,[v_2].$$

By defining

$$P = [v_1] \wedge [v_2] - [v_1 \wedge v_2]$$

we get a projector [12]) and, if we take into consideration propositions 1,5 and 4, we get

$$[Pv_1 \wedge Pv_2] = [P(v_1 \wedge v_2)] = P[v_1 \wedge v_2] = 0.$$

But this implies that $Pv_1 \wedge Pv_2 = \mathbf{0}$, whence, from the corollary of proposition 5, we have

$$[Pv_1]\,[Pv_2] = \mathbf{0}.$$

Hence

$$P = P([v_1]\,[v_2]) = (P[v_1])\,(P[v_2])$$
$$= [Pv_1]\,[Pv_2] = \mathbf{0};$$

therefore, we have

$$[v_1] \wedge [v_2] = [v_1 \wedge v_2].$$

The proposition is proved.

---

[12]) If $[a] \leqslant [b]$, then $[b] - [a] = [b - [a]b]$. Indeed, we have $[a^{\perp}]b = b - [a]b$; on the other hand, $[a^{\perp}][b] = [b] - [a][b] = [b] - [a]$. It follows that $[b] - [a] = [a^{\perp}][b] = [[a^{\perp}]b] = [b - [a]b]$.

Proposition 7. *If* $v_1 \perp v_2$, *then*

$$[v_1 + v_2] = [v_1] + [v_2].$$

Proof. For any elements v_1, v_2 the equality

$$[v_1] + [v_2] = [v_1] \vee [v_2] + [v_1] \wedge [v_2] \tag{20}$$

holds [13]). If $v_1 \perp v_2$, then

$$[v_1] \wedge [v_2] = [v_1]\ [v_2] = \mathbf{0}$$

and therefore, with proposition 6, we have

$$[v_1] + [v_2] = [v_1] \vee [v_2] = [|\, v_1 \,| + |\, v_2 \,|] = [|\, v_1 + v_2 \,|] = [v_1 + v_2].$$

Remark. If $|\, v_1 \,| \leqslant |\, v_2 \,|$, then we obviously have $[v_1] \leqslant [v_2]$. It is also easy to check that, (by taking into consideration propositions 4 and 5) the relation $[v_1] \leqslant [v_2]$ is equivalent to the equality $[v_2]\, v_1 = v_1$.

Definition 3. A sequence $\{x_n\}_{n \in \mathbb{N}}$ of elements belonging to a σ-complete vector lattice $\mathscr{X}$ is said to *converge individually* to an element x, if for any elements a, $b \in \mathscr{X}$ one has the relation

$$(x_n \wedge a) \vee b \overset{o}{\longrightarrow} (x \wedge a) \vee b.$$

One writes, in this case, $x_n \overset{\text{ind}}{\longrightarrow} x$.

One can show that $x_n \overset{\text{ind}}{\longrightarrow} x$ iff

$$(x \wedge a) = \overline{\lim}\, (x_n \wedge a),$$

and

$$x \vee a = \underline{\lim}\, (x_n \vee a),$$

for any $a \in \mathscr{X}$.

One can also prove the following proposition: In order that a sequence $\{x_n\}_{n \in \mathbb{N}}$ of elements of a σ-complete vector lattice $\mathscr{X}$ converge individually to an element x, it is necessary and sufficient that for any element $v \geqslant 0$ there should exists a sequence $\{v_m\}_{m \in \mathbb{N}}$ of elements of $\mathscr{X}$, such that $[v_m] \uparrow [v]$ and, moreover

$$[v_m]\, x = (o) \text{ - } \lim_n\, [v_m]\, x_n$$

for any $m \in \mathbb{N}$.

---

13) If I denotes the identity mapping, we have $[v_1] \vee [v_2] = [\{v_1, v_2\}^{\perp\perp}] = I - [\{v_1, v_2\}^{\perp}] = I - [v_1^{\perp}][v_2^{\perp}] = I - (I - [v_1])(I - [v_2]) = [v_1] + [v_2] - [v_1] \wedge [v_2^{\perp}]$, i.e., equality (20) is true.

3.3.5 Systems of components

Let $\mathscr{X}$ be a vector lattice.

Definition 1. Any total system of mutually orthogonal components $\{Q_j\}_{j\in J}$, is called a *decomposition* of the space $\mathscr{X}$.

If Q is any component of the space $\mathscr{X}$, then the system $\{Q, Q^{\perp}\}$ is a decomposition of $\mathscr{X}$. Any component thus determines a decomposition of the space. In this section we shall consider arbitrary decompositions.

Lemma 1. *If $\{Q_j\}_{j\in I}$ is a total system of components of the space $\mathscr{X}$, then $\mathscr{X}$ can be embedded into the product vector lattice $\mathscr{Y} = \prod_{j\in J} Q_j$.*

Proof. For any $x \in \mathscr{X}$, by putting

$$P(x) = \{[Q_j]\}_{j\in J} \tag{1}$$

we get a mapping of $\mathscr{X}$ into $\mathscr{Y}$. By taking into consideration the properties of the projections, we have

$$P(\alpha x_1 + \beta x_2) = \alpha P(x_1) + \beta P(x_2),$$

whereas $x \leqslant y$ implies $P(x) \leqslant P(y)$.

If [14]) $P(x) = \mathbf{0}$, then $[Q_j]\, x = \mathbf{0}$, for any $j \in J$, and this implies that $x \perp Q_j$, $\forall j \in J$. Since the system $\{Q_j\}_{j\in J}$ is total, it follows that $x = \mathbf{0}$. Consequently, the mapping P of $\mathscr{X}$ into $\mathscr{Y}$ is injective.

If $P(x) \geqslant \mathbf{0}$, i.e., if $[Q_j]\, x \geqslant \mathbf{0}$, for any $j \in J$, then $x \geqslant \mathbf{0}$. Indeed, if this be not true, then $x_- \neq \mathbf{0}$ and this implies that there would exist an index j_0, such that $[Q_{j_0}]\, x_- \neq \mathbf{0}$. In this case

$$([Q_{j_0}]\, x)_- = [Q_{j_0}]\, x_- \neq \mathbf{0},$$

which is a contradiction. From $P(x) \leqslant P(y)$ we then infer that $x \leqslant y$. The lemma is proved.

Definition 2. If $\{x_j\}_{j\in J}$ is a family of elements of the space $\mathscr{X}$ and if $\bigvee_{j\in J}(x_j)_+$ and $\bigvee_{J\in J}(x_j)_-$ exist, then the element

$$y = \bigvee_{j\in J}(x_j)_+ - \bigvee_{j\in J}(x_j)_-$$

is called the *association* of the family $\{x_j\}_{j\in J}$ and it is denoted by $y = \mathbf{S}_{j\in J} x_j$.

If $\{x_j\}_{j\in J}$is a family of positive elements, which admits of a least upper bound, then

$$\mathbf{S}_{j\in J} x_j = \bigvee_{j\in J} x_j.$$

If $x_1, x_2, \ldots, x_n$ are mutually orthogonal elements, then

$$\mathop{\mathbf{S}}_{j=1}^{n} x_j = \sum_{j=1}^{n} x_j.$$

---

[14]) We denote by $\mathbf{0}$ also the zero element of $\mathscr{Y}$.

Lemma 2. *If* $\{x_j\}_{j\in J}$ *is a family of mutually orthogonal elements and if* $\underset{j\in J}{\mathbf{S}}x_j$ *exists, then for any index* $j_0 \in J$ *the element* $\underset{\substack{j\in J\\ j\neq j_0}}{\mathbf{S}}x_j$ *exists, too, and the equality*

$$\underset{j\in J}{\mathbf{S}}\, x_j = x_{j_0} + \underset{j\neq j_0}{\mathbf{S}}\, x_j \tag{2}$$

holds.

Proof. Let us first consider the case in which all the elements of the family are positive. In this case we shall show that from the fact that the elements are mutually orthogonal, and from the existence of $\bigvee_{j\in J} x_j$ the existence of $\bigvee_{j\neq j_0} x_j$ follows and the equality

$$\bigvee_{j\in J} x_j = x_{j_0} + \bigvee_{j\neq j_0} x_j \tag{3}$$

holds. In order to do this, let us define

$$y = (\bigvee_{j\in J} x_j) - x_{j_0}.$$

If $j \neq j_0$, then

$$x_j + x_{j_0} = x_j \vee x_{j_0}$$

whence

$$x_j + x_{j_0} \leqslant \bigvee_{j\in J} x_j,$$

and this implies that

$$x_j \leqslant \bigvee_{j\in J} x_j - x_{j_0}.$$

It follows that $x_j \leqslant y$ for $j \neq j_0$.

Let z be an element, such that $x_j \leqslant z$, for any $j \neq j_0$. We have

$$x_j \vee x_{j_0} = x_j + x_{j_0} \leqslant z + x_{j_0},\ (j \neq j_0),$$

and this implies that

$$\bigvee_{j\in J} x_j \leqslant z + x_{j_0};$$

consequently, $y \leqslant z$. We have thus proved the equality

$$y = \bigvee_{j\neq j_0} x_j,$$

which implies (3).

Let us now consider the case in which the elements x_j are arbitrary. From what we have already shown, from the existence of the elements $\bigvee_{j\in J} (x_j)_+$, $\bigvee_{j\in J} (x_j)_-$, the existence of the elements $\bigvee_{j \neq j_0} (x_j)_+$, $\bigvee_{j \neq j_0} (x_j)_-$ follows, and, therefore, the element $\mathbf{S}_{j \neq j_0} x_j$ exists. Moreover, we have

$$\bigvee_{j \in J} (x_j)_+ = (x_{j_0})_+ + \bigvee_{j \neq j_0} (x_j)_+,$$

$$\bigvee_{j \in J} (x_j)_- = (x_{j_0})_- + \bigvee_{j \neq j_0} (x_j)_-.$$

By subtracting member by member, we immediately get formula (2).

Lemma 3. *If $\{Q_j\}_{j \in J}$ is a total family of components of the space $\mathscr{X}$, then any element $x \in \mathscr{X}$ can be represented as*

$$x = \mathbf{S}_{j \in J} [Q_j]\, x.$$

Proof. Let us consider the mapping P of $\mathscr{X}$ into $\prod_{j \in J} Q_j$, given by formula (1). As we have seen in the course of the proof of lemma 1, the mapping P is an isomorphism between the ordered vector spaces $\mathscr{X}$ and $P(\mathscr{X})$.

If x is a positive element of $\mathscr{X}$, then

$$\mathbf{0} \leqslant [Q_j]\, x \leqslant x, \ (\forall j \in J).$$

If y is an element, such that $[Q_j]\, x \leqslant y$, for any $j \in J$, then

$$[Q_j]\, x \leqslant [Q_j]\, y, \ (\forall j \in J),$$

i.e., $P(x) \leqslant P(y)$. Hence we infer that $x \leqslant y$. It follows that

$$x = \bigvee_{j \in J} [Q_j]\, x.$$

If x is an arbitrary element of $\mathscr{X}$, then

$$[Q_j]\, x_+ = ([Q_j]\, x)_+, \ [Q_j]\, x_- = ([Q_j]\, x)_-,$$

and

$$x_+ = \bigvee_{j \in J} [Q_j]\, x_+ = \bigvee_{j \in J} ([Q_j]\, x)_+$$

$$x_- = \bigvee_{j \in J} [Q_j]\, x_- = \bigvee_{j \in J} ([Q_j]\, x)_-.$$

It follows that

$$x = x_+ - x_- = \bigvee_{j \in J} ([Q_j]\, x)_+ - \bigvee_{j \in J} ([Q_j]\, x)_- = \mathbf{S}_{j \in J} [Q_j]\, x,$$

and the lemma is proved.

Lemma 4. *If* $\{Q_j\}_{j\in J}$ *is a family of mutually orthogonal components and if* $x = \mathop{\mathbf{S}}_{j\in J} x_j$, *where* $x_j \in Q_j$, *then* $x_j = [Q_j]\,x$.

Proof. Let j_0 be an arbitrary index of J. By virtue of lemma 2, from the existence of $\mathop{\mathbf{S}}_{j\in J} x_j$, the existence of $\mathop{\mathbf{S}}_{j\neq j_0} x_j$ follows and also the equality

$$x = \mathop{\mathbf{S}}_{j\in J} x_j = x_{j_0} + \mathop{\mathbf{S}}_{j\neq j_0} x_j.$$

Hence we infer that

$$[Q_{j_0}]\,x = [Q_{j_0}]\,(\mathop{\mathbf{S}}_{j\in J} x_j) = [Q_{j_0}]\,x_{j_0} + [Q_{j_0}]\,(\mathop{\mathbf{S}}_{j\neq j_0} x_j).$$

By taking into account the equality $[Q_{j_0}]\,x_{j_0} = x_{j_0}$ and the fact that

$$\mathop{\mathbf{S}}_{j\neq j_0} x_j \perp Q_{j_0},$$

we immediately infer that

$$[Q_{j_0}]\,x = x_{j_0}.$$

The lemma is proved.

From lemmas 1.3 and 4 we get the following

Theorem. *If* $\{Q_j\}_{j\in J}$ *is a decomposition of the vector lattice* $\mathscr{X}$, *then any element* $x \in \mathscr{X}$ *may be represented as* $x = \mathop{\mathbf{S}}_{j\in J} x_j$, *where* $x_j \in Q_j$, *iff* $x_j = [Q_j]\,x$.

3.4 Concrete spaces

3.4.1 Finitely dimensional spaces

The set $\mathbb{R}^m$ of all families, $x = \{\xi_1, \ldots, \xi_m\}$ of m real numbers, organized as a vector space in the usual way, and in which the "natural" order relation

$$\{\xi_1, \ldots, \xi_m\} \leqslant \{\eta_1, \ldots, \eta_m\} \Leftrightarrow \xi_i \leqslant \eta_i,\ i = 1, 2, \ldots, m,$$

is taken into consideration, is a complete vector lattice.

If $\{x_j\}_{j\in J}$ is a bounded from above family of elements of $\mathbb{R}^m$ and if $x_j = \{\xi_{j_1}, \ldots, \xi_{jm}\}$ then

$$\bigvee_{j\in J} x_j = \{\bigvee_{j\in J} \xi_{j_1}, \ldots, \bigvee_{j\in J} \xi_{jm}\}.$$

If the family is bounded from below, then

$$\bigwedge_{j\in J} x_j = \{\bigwedge_{j\in J} \xi_{j1}, \ldots, \bigwedge_{j\in J} \xi_{jm}\}.$$

If $x_n = \{\xi_{n1}, \dots, \xi_{nm}\}$ and $x = \{\xi_1, \dots, \xi_m\}$, then $x = (o)\text{-}\lim x_n$ iff $\xi_j = \lim_n \xi_{nj}$, $j = 1, 2, \dots, m$. Therefore, the equality

$$(o)\text{-}\lim x_n = \{\lim_n \xi_{n1}, \lim_n \xi_{n2}, \dots, \lim_n \xi_{nm}\}$$

holds.

Theorem. *Any m-dimensional Archimedean vector lattice is isomorphic to $\mathbb{R}^m$ (endowed with the natural order relations).*

Proof. Let $\mathscr{X}$ be an m-dimensional Archimedean vector lattice. Let us consider an arbitrary positive element $x > \mathbf{0}$. Then there exists a discrete element e, such that $\mathbf{0} < e \leqslant x$.

Indeed, if the element x is not discrete, then there exist orthogonal elements a, b, such that

$$\mathbf{0} < a \leqslant x, \quad \mathbf{0} < b \leqslant x.$$

If none of the elements a, b is discrete, then there exist orthogonal elements a_1, a_2, such that

$$\mathbf{0} < a_1 \leqslant a, \quad \mathbf{0} < a_2 \leqslant a$$

and orthogonal elements b_1, b_2, such that

$$\mathbf{0} < b_1 \leqslant b, \quad \mathbf{0} < b_2 \leqslant b.$$

If none of the elements a_1, a_2, b_1, b_2 is discrete, we can proceed. Since $\mathscr{X}$ is finitely dimensional, any system of non-zero mutually orthogonal elements is finite [15]). It follows that there exists such an element e.

If x is not discrete, then $\mathbf{0} < e < x$. There then exists a number $\lambda > 0$, such that $x - \lambda e \geqslant \mathbf{0}$ and $(x - \lambda e) \perp e$. Indeed, let

$$\lambda = \sup \{\mu;\ \mu e \leqslant x\},$$

and let us denote $v = x - \lambda e$, $z = v \wedge e$. If we had $z > \mathbf{0}$, then from $\mathbf{0} < z \leqslant e$ and from the fact that e is discrete, it would follow that there exists a number $\alpha > 0$, such that $z = \alpha e$ (Section 3.2.3, proposition 4). Consequently, $(\lambda + \alpha)\, e \leqslant x$, and this is a contradiction.

Let us now denote $x_1 = \lambda e$ and $y = x - x_1$. Hence $x = x_1 + y$, where x_1 is discrete, $x_1 > \mathbf{0}$ and $x_1 \perp y$.

If y is not discrete, then y can be written as $y = x_2 + w$, where x_2 is discrete, $x_2 > \mathbf{0}$, and $x_2 \perp w$. Since the space $\mathscr{X}$ is finitely dimensional, we infer that the element $\mathscr{X}$ can be represented as the sum of a finite system of discrete, positive, mutually orthogonal elements.

---

[15]) In a vector lattice, any m mutually orthogonal elements are linearly independent, as one can easily check. If $\mathscr{X}$ is an m-dimensional vector lattice, then any system of non-zero mutually orthogonal elements has at most m elements.

Let $e_1, e_2, \ldots, e_m$ be a system of m discrete positive, linearly independent elements [16]). These elements are then mutually orthogonal (Section 3.2.3, proposition 4, corollary). If $x \in \mathscr{X}$, and $x = \xi_1 e_1 + \ldots + \xi_m e_m$, then the mapping $x \mapsto (\xi_1, \ldots, \xi_m)$ determines an isomorphism between the ordered vector spaces $\mathscr{X}$ and $\mathbb{R}^m$.

3.4.2 Spaces of real sequences

Let (s) be the set of all real sequences. If $x, y \in (s)$ and, namely, $x = \{\xi_n\}_{n \in \mathbb{N}}$, $y = \{\eta_n\}_{n \in \mathbb{N}}$, we shall define the ordered vector space structure by the formulae

$$x + y = \{\xi_n + \eta_n\}_{n \in \mathbb{N}}$$

$$\alpha x = \{\alpha \xi_n\}_{n \in \mathbb{N}}$$

$$x \leqslant y \Leftrightarrow \xi_n \leqslant \eta_n, \ (\forall n \in \mathbb{N}).$$

The set (s) becomes, in this case, a complete vector lattice.

If $\{x_j\}_{j \in J}$ is a bounded from above family of elements of (s), and if $x_j = \{\xi_{jn}\}_{n \in \mathbb{N}}$, then

$$\bigvee_{j \in J} x_j = \{\bigvee_{j \in J} \xi_{jn}\}_{n \in \mathbb{N}}.$$

If the system is bounded from below, then

$$\bigwedge_{j \in J} x_j = \{\bigwedge_{j \in J} \xi_{jn}\}_{n \in \mathbb{N}}.$$

The order convergence of a sequence $\{x_n\}_{n \in \mathbb{N}}$ (where $x_n = \{\xi_{nj}\}_{j \in \mathbb{N}}$) has the following meaning

$$(o)\text{-}\lim_n x_n = \{\lim_n \xi_{nj}\}_{j \in \mathbb{N}}.$$

The subset (m) of all bounded sequences from (s) is a normal subspace of (s).

Let p be a real number, such that $p \geqslant 1$. We shall denote by (l^p) the set of all sequences $x = \{\xi_n\}_{n \in \mathbb{N}}$ of real numbers, such that the series $\sum_{n=1}^{\infty} |\xi_n|^p$ be convergent. As a subset of the space (s), the set (l^p) is a normal subspace.

3.4.3 Spaces of real functions

(1) Let T be any set and let $F(T)$ be the set of all real functions, defined on T. We shall consider in $F(T)$ the usual vector space operations, defined by

$$(x + y)(t) = x(t) + y(t), \qquad t \in T,$$

$$(\alpha x)(t) = \alpha x(t), \qquad t \in T,$$

[16]) The fact that the elements e_i can be assumed to be discrete (and positive) is an immediate consequence of what we have already shown.

for any x, $y \in F(T)$ and any $t \in T$. We shall also consider the following order relation

$$x \leqslant y \Leftrightarrow x(t) \leqslant y(t), \quad t \in T. \tag{1}$$

The set $F(T)$, endowed with this structure, is a complete vector lattice.

If $\{x_j\}_{j \in J}$ is a bounded from above family of elements of $F(T)$, then

$$(\bigvee_{j \in J} x_j)(t) = \bigvee_{j \in J} x_j(t).$$

If the family is bounded from below, then

$$(\bigwedge_{j \in J} x_j)(t) = \bigwedge_{j \in J} x_j(t).$$

The order convergence has the following meaning

$$((o)\text{-lim } x_n)(t) = \lim_n x_n(t).$$

If T is a topological space, then the set $C(T)$ of all continuous real functions, defined on T, is a vector sublattice of $F(T)$.

(2) Let $T = [\alpha, \beta]$ be an interval of the real line and let us denote by $V(T)$ the set of all real functions, of bounded variation on T. Endowed with the same operations as in $F(T)$, the space $V(T)$ is a vector space. We shall define in $V(T)$ the relation $x \leqslant y$ by the following two conditions:

(i) $x(\alpha) \leqslant y(\alpha)$;

(ii) $x(t) - x(s) \leqslant y(t) - y(s)$, for any s, $t \in [\alpha, \beta]$, such that $s \leqslant t$.

A positive element of $V(T)$ is, therefore, an increasing, positive function.

It is easy to check that T is an ordered vector space. Since any function on T, which is of bounded variation, is the difference of two increasing, positive functions, we infer that $V(T)$ is a directed space. If A is a bounded from above set of elements of $V(T)$, and if B is the set of all upper bounds of A, then the function defined by the formula

$$z(t) = \inf_{y \in B} y(t)$$

is the least upper bound of A. This fact is easy to check. It follows that $V(T)$ is a complete vector space.

In the space $V(T)$ the absolute value of any element is given by the formula

$$|x|(t) = |x(\alpha)| + \operatorname*{var}_{\alpha}^{t} x(s).$$

The subset $CV(T)$ of the space $V(T)$, consisting of all the continuous functions, is a vector sublattice. This fact is a consequence of the fact that if x is continuous and of bounded variation, then the function $y(t) = \operatorname*{var}_{\alpha}^{t} x(s)$ is increasing and continuous. It follows that if $x \in CV(T)$, then $|x| \in CV(T)$, where the absolute value is computed in $V(T)$.

Let $AC(T)$ be the set of all (real) absolutely continuous functions, defined on T, which vanish at α. It is obvious that $AC(T)$ is a vector subspace of $V(T)$. If $x \in V(T)$, $y \in AC(T)$ and $|x| \leqslant |y|$, then

$$\operatorname*{var}_{t_1}^{t_2} x(t) \leqslant \operatorname*{var}_{t_1}^{t_2} y(t),$$

for any $t_1, t_2 \in T$, such that $t_1 < t_2$. Hence we easily infer that x is absolutely continuous, i.e., $x \in AC(T)$. It follows that $AC(T)$ is a normal subspace of the space $V(T)$. One can also prove that if $E \subset AC(T)$, and if $y = \sup E$ in the space $V(T)$, then $y \in AC(T)$. This fact implies that $AC(T)$ is a component of the space $V(T)$.

(3) Let us consider a real number $p > 1$. A real function x, defined on an interval T, is said to be of bounded p-variation on T, if there exists a number $\lambda > 0$, such that

$$\sum_{i=1}^{n} \frac{|x(t_i) - x(t_{i-1})|^p}{(t_i - t_{i-1})^{p-1}} \leqslant \lambda, \tag{2}$$

for any system of points $t_i \in T$, which determine a partition of T ($\alpha = t_0 < t_1 < \ldots < t_n = b$).

If σ is the supremum of the sums which appear in (2), then the number $\sqrt[p]{\sigma}$ is called the total p-variation of x and it is denoted by $(p)\text{-}\operatorname*{var}_{t \in T} x(t)$.

Any function of bounded p-variation is absolutely continuous.

Let us denote by $V_0^p(T)$ the set of all functions of bounded p-variation, defined on the interval $T = [\alpha, \beta]$, which vanish at α.

One can show that $V_0^p(T)$ is a normal subspace of the space $AC(T)$.

3.4.4 Spaces of analytic functions

Let us denote by $\mathscr{A}$ the set of all complex functions f, defined on the open disk $|\zeta| < 1$ of the complex plane [17]), which can be represented as the sum of an absolutely convergent series

$$f(\zeta) = \sum_{n=0}^{\infty} \alpha_n \zeta^n. \tag{1}$$

With the usual operations, the set $\mathscr{A}$ becomes a vector space. If f is of the form (1), whereas

$$g(\zeta) = \sum_{n=0}^{\infty} \beta_n \zeta^n$$

---

[17]) We denote by ζ an arbitrary point of the complex plane $\mathbf{C}$. We also denote by re ζ the real, and by im ζ the imaginary parts of ζ.

we shall define the relation $f \leqslant g$ by re $\alpha_n \leqslant$ re β_n and im $\alpha_n \leqslant$ im β_n, for any n. It is easy to check that $\mathscr{A}$ becomes an ordered vector space.

If $f \in \mathscr{A}$ and if f is given by (1), then by defining

$$f_1(\zeta) = \sum_{n=0}^{\infty} (|\text{re}\, \alpha_n| + i\,|\text{im}\, \alpha_n|)\, \zeta^n,$$

we have $f_1 \in \mathscr{A}$, $f_1 \geqslant \mathbf{0}$ and $f_1 \geqslant f$. It follows that $\mathscr{A}$ is a directed space. Let now $\{f_j\}_{j \in J}$ be a bounded from above family of elements of $\mathscr{A}$, given by

$$f_j(\zeta) = \sum_{n=0}^{\infty} \alpha_{jn} \zeta^n.$$

If we define

$$\gamma_n = (\sup_{j \in J} \text{re}\, \alpha_{jn}) + i(\sup_{j \in J} \text{im}\, \alpha_{jn}),$$

then the power series, having the numbers γ_n as coefficients, is absolutely convergent for $|\zeta| < 1$ and, by defining

$$f(\zeta) = \sum_{n=0}^{\infty} \gamma_n \zeta^n,$$

we have $f \in \mathscr{A}$ and $f = \bigvee_{j \in J} f_j$.

It follows that $\mathscr{A}$ is a complete vector lattice.

The order convergence in the space $\mathscr{A}$ is equivalent to the uniform convergence (of the considered sequence of functions) on any disk $|\zeta| \leqslant \rho < 1$. Indeed, let $f_m \in \mathscr{A}$ be given by

$$f_m(\zeta) = \sum_{n=0}^{\infty} \alpha_{mn} \zeta^n.$$

If $f_m \xrightarrow{o} 0$ in $\mathscr{A}$, then, by virtue of what we have already shown, $\lim \alpha_{mn} = 0$ for any n, and this implies that $f_m(\zeta) \to 0$ uniformly for $|\zeta| \leqslant \rho < 1$.

Conversely, let us assume that $f_m(\zeta) \to 0$ uniformly on any disk $|\zeta| \leqslant \rho < 1$. If we define

$$\mu(\rho) = \sup_{\substack{|\zeta| \leqslant \rho \\ m \in \mathbb{N}}} |f_m(\zeta)|,$$

then (by using the well-known inequalities of Cauchy and the fact that $\rho < 1$), we get

$$|\alpha_{mn}| \leqslant \frac{\mu(\rho)}{\rho^{n+1}}$$

for any m and n.

Let us now define $g_1 = \underline{\lim} f_n$ and $g_2 = \overline{\lim} f_n$ (in the space $\mathscr{A}$) and let

$$g_j(\zeta) = \sum_{n=0}^{\infty} \beta_n^j \zeta^n, \quad (j = 1,2).$$

Then

$$|\beta_n^j| \leqslant \frac{2\,\mu(\rho)}{\rho^{n+1}}, \quad (j = 1,2). \tag{2}$$

Let k be a natural number and let us define

$$\mu_k(\rho) = \sup_{\substack{|\zeta| \leqslant \rho \\ m \geqslant k}} |f_m(\zeta)|.$$

In (2) the number $\mu(\rho)$ can be replaced by $\mu_k(\rho)$. Since $\mu_k(\rho) \to 0$, it follows that $\beta_n^j = 0, j = 1, 2, n \in \mathbb{N}$. It follows that $g_1 = g_2 = \mathbf{0}$, hence $f_n \overset{o}{\to} \mathbf{0}$.

Bibliographical comments

Ordered vector spaces have been considered for the first time by F. Riesz at the International Congress of Mathematicians, held at Bologna in 1928. F. Riesz [1] considers as an ordered vector space the set of all linear functionals defined on a space of real continuous functions. In another paper (F. Riesz [2]) he considers ordered vector spaces of functionals defined on an abstract set.

The vector lattices are also called Riesz spaces by N. Bourbaki [4].

An axiomatic presentation of the theory of ordered vector spaces of various types has been made by L. V. Kantorovitch (see [1], [2]) in 1935—1936, and independently, by H. Freudenthal [1], in 1936. The theory of ordered vector spaces develops from these papers, which have been followed by many others, due to L. V. Kantorovitch, G. Birkhoff, H. Nakano, B. Z. Vulich, A. G. Pinsker, etc.

L. V. Kantorovitch and his co-workers mainly considered the complete vector lattices, which are also called K-spaces. The majority of the results obtained by them during the period 1935—1950 are presented in the book published in 1950 by L. V. Kantorovitch, B. Z. Vulich and A. G. Pinsker [1].

H. Nakano also published, in 1950, two books on ordered vector spaces (see H. Nakano, [10], [11]).

We also mention the author's book, published in 1959 (R. Cristescu, [15]) and the book of B. Z. Vulich [16], published in 1961.

The arbitrary ordered vector spaces have been studied by G. Birkhoff [4], whereas the directed spaces were considered by D. Wassilkoff [1].

In a directed vector space $\mathscr{X}$ an element $u > 0$ is said to be *axial* if for any $x \in \mathscr{X}$ there exists a real number $\alpha > 0$ such that $u > \alpha x$.

The spaces in which axial elements exist have been studied by D. Wassilkoff [1] and by N. Popa [1].

The ordered groups were studied by G. Birkhoff [6], A. H. Clifford [1], E. P. Shimbireva [1], F. Šik [1], P. Jaffard [1], R. Cristescu [16]. One can also see the expository paper by L. V. Kantorovitch, B. Z. Vulich and A. G. Pinsker [2].

The topology of the (o)-boundedness in ordered vector spaces was studied by I. Namioka [1] and H. H. Schaefer [2]. Concerning other topologies, generated by the order relation, one can see G. Birkhoff [7], L. V. Kantorovitch, B. Z. Vulich, A. G. Pinsker [1], B. Z. Vulich [16], A. L. Peressini [1], A. V. Potepun [1], etc.

The property of decomposition of the elements (Section 3.1.1, proposition 3) is also known as the Riesz decomposition property, because F. Riesz [2] used it for the first time as an axiom for certain semi-groups.

The Riesz decomposition property is useful for establishing some results, and it can sometimes replace the condition that the space be a lattice (for instance, this is the case with the operator theory on vector lattices).

The infinite distributivity laws in vector lattices were established by H. Freudenthal [1] (in the countable case) and by B. Z. Vulich [2] (for the general case). G. Birkhoff remarked that they still hold for the non-commutative groups. Concerning the infinite distributivity laws, one can also see H. Nakano [10].

The convergence with regulator in vector lattices was introduced by L. V. Kantorovitch [13].

The individual convergence was introduced by H. Nakano [9].

The notion of association of a family of elements was introduced for the first time in the monograph by L. V. Kantorovitch, B. Z. Vulich and A. G. Pinsker [1], for the case of complete vector spaces, and under the additional condition that the elements of the family be mutually orthogonal. The general definition given in this chapter was introduced by F. Šik [1].

The extension of the ordered vector spaces was studied by A. I. Iudin [2], D. A. Wassilkoff [1], G. Birkhoff [7].

The quasi-σ-complete spaces were studied by C. D. Aliprantis and E. Langford [1].

The notion of projection was introduced by H. Freudenthal [1]. Various definitions have been given by F. Riesz, H. Nakano, F. Maeda, S. Bochner and R. S. Phillips. The principal projections in arbitrary vector lattices were thoroughly studied by H. Nakano [10].

If v is an element of a vector lattice and if $v^{\perp\perp}$ is a component, then one can define the projection $[v]$ as in the case of the σ-complete vector lattices. One can prove (see B. C. Anderson and H. Nakano [1], H. Nakano [18]) that if in a vector lattice $\mathscr{X}$ the set $v^{\perp\perp}$ is a component, for any $v \in \mathscr{X}$, and if any (ρ)-Cauchy sequence (with the natural definition) is (o)-convergent, then $\mathscr{X}$ is σ-complete.

The theorem on decomposition (Section 3.3.5) was established by A. G. Pinsker for the case of the complete vector lattices (see L. V. Kantorovitch, B. Z. Vulich, A. G. Pinsker [1]). This theorem was generalized by F. Šik [1], for lattice ordered groups, and by R. Cristescu [16], for the directed groups.

The theorem on the representation of the finitely dimensional Archimedean vector lattices (Section 3.4.1) was first proved by A. I. Iudin [1].

D. Wassilkoff [1] shows that any Archimedean directed ordered vector space is isomorphic to a space of real functions (on a certain set T), endowed with the natural order.

C. Goffman [2] obtains representations of the complete vector spaces by Caratheodory function spaces.

Spaces of analytic functions were considered by A. I. Fet [1] in connection with their applications to the Cauchy-Kowalewska problem.

4

Vector lattices with special properties

In this chapter vector lattices are considered which satisfy some additional conditions, such as the existence of total elements, the condition that (o)-convergence coincide with (ρ)-convergence, etc. In one of its sections lattice ordered algebras are studied.

4.1 Vector lattices with unit element

4.1.1 The notion of space with unit

Let $\mathscr{X}$ be a vector lattice. In accordance with the definition of a total set (Section 3.3.2), an element $c \in \mathscr{X}$ is said to be *total* if $x \perp c$ implies $x = \mathbf{0}$.

Definition 1. A vector lattice is said to be a *space with unit* if it has total elements and if a positive total element u is chosen in it. The element u is called a *unit element.*

Definition 2. If $\mathscr{X}$ is a space with unit, any element $e \in \mathscr{X}$ for which $e \wedge (u - e) = \mathbf{0}$ is called a *unitary element*. The set of all unitary elements of $\mathscr{X}$ is called the *base* of the space $\mathscr{X}$.

In this chapter we shall denote by $\mathscr{B}(\mathscr{X})$ the base of a space $\mathscr{X}$, with unit $\boldsymbol{u}$.

Proposition 1. *In a space with unit, the unitary elements have the following properties.*

(i) *If* $\{e_j\}_{j \in J}$ *is a family of unitary elements and if there exists* $e = \bigvee_{j \in J} e_j$ *(or* $e = \bigwedge_{j \in J} e_j$*), then the element* e *is unitary.*

(ii) *if* $e = e_1 + e_2$ *and if the elements* e_1, e_2 *are unitary, then the element* e *is unitary, if and only if,* $e_1 \perp e_2$.

(iii) *any projection of the unit element is a unitary element.*

Proof. (i) If

$$e = \bigvee_{j \in J} e_j,$$

then

$$\mathbf{0} \leqslant u - e \leqslant u - e_j, \quad (\forall j \in J)$$

and from the equality

$$e_j \wedge (u - e_j) = \mathbf{0}$$

we infer that

$$e_j \wedge (u - e) = \mathbf{0}.$$

By using the distributivity laws (Section 3.2.1, proposition 2) we get

$$e \wedge (u - e) = (\bigvee_{j \in J} e_j) \wedge (u - e) = \bigvee_{j \in J} (e_j \wedge (u - e)) = \mathbf{0}$$

and, consequently, e is a unitary element.

If

$$e = \bigwedge_{j \in J} e_j$$

then

$$u - e = \bigvee_{j \in J} (u - e_j)$$

and, from what we have already proved, we infer that $u - e$ is a unitary element; this shows that e is unitary too.

(ii) If $e_1 \perp e_2$, then

$$e = e_1 + e_2 = e_1 \vee e_2$$

and from (i) we infer that the element e is unitary.

Conversely, if the element $e_1 + e_2$ is unitary, then $e_1 + e_2 \leqslant u$; hence $e_1 \leqslant u - e_2$ and this implies that $e_1 \perp e_2$.

(iii) If Q is a component of $\mathscr{X}$, then

$$u = [Q]\, u + [Q^{\perp}]\, u;$$

from the relations

$$[Q]\, u \perp [Q^{\perp}]\, u$$

and

$$[Q^{\perp}]\, u = u - [Q]\, u$$

it follows that the element $[Q]\, u$ is unitary.

Remark. The base of a space with unit is a Boolean lattice. Indeed, the set $\mathscr{B}(\mathscr{X})$ with the order relation induced by that of $\mathscr{X}$ has $\mathbf{0}$ as smallest element and u as greatest element. From proposition 1 we infer that if $e_1, e_2 \in \mathscr{B}(\mathscr{X})$, then $e_1 \vee e_2 \in \mathscr{B}(\mathscr{X})$ and $e_1 \wedge e_2 \in \mathscr{B}(\mathscr{X})$. The set $\mathscr{B}(\mathscr{X})$ is distributive lattice because the distributivity laws hold in $\mathscr{X}$. For any element $e \in \mathscr{B}(\mathscr{X})$ the element $u - e$ plays the role of the complement of e in the Boolean lattice $\mathscr{B}(\mathscr{X})$.

Definition 3. If $\mathscr{X}$ is a space with unit, any element x which can be represented in the following form

$$x = \sum_{i=1}^{n} \alpha_i e_i,$$

where e_i are unitary elements, whereas α_i are real numbers, is called a *simple element.*

The set of the simple elements of a space with unit is a vector sublattice.

Examples. (1) In the space (s) (Section 3.4.2) any element $x = \{\xi_n\}_{n \in \mathbb{N}}$, such that $\xi_n \neq 0$ for any $n \in \mathbb{N}$, is total. The element $u = \{\alpha_n\}_{n \in \mathbb{N}}$ such that $\alpha_n = 1$, for any $n \in \mathbb{N}$, is usually chosen to be the unit element of the space (s).

In this case, an element $e = \{\lambda_n\}_{n \in \mathbb{N}}$ is unitary iff $\lambda_n \in \{0,1\}$, for any $n \in \mathbb{N}$.

(2) In the space $F(T)$ (Section 3.4.3) any element x, such that $x(t) \neq 0$ for any $t \in T$, is total. The element u, such that $u(t) = 1$, for any $t \in T$, is usually chosen to be the unit element. In this case the base of the space $F(T)$ consists of all the characteristic functions of all the subsets of T.

(3) Let us consider the set of all the Lebesgue measurable, real functions, defined on an interval T. In this set we consider the following equivalence relation: $x \sim y$ iff $x(t) = y(t)$ almost everywhere. Let $S(T)$ be the set of all the equivalence classes. Let us denote by $\dot{x}$ the class determined by an element x. The operations $\dot{x} + \dot{y}$ and $\alpha\dot{x}$ in $S(T)$ can be defined in a natural manner. We shall write $\dot{x} \leqslant \dot{y}$ iff $x(t) \leqslant y(t)$ almost everywhere. It is obvious that $S(T)$ becomes a vector lattice [1]). If $\dot{x} \in S(T)$ and $x(t) \neq 0$ almost everywhere, then $\dot{x}$ is a total element. The element $\dot{u}$, such that $u(t) \equiv 1$, is usually chosen to be the unit element. In this case an element $\dot{e} \in S(T)$ is unitary iff e is equal almost everywhere to the characteristic function of a measurable subset of T.

4.1.2 σ-complete vector lattices with unit

Proposition 1. *For any σ-complete vector lattice there exists a decomposition into spaces with unit.*

Proof. Let $\mathscr{X}$ be a σ-complete vector lattice and let $\mathscr{E}$ be the set of all the subsets of $\mathscr{X}$ consisting of (non-zero) positive, mutually orthogonal elements. The set $\mathscr{E}$ is not empty, since for any element $x_0 > \mathbf{0}$, which belongs to $\mathscr{X}$, we have $\{x_0\} \in \mathscr{E}$.

The set $\mathscr{E}$ obviously satisfies the conditions in order to apply Zorn's lemma and, consequently, it has a maximal element E. Any element $x \in E$ is total in the component Q_x generated by x, and one can choose it to be the unit element in this component. The system $\{Q_x\}_{x \in E}$ gives a decomposition of the space $\mathscr{X}$. The proposition is proved.

---

1) In Section 6.4.1, we shall show that $S(T)$ is even a complete vector lattice.

Corollary. *Any complete vector lattice is isomorphic to a total, normal subspace of a complete vector lattice with unit.*

Indeed, with the notations used in the proof of proposition 1, let us define $\mathscr{Y} = \prod_{x \in E} Q_x$.

Then the space $\mathscr{X}$ is isomorphic to a subspace $\mathscr{Y}_0$ of $\mathscr{Y}$. It remains only to remark that $\mathscr{Y}_0$ is a total, normal subspace of $\mathscr{Y}$ and that the element of $\mathscr{Y}$ given by E is total in $\mathscr{Y}$.

In the sequel of this section $\mathscr{X}$ will stand for a σ-complete vector lattice with unit.

Remarks. (1) If Q is a component of $\mathscr{X}$ and if $e = [Q]u$, then $[e] = [Q]$ (see Chapter III, § 3.4, proposition 5). It follows that any projector in $\mathscr{X}$ is principal and it is defined by a unitary element. In particular, when the unit is changed, the new base is isomorphic to the first one.

(2) As we have seen in the preceding section, any projector of the unit element is a unitary element. Conversely, if $e \in \mathscr{B}(\mathscr{X})$, then the element e is a projection of the unit element, because $e = [e]u$.

(3) If $e_1, e_2 \in \mathscr{B}(\mathscr{X})$, then

$$[e_1]e_2 = e_1 \wedge e_2.$$

Indeed, by taking into account the properties of the principal projectors, we get the following equalities

$$e_1 \wedge e_2 = [e_1 \wedge e_2]\, u = [e_1]\, [e_2]u = [e_1]e_2.$$

(4) If $\{e_j\}_{j \in J}$ is a system of mutually orthogonal elements of $\mathscr{X}$, if $e = \bigvee_{j \in J} e_j$ and if $e \in \mathscr{B}(\mathscr{X})$, then $e_j \in \mathscr{B}(\mathscr{X})$, for any $j \in J$.

Indeed, if $h \in J$, we have

$$[e_h]u = [e_h][e]u = [e_h]e = \bigvee_{j \in J} [e_h]e_j = e_h.$$

It follows that e_h is a unitary element.

Proposition 2. *If $[e_n]_{n \in \mathbb{N}}$ is a sequence of unitary elements of the σ-complete vector lattice $\mathscr{X}$ and if $e_n \downarrow \mathbf{0}$, then $[e_n]x \xrightarrow{o} \mathbf{0}$, for any $x \in \mathscr{X}$.*

Proof. Let us first assume that $x \geqslant \mathbf{0}$ and let $a = \bigwedge_{n \in \mathbb{N}} [e_n]x$. We have

$$[a]u \leqslant [[e_n]x]u = ([e_n][x])u = e_n \wedge [x]u$$

for any n. It follows that $a = \mathbf{0}$ and, consequently, $[e_n]x \downarrow \mathbf{0}$.

If x is arbitrary, then

$$[e_n]x = [e_n]x_+ - [e_n]x_- \xrightarrow{o} \mathbf{0}.$$

Proposition 3. *For any element $x > \mathbf{0}$ of the σ-complete vector lattice $\mathscr{X}$ there exists a unitary element $e > \mathbf{0}$ and a number $\alpha > \mathbf{0}$, such that $x > \alpha e$.*

Proof. Since $\mathscr{X}$ is Archimedean, there exists a number $\alpha > 0$ such that $x \nleqslant \alpha u$. The element e defined by

$$e = [(x - \alpha u)_+]u$$

is unitary and different from zero. On the other hand, we have

$$[(x - \alpha u)_+]\,(x - \alpha u) = (x - \alpha u)_+ > \mathbf{0}$$

and, consequently,

$$x \geqslant [e]x = [(x - \alpha u)_+]\,x > [(x - \alpha u)_+]\,(\alpha u)$$

$$= [e]\alpha\, u = \alpha e.$$

The proof is complete.

In the following theorem $\mathscr{S}(\mathscr{X})$ stands for the set of all the simple elements of a space with unit.

Theorem. *If $\mathscr{X}$ is a complete vector lattice with unit, then for any element $x > \mathbf{0}$ of $\mathscr{X}$ the following equality holds*

$$x = \sup\{y; \mathbf{0} \leqslant y \leqslant x; \quad y \in \mathscr{S}(\mathscr{X})\}. \tag{1}$$

Proof. Let us denote

$$S(x) = \{y; \mathbf{0} \leqslant y \leqslant x; \quad y \in \mathscr{S}(\mathscr{X})\},$$

and let

$$z = \sup S(x).$$

Obviously, $z \leqslant x$. Let us assume that $z < x$ and, consequently, $x - z > \mathbf{0}$. With proposition 3 we infer that there exists a unitary element $e > \mathbf{0}$ and real number $\alpha > 0$, such that $x - z > \alpha e$. If $y \in S(x)$, then $y + \alpha e \in \mathscr{S}(\mathscr{X})$ and from

$$y + \alpha e \leqslant z + \alpha e < x$$

we get $y + \alpha e \in S(x)$. We inductively get that $y + ne \in S(x)$ for any $n \in \mathbb{N}$. But this implies that $ne \leqslant x - y$, for any $n \in \mathbb{N}$, and this contradicts the fact that $\mathscr{X}$ is Archimedean. It follows that $z = x$ and equality (1) is proved.

4.1.3 Integral representation

Let $\mathscr{X}$ be a σ-complete vector lattice with unit and let $\mathscr{B}(\mathscr{X})$ be the base of this space.

Let φ be a real-valued function defined on $\mathbb{R}$, and g an $\mathscr{X}$-valued function, defined on $\mathbb{R}$.

Let us consider a partition Δ of the real axis, given by the points λ_i ($i = 0, \pm 1, \pm 2, \ldots$), such that $\lambda_i < \lambda_{i+1}$ and $\lambda_{i+1} - \lambda_i \leqslant \varepsilon$ (for a given ε). Let us assume that for any such partition and any choice of the intermediate points γ_i(i.e., $\lambda_i \leqslant \gamma_i \leqslant \lambda_{i+1}$) the following sum

$$s_\Delta = (o) \text{-} \sum_{-\infty}^{+\infty} \varphi(\gamma_i)\,(g(\lambda_{i+1}) - g(\lambda_i)) \tag{1}$$

exists [2]). Let $\nu(\Delta)$ be the norm of the partition Δ, i.e.

$$\nu(\Delta) = \sup\,(\lambda_{i+1} - \lambda_i).$$

For any sequence $\{\Delta_n\}_{n \in \mathbb{N}}$ of partitions, such that $\nu(\Delta_n) \to 0$, we shall consider a sequence of elements of the form (1), where the intermediate points are chosen arbitrarily for each partition Δ_n.

If for any sequence $\{\Delta_n\}_{n \in \mathbb{N}}$, such that $\nu(\Delta_n) \to 0$, the sequence $\{s_{\Delta_n}\}_{n \in \mathbb{N}}$ is (o)-convergent to a given element $x \in \mathscr{X}$, which is independent of the chosen sequence of partitions, one writes

$$x = \int_{-\infty}^{+\infty} \varphi(\lambda)\,\mathrm{d}g(\lambda).$$

Definition. For any element $x \in \mathscr{X}$, the function $g^x \colon \mathbb{R} \to \mathscr{B}(\mathscr{X})$, defined by the formula

$$g^x(\lambda) = [(\lambda u - x)_+]u,$$

is called the *spectral function* of x.

It is obvious that the spectral function of an element x is an increasing function.

For any λ one has the inequality

$$[g^x(\lambda)]x \leqslant \lambda g^x(\lambda). \tag{2}$$

Indeed, we have

$$\lambda g^x(\lambda) - [g^x(\lambda)]x = [g^x(\lambda)]\,(\lambda u - x) = (\lambda u - x)_+ \geqslant \mathbf{0},$$

whence (2) immediately follows.

---

[2]) For a system $\{x_i; \mathrm{i} = 0, \pm 1, \pm 2, \ldots\}$ of elements of a σ-complete vector lattice one writes.

$$(o)\text{-}\sum_{-\infty}^{+\infty} x_i = (o)\text{-}\sum_{i=0}^{\infty} x_i + (o)\text{-}\sum_{i=0}^{\infty} x_{-i},$$

if the series in the right-hand member are converging.

With the help of (2) we can prove the following equality

$$\mathbf{0} = \bigwedge_{\lambda \in \mathbb{R}} g^x(\lambda). \tag{3}$$

Indeed, let $\lambda_n \downarrow -\infty$ and let us define

$$e = \bigwedge_{n \in \mathbb{N}} g^x(\lambda_n).$$

By taking into account equality (2), we get

$$[e]x = [e]\,[g^x(\lambda_n)]x \leqslant \lambda_n[e]g^x(\lambda_n) = \lambda_n e;$$

it follows that

$$-\lambda_n e \leqslant -[e]x, \quad \forall\, n \in \mathbb{N}.$$

Since $-\lambda_n \uparrow +\infty$ and $\mathscr{X}$ is Archimedean, it follows that $e = \mathbf{0}$. Equality (3) now immediately follows.

Similarly, one can prove the equality

$$u = \bigvee_{\lambda \in \mathbb{R}} g^x(\lambda). \tag{4}$$

Let us now state that if $\alpha, \beta \in \mathbb{R}$ and $\alpha < \beta$, then

$$\alpha(g^x(\beta) - g^x(\alpha)) \leqslant ([g^x(\beta)] - [g^x(\alpha)])\, x \leqslant \beta(g^x(\beta) - g^x(\alpha)). \tag{5}$$

Indeed, by using (2), in which we make $\lambda = \beta$, we get

$$\begin{aligned}([g^x(\beta)] - [g^x(\alpha)])x &= ([g^x(\beta)] - [g^x(\alpha)])\,[g^x(\beta)]x \\ &\leqslant \beta([g^x(\beta)] - [g^x(\alpha)])(g^x(\beta)) = \beta([g^x(\beta)]\,g^x(\beta) - [g^x(\alpha)]\,g(\beta)) \\ &= \beta(g^x(\beta) - g^x(\alpha) \wedge g^x(\beta)) = \beta(g^x(\beta) - g^x(\alpha)),\end{aligned}$$

and this shows that the second inequality in (5) is true. The first one can be proved in a similar manner.

Theorem. *Any element x of a σ-complete vector lattice with unit can be represented in the form*

$$x = \int_{-\infty}^{+\infty} \lambda \mathrm{d} g^x(\lambda). \tag{6}$$

Proof. Let us consider a partition Δ of $\mathbb{R}$ given by the points $\lambda_i (i = 0, \pm 1, \pm 2, \ldots)$ such that $\lambda_{i+1} - \lambda_i \leqslant \varepsilon$. Since the function g^x is increasing, from equalities (3) and (4) one gets

$$u = (o)\text{-}\sum_{-\infty}^{\infty} (g^x(\lambda_{i+1}) - g^x(\lambda_i)).$$

Hence we get [3])

$$x = (o)\text{-}\sum_{-\infty}^{+\infty} [g^x(\lambda_{i+1}) - g^x(\lambda_i)]x$$

On the other hand, by using (5) (in which we make $\alpha = \lambda_i$, $\beta = \lambda_{i+1}$) we get

$$|\gamma_i(g^x(\lambda_{i+1}) - g^x(\lambda_i)) - [g^x(\lambda_{i+1}) - g^x(\lambda_i)]x| \leqslant \varepsilon(g^x(\lambda_{i+1}) - g^x(\lambda_i)).$$

if $\lambda_i \leqslant \gamma_i \leqslant \lambda_{i+1}$.

Hence we infer that the following element

$$s_\Delta = (o)\text{-}\sum_{-\infty}^{\infty} \gamma_i(g^x(\lambda_{i+1}) - g^x(\lambda_i)) \tag{7}$$

exists and that we have the following relations

$$|s_\Delta - x| = \left|\sum_{-\infty}^{\infty} \gamma_i(g^x(\lambda_{i+1}) - g^x(\lambda_i)) - \sum_{-\infty}^{\infty} [g^x(\lambda_{i+1}) - g^x(\lambda_i)]x\right|$$

$$\leqslant \sum_{-\infty}^{+\infty} \varepsilon(g^x(\lambda_{i+1}) - g^x(\lambda_i)) = \varepsilon u.$$

It follows that if $\nu(\Delta) \leqslant \varepsilon$, then

$$|s_\Delta - x| \leqslant \varepsilon u. \tag{8}$$

The proof is complete.

Remarks. (1) From (8) we infer that if $\nu(\Delta_n) \to 0$, then $s_{\Delta_n} \xrightarrow{\rho} x$. The element u plays the role of a convergence regulator.

(2) If $x, y \in \mathscr{X}$ and $g^x(\lambda) = g^y(\lambda)$ for any rational λ, then $x = y$.

This is an immediate consequence of the preceding theorem.

(3) If $x \in \mathscr{X}$ and if there exists a number ξ such that $|x| \leqslant \xi u$, then $g^x(\lambda) = \mathbf{0}$ for $\lambda \leqslant -\xi$ and $g^x(\lambda) = u$ for $\lambda > \xi$.

---

[3]) By defining $e_i = g^x(\lambda_{i+1}) - g^x(x_i)$, we have $u = \bigvee e_i$ and, consequently, for $x \geqslant \mathbf{0}$, we have $x = \bigvee [e_i]x$. For an arbitrary x, by taking into account the fact that the elements e_i are mutually orthogonal, it follows that

$$x = S[e_i]x = \sum [e_i]x.$$

Indeed, if $\lambda \leqslant -\xi$, then

$$\lambda u - x \leqslant \lambda u + \xi u \leqslant \mathbf{0}$$

and, therefore, $(\lambda u - x)_+ = \mathbf{0}$. It follows that $g^x(\lambda) = \mathbf{0}$.

If $\lambda > \xi$, then $\lambda u - x > \mathbf{0}$ and, therefore,

$$(\lambda u - x)_+ = \lambda u - x \geqslant (\lambda - \xi)u,$$

which implies that $g^x(\lambda) = u$.

In this case, in formula (7), the series reduces to a finite sum and, therefore, the integral in (6) reduces to an integral on a finite interval.

In particular, there exists a sequence $\{s_n\}_{n \in \mathbb{N}}$ of simple elements, which (ρ)-converges to x with respect to the element u, taken as a convergence regulator.

4.1.4 Spaces with strong unit and spaces with M-unit

Definition 1. If $\mathscr{X}$ is a vector lattice with unit, any element $x \in \mathscr{Y}$, such that there exists a $\xi \in \mathbb{R}$, for which $|x| \leqslant \xi u$, is called a *bounded element*.

The set $\mathscr{M}(\mathscr{X})$ of the bounded elements of $\mathscr{X}$ is a normal, total subspace of $\mathscr{X}$.

Definition 2. A space with unit, in which any element is bounded, is called a *space with strong unit*.

Remarks. (1) Let $\mathscr{X}$ be a vector lattice in which there exists an element u with the following property: for any element $x \in \mathscr{X}$ there is a $\xi \in \mathbb{R}$ such that $|x| \leqslant \xi u$.

Then u is total in $\mathscr{X}$ and, by choosing u as a unit element, $\mathscr{X}$ becomes a space with strong unit.

(2) If $\mathscr{X}$ is a vector lattice, then, for any element $v > \mathbf{0}$ of $\mathscr{X}$, the set E_v which consists of all the elements $x \in \mathscr{X}$, such that there exists a $\xi \in \mathbb{R}$ for which $|x| \leqslant \xi v$, is a space with strong unit (if v is chosen to be the unit). The space E_v is the smallest normal subspace of $\mathscr{X}$ which contains v.

Definition 3. Any σ-complete vector lattice $\mathscr{X}$ which has a unit u, and which contains a total set E of unitary elements such that the following property holds:

for any $x \in \mathscr{X}$ and $e \in E$ there exists a $\xi \in \mathbb{R}$ such that

$$[e]|x| \leqslant \xi u \tag{1}$$

is called a *space with* M-*unit.*

Inequality (1) is obviously equivalent to the following one

$$[e]\,|x| \leqslant \xi e. \tag{2}$$

Proposition 1. *The product of any family of σ-complete vector lattices with strong unit is a space with M-unit.*

Proof. Let $\{\mathscr{X}_j\}_{j\in J}$ be a family of σ-complete vector lattices with strong unit. For any $j \in J$ let us denote by u_j the unit element in $\mathscr{X}_j$. Let us consider the product σ-complete vector lattice

$$\mathscr{X} = \prod_{j\in J} \mathscr{X}_j$$

Let us denote $u = \{u_j\}_{j\in J}$. It is easy to show that u is a total element of $\mathscr{X}$. We shall choose u as unit element in $\mathscr{X}$.

It is also easy to check that if we define [4]) $e_j = \{\delta_{jh}u_h\}_{h\in J}$ then we get unitary elements in $\mathscr{X}$, whereas the system $\{e_j\}_{j\in J}$ is total in $\mathscr{X}$.

If $x = \{x_j\}_{j\in J} \in \mathscr{X}$ and

$$|x_j| \leqslant \lambda_j u_j,$$

then

$$[e_j]|x| = \{\delta_{jh}|x_h|\}_{h\in J} \leqslant \{\delta_{jh}\lambda_h u_h\}_{h\in J} = \lambda_j e_j.$$

The proof is complete.

Proposition 2. *Any space with M-unit can be embedded into a product of spaces with strong unit.*

Proof. It is enough to remark that (with the notations of the definition 3) $\{[e]\mathscr{X}\}_{e\in E}$ is a total system of components with strong unit (if e is chosen to be the unit in $[e]\mathscr{X}$). The proposition then follows from what we have shown in Section 3.3.5 (lemma 1).

4.2 Regular spaces, spaces of countable type, maximal spaces

4.2.1 Regular and σ-regular spaces

Definition 1. A σ-complete vector lattice, in which the (o)-convergence and the (ρ)-convergence of the sequences are equivalent to one another, is called a *regular* space.

Proposition 1. *A σ-complete vector lattice $\mathscr{X}$ is a regular space iff for any sequence $\{x_n\}_{n\in\mathbb{N}}$ of elements of $\mathscr{X}$, which is (o)-convergent to* $\mathbf{0}$, *there exists a sequence $\{\alpha_n\}_{n\in\mathbb{N}}$ of positive real numbers such that $\alpha_n \to +\infty$ and $\alpha_n x_n \xrightarrow{o} \mathbf{0}$.*

Proof. Let us assume that $\mathscr{X}$ is regular. Let $x_n \xrightarrow{o} \mathbf{0}$ and let r be a convergence regulator of the sequence. Then there exists an increasing sequence $\{n_k\}_{k\in\mathbb{N}}$ of natural numbers, such that

$$|x_n| \leqslant \frac{1}{k} r \qquad (n \geqslant n_k).$$

---

[4]) Here δ_{jh} stands for the Kronecker symbol.

Let us define $\lambda_n = k$ for $n_k \leqslant n < n_{k+1}$; then we have $\sqrt{\lambda_n} \to +\infty$ and

$$|\sqrt{\lambda_n}\, x_n| \leqslant \frac{1}{\sqrt{\lambda_n}} r, \qquad (n \geqslant n_1);$$

consequently, $\sqrt{\lambda_n}\, x_n \xrightarrow{o} \mathbf{0}$.

Conversely, let us now assume that $\mathscr{X}$ satisfies the condition from the statement of the proposition. Let $x_n \xrightarrow{o} x$ and let $\{\alpha_n\}_{n\in\mathbb{N}}$ be a sequence of positive numbers, such that $\alpha_n \to +\infty$ and $\alpha_n|x_n - x| \xrightarrow{o} \mathbf{0}$. If r is an upper bound of the sequence $\{\alpha_n|x_n - x_n|\}_{n\in\mathbb{N}}$, then

$$|x_n - x| \leqslant \frac{1}{\alpha_n} r$$

and this shows that $x_n \xrightarrow{o} x$.

Definition 2. Any σ-complete vector lattice in which any countable set of (o)-convergent sequences has a common convergence regulator is called a *σ-regular space.*

Proposition 2. *A σ-complete vector lattice $\mathscr{X}$ is a σ-regular space iff it satisfies the following condition: if (o)-$\lim\limits_m x_{nm} = x_n$ and (o)-$\lim x_n = x$, then there exists an increasing sequence $\{m_n\}_{n\in\mathbb{N}}$ of natural numbers, such that $x = (o)$-$\lim x_{nm_n}$.*

Proof. Let us assume that $\mathscr{X}$ is a σ-regular space. Let (o)-$\lim x_{nm} = x_n$. Let r be the common convergence regulator of the sequences $\{x_{nm}\}_{m\in\mathbb{N}}$. If $\{\varepsilon_n\}_{n\in\mathbb{N}}$ is a sequence of positive numbers, such that $\varepsilon_n \downarrow 0$, then there exists an increasing sequence $\{m_n\}_{n\in\mathbb{N}}$ of natural numbers, such that

$$|x_{nm_n} - x_n| \leqslant \varepsilon_n r.$$

From

$$|x_{nm_n} - x| \leqslant |x_{nm_n} - x_n| + |x_n - x| \leqslant \varepsilon_n r + |x_n - x|$$

and from $x_n \xrightarrow{o} x$ it follows that $x_{nm_n} \xrightarrow{o} x$.

Conversely, let us assume that the space $\mathscr{X}$ satisfies the condition from the statement of the proposition.

We shall first show that the space is regular; more precisely, it has the property given in proposition 1. Let $x_n \xrightarrow{o} 0$, and define

$$y_m = \bigvee_{i\geqslant m} |x_i|.$$

We then have $y_m \downarrow \mathbf{0}$ and, therefore, for any natural number n we have (o)-$\lim\limits_m n y_m = \mathbf{0}$.

It follows that there exists an increasing sequence $\{m_n\}_{n\in\mathbb{N}}$ of natural numbers, such that (o)-$\lim\limits_n n y_{m_n} = \mathbf{0}$. If we define $\lambda_m = n$ for $m_n \leqslant m < m_{n+1}$, it follows that

$\lambda_m y_m \xrightarrow{o} \mathbf{0}$ and $\lambda_m \to +\infty$. But this implies that $\lambda_m x_m \xrightarrow{o} 0$ and, if we take into account proposition 1, the space $\mathscr{X}$ is regular.

Let us now show that for any countable system of (o)-convergent sequences there exists a common convergence regulator. Let then (o)-lim $x_{nm} = x_n (n \in \mathbb{N})$ and let r_n be the convergence regulator of the sequence $\{x_{nm}\}_{m\in\mathbb{N}}$. For any number $\eta > 0$ there exists a natural number $k = k(\eta, n)$, such that

$$|x_{nm} - x_n| \leqslant \eta r_n, \ (m \geqslant k(\eta, n)). \tag{1}$$

On the other hand, there exists a sequence $\{\lambda_n\}_{n\in\mathbb{N}}$ of strictly positive numbers, such that the sequence $\{\lambda_n r_n\}_{n\in\mathbb{N}}$ be bounded. Indeed, for any $n \in \mathbb{N}$ we have $(o)\text{-}\lim\limits_m \frac{1}{m} r_n = \mathbf{0}$; consequently, there exists an increasing sequence $\{m_n\}_{n\in\mathbb{N}}$ of natural numbers, such that $\frac{1}{m_n} r_n \xrightarrow{o} \mathbf{0}$.

Let then $\{\lambda_n\}_{n\in\mathbb{N}}$ and $r \in \mathscr{X}$ be such that $\lambda_n r_n \leqslant r$, for any n. Let us fix a natural number n and take $\eta = \varepsilon\lambda_n$. We then have, if we take (1) into account,

$$|x_{nm} - x_n| \leqslant \varepsilon r, \ (m \geqslant k_0(\varepsilon, n)),$$

where $k_0(\varepsilon, n) = k(\varepsilon\lambda_n, n)$. The proof is complete.

Proposition 3. *A regular space $\mathscr{X}$ is σ-regular iff for any sequence $\{x_n\}_{n\in\mathbb{N}}$ of elements of $\mathscr{X}$, there exists a sequence $\{\lambda_n\}_{n\in\mathbb{N}}$ of real, non-zero numbers, such that $\lambda_n x_n \xrightarrow{o} \mathbf{0}$.*

Proof. Let us assume that $\mathscr{X}$ is a σ-regular space. If $\{x_n\}_{n\in\mathbb{N}}$ is any sequence of elements of $\mathscr{X}$, then $(o)\text{-}\lim\limits_m \frac{1}{m} x_n = \mathbf{0}$, for any n. Taking into account proposition 2, it follows that there exists an increasing sequence $\{m_n\}_{n\in\mathbb{N}}$ of natural numbers, such that $(o)\text{-}\lim\limits_n \frac{1}{m_n} x_n = \mathbf{0}$.

Conversely, let us assume that $\mathscr{X}$ is a regular space, which has the property from the statement of the proposition. If $(o)\text{-}\lim\limits_m x_{nm} = x_n$ $(n \in \mathbb{N})$, then there exists $r_n \in \mathscr{X}$, whereas for any $\eta > 0$ there exists a natural number $k(\eta, n)$, such that

$$|x_{nm} - x_n| \leqslant \eta r_n (m \geqslant k(\eta, n)).$$

By hypothesis, there exists a sequence $\{\lambda_n\}_{n\in\mathbb{N}}$ of strictly positive real numbers, such that $\lambda_n r_n \xrightarrow{o} \mathbf{0}$.

In particular, there exists an $r \in \mathscr{X}$, such that $\lambda_n r_n \leqslant r$ (n $\in \mathbb{N}$), and it is now easy to show that r is a common convergence regulator for the sequences $\{x_{nm}\}_{m\in\mathbb{N}}$.

4.2.2 Spaces of countable type and spaces of bounded type

Definition 1. A vector lattice $\mathscr{X}$ is called a space of (o)-*countable type* if any bounded set of non-zero (mutually) orthogonal elements of $\mathscr{X}$ is at most countable.

Proposition 1. *If $\mathscr{X}$ is an Archimedean vector lattice of* (o)-*countable type, then its Dedekind extension $\widetilde{\mathscr{X}}$ is a space of* (o)-*countable type.*

Proof. Let $\{y_j\}_{j\in J}$ be a bounded system of non-zero mutually orthogonal elements of $\widetilde{\mathscr{X}}$. We can obviously assume that the elements y_j are positive. Let

$$\mathbf{0} < y_j \leqslant y \ (\forall j \in J).$$

Let $\widetilde{\mathscr{X}}_j$ be the component generated by y_j in the space $\widetilde{\mathscr{X}}$. For any $j \in J$ there exists an x_j, such that

$$x_j \in \widetilde{\mathscr{X}}_j \cap \mathscr{X},$$

and $\mathbf{0} < x_j \leqslant y_j$ ($j \in J$). If $j_1 \neq j_2$, then $y_{j_1} \neq y_{j_2}$; hence

$$\widetilde{\mathscr{X}}_{j_1} \perp \widetilde{\mathscr{X}}_{j_2}.$$

In particular, we get $x_{j_1} \perp x_{j_2}$.

On the other hand, there exists a $z \in \mathscr{X}$, such that $y \leqslant z$; consequently

$$\mathbf{0} < x_j \leqslant z \qquad (j \in J).$$

The system $\{x_j\}_{j\in J}$ is, therefore, bounded and it consists of non-zero, mutually orthogonal elements. Since the space $\mathscr{X}$ is of (o)-countable type, the system $\{x_j\}_{j\in J}$ is at most countable. Since the correspondence $x_j \mapsto y_j$ is one-to-one, the system $\{y_j\}_{j\in J}$ is also at most countable. It follows that $\widetilde{\mathscr{X}}$ is of (o)-countable type.

Proposition 2. *Let $\mathscr{X}$ be a complete vector lattice of* (o)-*countable type. If A is a bounded from above* (*or from below*) *set of elements from $\mathscr{X}$, then there exists an almost countable set $A' \subset A$, such that* $\sup A' = \sup A$, ($\inf A' = \inf A$, *respectively*).

Proof. Let us first assume that $\mathscr{X}$ is a space with unit and let $\mathscr{B}(\mathscr{X})$ be its base. Let A be an (infinite) set of unitary elements.

Let $\mathscr{E}$ the set of all the subsets $E \subset \mathscr{B}(\mathscr{X})$, such that each E consists of non-zero mutually orthogonal elements, and such that each element of E be bounded from above by an element of A. It is easy to verify that the set $\mathscr{E}$, ordered by inclusion, is inductive. With Zorn's lemma, there exists a maximal element $E_0 \in \mathscr{E}$, and let $e_0 = \sup E_0$. Then $e_0 = \sup A$. Indeed, if this be not true, then it would exist an element $e_1 \in A$ such that $e_1 \nleqslant e_0$. Let us define

$$e_2 = e_1 \wedge (u - e_0),$$

(where u is the unit element); then $e_2 \neq \mathbf{0}$ and, if we denote

$$E_1 = E_0 \cup \{e_2\},$$

we get $E_1 \in \mathscr{E}$, which contradicts the maximality of E_0.

Since the set E_0 is at most countable, and since for any element $e \in E_0$ there exists an $e_1 \in A$ such that $e \leqslant e_1$, it follows that there exists a subset A' of A, which is at most countable and for which $\sup A' \geqslant e_0$.

Since $e_0 = \sup A$, it follows that $\sup A' = \sup A$.

In a similar way one can consider the case of the greatest lower bound.

Now let A be a bounded from below set of arbitrary elements of $\mathscr{X}$ (the assumption that $\mathscr{X}$ has a unit element is maintained). If $a = \inf A$, then, by taking into account the notations from Section 4.1.3, we get easily that [5])

$$g^a(\lambda) = \sup_{x \in A} g^x(\lambda), \qquad \forall\, \lambda \in \mathbb{R},$$

Let us arrange the set of all rational numbers as a sequence $\{\lambda_n\}_{n \in \mathbb{N}}$. From what we have shown in the first part of the proof, for any $n \in \mathbb{N}$ there exists a sequence $\{x_{nm}\}_{m \in \mathbb{N}}$ of elements of A, such that

$$g^a(\lambda_n) = \sup_{m \in \mathbb{N}} g^{x_{nm}}(\lambda_n), \qquad \forall\, n \in \mathbb{N}.$$

If

$$b = \inf_{n,m \in \mathbb{N}} x_{nm},$$

then $a \leqslant b$ and, therefore,

$$g^a(\lambda_n) \geqslant g^b(\lambda_n),$$

for any $n \in \mathbb{N}$. On the other hand, the opposite inequality holds too

$$g^b(\lambda_n) = \sup_{k,m \in \mathbb{N}} g^{x_{km}}(\lambda_n) \geqslant \sup_{m \in \mathbb{N}} g^{x_{nm}}(\lambda_n) = g^a(\lambda_n)$$

and, consequently,

$$g^a(\lambda_n) = g^b(\lambda_n), \qquad n \in \mathbb{N}.$$

By taking into account the remark (2) made after the theorem from Section 4.1.3, it follows that $a = b$. We have thus shown that there exists an at most countable subset A' of A, such that $\inf A' = \inf A$.

Let us now drop the assumption that $\mathscr{X}$ is a space with unit. Let A be a bounded from below set of elements of $\mathscr{X}$, and let x_0 be any element of A. Let us denote

$$A_0 = \{x \wedge x_0;\ x \in A\};$$

then the set A_0 is bounded. If $|y| \leqslant c$ for any $y \in A$, then A_0 is a subset of the component $\mathscr{X}_0$ generated by c.

---

[5]) From $\lambda u - a = \sup\limits_{x \in A} (\lambda u - x)$ it follows that $(\lambda u - a)_+ = \sup\limits_{x \in A} (\lambda u - x)_+$ and, also, that $g^a(\lambda) = [(\lambda u - a)_+]u = \sup\limits_{n \in \mathbb{N}} (n(\lambda u - a)_+ \wedge u) = \sup\limits_{x \in A} \sup\limits_{n \in \mathbb{N}} (n(\lambda u - x)_+ \wedge u) = \sup\limits_{x \in A} g^x(\lambda)$.

Since the element c may be chosen to be a unit element in $\mathscr{X}_0$, taking into account what we have already shown, there exists an at most countable set $A' \subset A$, such that

$$\inf A' = \inf_{x \in A'} x \wedge x_0 = \inf A_0 = \inf A.$$

Finally, if A is a bounded from above subset of $\mathscr{X}$, then we shall consider the set $-A = \{-x; x \in A\}$. The proof is complete.

Corollary. *If $\{x_\delta\}_{\delta \in \Delta}$ is a generalized sequence of elements of a complete vector lattice of (o)-countable type, and if $x_\delta \xrightarrow[\delta \in \Delta]{\omega} x$, then there exists an increasing sequence of indices $\{\delta_n\}_{n \in \mathbb{N}}$, such that $x_{j_n} \xrightarrow{o} x$.*

The proof can easily be made.

Definition 2. A subset A of an Archimedean vector lattice is called *(o)-annihilating* if for any sequence $\{x_n\}_{n \in \mathbb{N}}$ of elements of A and any sequence $\{\alpha_n\}_{n \in \mathbb{N}}$ of real numbers, which converges to zero, one has $\alpha_n x_n \xrightarrow{o} \mathbf{0}$.

If $\mathscr{X}$ is an Archimedean vector lattice and if A is a bounded subset of A, then A is (o)-annihilating.

The converse is not always true.

Definition 3. Any Archimedean vector lattice, in which any (o)-annihilating set of elements is bounded, is called a space *of bounded type.*

From the definition it follows that in any space of bounded type, a set A is bounded iff any sequence of elements of A is bounded.

Let $\mathscr{X}$ be a space of bounded type and of (o)-countable type. If A is an uncountable set of non-zero elements of $\mathscr{X}$ and if the elements of A are mutually orthogonal, then the set A contains an unbounded countable subset. The proof is easy.

4.2.3 Maximal spaces

Definition 1. Any complete vector lattice, in which any set of mutually orthogonal elements is bounded is called a *maximal space.*

Proposition 1. *In any maximal space there exist total elements.*

Proof. Let $\mathscr{X}$ be a maximal space. Let $\{\mathscr{X}_j\}_{j \in J}$ be a decomposition of $\mathscr{X}$ into spaces with unit (see Section 4.1.2). Let us denote by u_j the unit element of $\mathscr{X}_j$. Since the elements u_j are mutually orthogonal, they form a bounded set. It follows that there exists $u = \bigvee_{j \in J} u_j$. The element u is total in $\mathscr{X}$. Indeed, if $x \perp u$, then $x \perp u_j$, and this shows that $x \perp \mathscr{X}_j$ (for any $j \in J$). But this implies that $x = 0$, because the system $\{\mathscr{X}_j\}_{j \in J}$ is total.

Definition 2. Let $\mathscr{X}$ and $\mathscr{Y}$ be two vector lattices. The space $\mathscr{Y}$ is called an *enlargement* of $\mathscr{X}$ if $\mathscr{X}$ is isomorphic to a normal total subspace of $\mathscr{Y}$. If $\check{\mathscr{X}}$ is an enlargement of any enlargement $\mathscr{Y}$ of $\mathscr{X}$, then $\check{\mathscr{X}}$ is called the *maximal enlargement* of $\mathscr{X}$.

Proposition 2. *For any complete vector lattice $\mathscr{X}$, the maximal enlargement $\check{\mathscr{X}}$ of $\mathscr{X}$ exists, it is unique (up to an isomorphism) and it is a maximal space.*

Proof. Let $\mathscr{P}$ be the set of all the subsets E of $\mathscr{X}$, consisting of positive elements and satisfying the following conditions.

(i) if $x \in E$ and $\mathbf{0} \leqslant x' \leqslant x$, then $x' \in E$;

(ii) if A is a bounded set of elements of E, then $\sup A \in E$.

(iii) for any $x > \mathbf{0}$ belonging to $\mathscr{X}$, there exists a natural number n, such that $nx \notin E$.

We shall order the set $\mathscr{P}$ by inclusion. It is not difficult to check that $\mathscr{P}$ is a relatively complete lattice. For example, let $\{E_j\}_{j\in J}$ be a bounded from above set of elements of $\mathscr{P}$, such that $E_j \subset E$, where $E \in \mathscr{P}$.

Let $B = \bigcup_{j\in J} E_j$ and denote by C the set formed of the elements of B and of the least upper bounds of all bounded from above subsets of B. Then $C \in \mathscr{P}$, $C \subset E$ and $E_j \subset C$, for any $j \in J$. It follows that $C = \vee E_j$ in $\mathscr{P}$.

The set $\mathscr{P}$ becomes a commutative semi-group if addition is defined in the following way

$$E_1 + E_2 = \{x + y;\, x \in E_1, y \in E_2\}.$$

The associativity and commutativity laws for addition, as well as the implication

$$E_1 + E = E_2 + E \Rightarrow E_1 = E_2$$

are easy to prove.

The set $\{\mathbf{0}\}$, formed of the zero element of $\mathscr{X}$, belongs to $\mathscr{P}$, and if E, E_1, E_2 belong to $\mathscr{P}$, then

$$E + \{\mathbf{0}\} = E,$$

$$E_1 + E_2 = \{\mathbf{0}\} \Rightarrow E_1 = E_2 = \{0\}.$$

It is easy to check that if $E_1, E_2 \in \mathscr{P}$, then $E_1 \subset E_2$ iff there exists an $E \in \mathscr{P}$, such that $E + E_1 = E_2$.

For any number $\alpha \geqslant 0$ and any $E \in \mathscr{P}$, we shall define

$$\alpha E = \{\alpha x;\, x \in E\}.$$

Let now $\mathscr{Y} = \mathscr{P} \times \mathscr{P}$ and define on $\mathscr{Y}$ the following equivalence relation

$$(E_1, E_2) = (E', E'') \Leftrightarrow E_1 + E'' = E_2 + E'.$$

We shall denote by $\check{\mathscr{X}}$ the set of all the equivalence classes.

The class represented by (E_1, E_2) will be denoted also by (E_1, E_2), in order to avoid unnecessary complications.

We shall define on $\check{\mathscr{X}}$ the algebraic operations as follows

$$(E_1, E_2) + (E_3, E_4) = (E_1 + E_3, E_2 + E_4),$$

$$\alpha(E_1, E_2) = \begin{cases} (\alpha E_1, \alpha E_2), & \text{if } \alpha \geqslant 0, \\ (-\alpha E_2, -\alpha E_1), & \text{if } \alpha < 0. \end{cases}$$

The element $\mathbf{0} = (\{\mathbf{0}\}, \{\mathbf{0}\})$ of $\check{\mathscr{X}}$ plays the role of zero.

We shall define $(E_1, E_2) \geqslant \mathbf{0}$ if there exists an $E \in \mathscr{P}$, such that $(E_1, E_2) = (E, \{\mathbf{0}\})$. The inequality $(E_1, E_2) \leqslant (E_3, E_4)$ is defined to mean that $(E_3, E_4) - (E_1, E_2) \geqslant \mathbf{0}$.

The set $\check{\mathscr{X}}$, endowed with this structure, is a complete vector lattice in which the set of the positive elements is isomorphic to $\mathscr{P}$.

Let us now remark that for any positive element $x \geqslant \mathbf{0}$ we have $[\mathbf{0}, x] \in \mathscr{P}$. We infer that $\mathscr{X}$ is isomorphic to a normal and total subspace of $\check{\mathscr{X}}$; in other words, $\check{\mathscr{X}}$ is an enlargement of $\mathscr{X}$. Taking into account the construction of $\check{\mathscr{X}}$, we infer that it is the maximal enlargement of $\mathscr{X}$, and that any maximal enlargement of $\mathscr{X}$ is isomorphic to $\check{\mathscr{X}}$.

Let G be a set of mutually orthogonal positive elements of $\check{\mathscr{X}}$. Let H be the set obtained from G by forming its union with the set of all the suprema of the bounded sets of G. Let

$$E = \bigcup_{y \in H} [\mathbf{0}, y].$$

It follows that E has the properties (i), (ii), (iii) by which the set $\mathscr{P}$ was defined. It follows that the set G is bounded in $\check{\mathscr{X}}$ and, hence, $\check{\mathscr{X}}$ is a maximal space.

Remarks. (1) One can prove that any maximal space is isomorphic to its maximal enlargement.

(2) One can prove that in a complete vector lattice $\mathscr{X}$, the individual convergence of the sequences coincides with the (o)-convergence in the maximal enlargement $\check{\mathscr{X}}$; i.e., $x_n \xrightarrow{\text{ind}} x$ in $\mathscr{X}$ iff $x_n \xrightarrow{o} x$ in $\check{\mathscr{X}}$.

4.3 Lattice ordered algebras

4.3.1 The notions of lattice ordered quasi-algebra and of lattice ordered algebras

Any vector lattice $\mathscr{X}$, in which for some pairs of elements x, y a multiplication xy is defined (that is, a mapping $(x, y) \mapsto xy$ of a subset of $\mathscr{X} \times \mathscr{X}$ into $\mathscr{X}$) is called a *lattice ordered quasi-algebra*, provided this multiplication is subject to the following conditions:

(i) if xy exists, and if $|x'| \leqslant |x|$ and $|y'| \leqslant |y|$, then $x'y'$ exists too; for any x, the products $x\mathbf{0}$ and $\mathbf{0}x$ exist.

(ii) if xy, $(xy)z$ and yz exist, then $x(yz)$ exists too and the equality $x(yz) = (xy)z$ holds.

(iii) if x_1y and x_2y exist, then $(x_1 + x_2)y$ exists too, and the equality

$$(x_1 + x_2)\, y = x_1y + x_2y$$

holds; if xy_1 and xy_2 exist, then $x(y_1 + y_2)$ exists too, and the equality

$$x(y_1 + y_2) = xy_1 + xy_2$$

holds;

(iv) if xy exists, then for any real α the equalities

$$\alpha(xy) = (\alpha x)y = x(\alpha y)$$

hold [6]);

(v) if xy exists and if $x \geqslant \mathbf{0}$, $y \geqslant \mathbf{0}$, then $xy \geqslant 0$.

Any lattice ordered quasi-algebra, for which the following condition holds:

(vi) if xy exists, then yx exists too and the equality

$$xy = xy$$

holds;

is called a *commutative lattice ordered quasi-algebra.*

If a lattice ordered quasi-algebra has an element v such that xv and vx exist for any x and if the equalities $xv = vx = x$ hold, then v is called the *neutral element*.

The following properties of the multiplication in a lattice ordered quasi-algebra are easy to check

(1) $x\mathbf{0} = \mathbf{0}x = \mathbf{0}$, for any x;

(2) if $x \leqslant y$ and $z \geqslant \mathbf{0}$ and if xz and yz exist, then $xz \leqslant yz$; if zx and zy exist, then $zx \leqslant zy$;

(3) if xy exists and $\mathbf{0} \leqslant x' \leqslant x$, $\mathbf{0} \leqslant y' \leqslant y$, then $x'y'$ exists and $x'y' \leqslant xy$;

(4) if xy exists, then $|x|\,|y|$ exists and

$$|xy| \leqslant |x|\,|y|.$$

If $\mathscr{X}$ is a lattice ordered quasi-algebra and if the multiplication is defined for any pair of elements, then $\mathscr{X}$ is called a *lattice ordered algebra.*

The notion of commutative lattice ordered algebra has an obvious meaning.

4.3.2 The multiplication of the elements of a σ-complete vector lattice with unit

Let $\mathscr{X}$ be a σ-complete vector lattice with unit. Let u be the unit element of $\mathscr{X}$, $\mathscr{B}(\mathscr{X})$ the base of $\mathscr{X}$, $\mathscr{S}(\mathscr{X})$ the set of the simple elements of $\mathscr{X}$, $\mathscr{M}(\mathscr{X})$ the set of the bounded elements of $\mathscr{X}$.

If $x, y \in \mathscr{S}(\mathscr{X})$ and specifically

$$x = \sum_{i=1}^{m} \alpha_i e_i, \quad y = \sum_{j=1}^{n} \beta_j e'_j,$$

---

[6]) The existence of $(\alpha x)y$ and of $x(\alpha y)$ follows from (i) and (iii). If $\mathscr{X}$ is Archimedean, then the equalities

$$\alpha(xy) = (\alpha x)y = x(\alpha y)$$

follow too.

we shall define the product of the elements x and y to be the element xy defined by

$$xy = \sum_{i,j} \alpha_i \beta_j e_i \wedge e_j'.$$

In particular, for $e_1, e_2 \in \mathscr{B}(\mathscr{X})$ one has

$$e_1 e_2 = e_1 \wedge e_2.$$

The product of two simple elements does not depend on their representation.

It is easy to check that $\mathscr{S}(\mathscr{X})$ is a commutative lattice ordered algebra, in which u is the neutral element.

Let $x, y \in \mathscr{M}(\mathscr{X})$. There then exist (see Section 4.1.3, remarks) two sequences of simple elements $\{x_n\}_{n\in\mathbb{N}}$ and $\{y_n\}_{n\in\mathbb{N}}$, such that $x_n \xrightarrow{o} x$ and $y_n \xrightarrow{o} y$ in $\mathscr{M}(\mathscr{X})$. In other words, $x_n \xrightarrow{o} x$ in $\mathscr{X}$ and there exists a $\xi \in \mathbb{R}$, such that $|x_n| \leqslant \xi u$ for any n; analogously, $y_n \xrightarrow{o} y$ in $\mathscr{X}$ and there exists an $\eta \in \mathbb{R}$, such that $|y_n| \leqslant \eta u$. Since for any simple elements a, b one has $|ab| = |a|\,|b|$, it follows that

$$|x_n y_n - x_m y_m| \leqslant |x_n y_n - x_n y_m| + |x_n y_m - x_m y_m|$$

$$= |x_n|\,|x_n - y_m| + |x_n - x_m|\,|y_m| \leqslant \xi |y_n - y_m| + \eta\,|x_n - x_m|$$

and therefore, (see Section 3.2.6, proposition 1) the limit $z = (o)\text{-}\lim x_n y_n$ exists. In a similar manner one shows that the limit z does not depend on the choice of the sequences $\{x_n\}_{n\in\mathbb{N}}$ and $\{y_n\}_{n\in\mathbb{N}}$.

We shall therefore define the product of the elements x and y by putting $xy = (o)\text{-}\lim\limits_{n} x_n y_n$.

If multiplication is defined on $\mathscr{M}(\mathscr{X})$ in this manner, then this space becomes a commutative lattice ordered algebra in which u is the neutral element. It is also easy to check that if $x_1 \perp x_2$, then $x_1 x_1 \perp x_2 y_2$, for any y_1 and y_2.

If $x_n \xrightarrow{o} x$ and $y_n \xrightarrow{o} y$ in $\mathscr{M}(\mathscr{X})$, then there exists an $\eta \in \mathbb{R}$ such that $|y_n| \leqslant \eta u$. Let us suppose that $|x| \leqslant \xi u$. Then

$$|xy - x_n y_n| \leqslant |x|\,|y - y_n| + |x - x_n|\,|y_n| \leqslant \xi\,|y - y_n| + \eta\,|x - x_n|,$$

whence it follows that $x_n y_n \xrightarrow{o} xy$.

Let x and y be arbitrary positive elements of $\mathscr{X}$. If we define $x_n = x \wedge (nu)$ and $y_n = y \wedge (nu)$, we get bounded elements and $x_n \xrightarrow{o} x$ and $y_n \xrightarrow{o} y$. If the sequence $\{x_n y_n\}_{n\in\mathbb{N}}$ is bounded, then we shall define $xy = (o)\text{-}\lim x_n y_n$.

If x and y are arbitrary elements of the space $\mathscr{X}$, we shall define

$$xy = x_+ y_+ - x_+ y_- - x_- y_+ + x_- y_- \tag{1}$$

if each of the products in the right-hand member of this equality exists.

One can show that $\mathscr{X}$ is a commutative lattice orderd quasialgebra, in which u is the neutral element.

Conditions (i), (v) and (vi) from Section 4.3.1 are obviously satisfied. Let us verify condition (iii).

In virtue of formula (1) it is sufficient to consider positive elements.

Let x', x'', y be positive elements, such that $x'y$ and $x''y$ exist. Let $x'_n = x' \wedge (nu)$ and $x''_n = x'' \wedge (nu)$, $y_n = y \wedge (nu)$. We have

$$(x'_n + x''_n)\, y_n = x'_n\, y_n + x''_n\, y_n.$$

On the other hand we have

$$(x' + x'') \wedge (nu) \leqslant (x' \wedge nu) + (x'' \wedge nu) = x'_n + x''_n;$$

consequently we have

$$((x' + x'') \wedge nu)\, y_n \leqslant x'_n\, y_n + x''_n\, y_n.$$

It consequently follows that $(x' + x'')y$ exists and we have

$$(x' + x'')\, y \leqslant x'y + x''\, y.$$

The opposite inequality follows from the relations

$$x'_n\, y_n + x''_n\, y_n = (x'_n + x''_n)\, y_n \leqslant (x' + x'')y.$$

It follows that condition (iii) is satisfied.

Condition (iv) follows from the fact that the space $\mathscr{X}$ is Archimedean. Also it is easy to check condition (ii).

It is obvious that u is the neutral element.

It is also easy to check that from $x_1 \wedge x_2 = \mathbf{0}$ it follows that $x_1 y \wedge x_2 y = \mathbf{0}$ (for any $y \geqslant \mathbf{0}$ such that $x_1 y$ and $x_2 y$ exist) and that the relation $x_1 \perp x_2$ is equivalent to $x_1 x_2 = \mathbf{0}$[7]).

If xy exists, then

$$[xy] = [x]\,[y]. \tag{2}$$

[7]) If $x_1 \perp x_2$, then $x_1 x_2 = ([x_1]u)x_1\ ([x_2]u)x_2 = \mathbf{0}$. If $x \neq \mathbf{0}$, then $x^2 = |x|\ |x| > \mathbf{0}$ (see Section 4.1.2, proposition 3). It is now easy to infer that from $x_1 x_2 = \mathbf{0}$ it follows that $x_1 \perp x_2$.

Indeed, let $e_x = [x]u$ and $e_y = [y]u$. From $x \perp (u - e_x)$ it follows that $x(u - e_x) = \mathbf{0}$; therefore, we have

$$(xy)(u - e_x) = y(x(u - e_x)) = \mathbf{0}$$

and this means that

$$xy \perp (u - e_x).$$

By putting $e_{xy} = [xy]u$, it follows that $e_{xy} \leqslant e_x$. If the role of the elements x and y is exchanged, we get that $e_{xy} \leqslant e_y$. It follows that $e_{xy} \leqslant e_x \wedge e_y$. Let us now define

$$e = e_x \wedge e_y - e_{xy};$$

then $e \perp e_{xy}$ and, therefore, $e \perp xy$; it follows that $e(xy) = \mathbf{0}$, i.e. $(ex)y = \mathbf{0}$ and this shows that $ex \perp y$.

If $e \neq \mathbf{0}$, then from $e \leqslant e_x$ it follows that $ex \neq \mathbf{0}$. On the other hand, from $e \leqslant e_y$ it follows that $[e] \leqslant [y]$; since $ex = [e]x$, the element ex cannot be orthogonal to y, and this is a contradiction.

It follows that $e = \mathbf{0}$, i.e.,

$$e_{xy} = e_x \wedge e_y = e_x e_y$$

and equality (2) is proved.

If x is an element of $\mathscr{X}$ and if n is a natural number exceeding 1, the element x^n is defined by induction on n (i.e., one defines $x^n = x^{n-1}x$) but it generally does not exist for any x. If x^n exists and if $n = n_1 + n_2$, then $x^n = x^{n_1} x^{n_2}$. Also it is easy to prove by induction the following equalities

$$x^n = (x_+)^n + (-1)^n (x_-)^n,$$

$$|x^n| = (x_+)^n + (x_-)^n.$$

The equality $x = x^2$ holds iff the element x is unitary.

Indeed, if e is unitary, then $e^2 = e \wedge e = e$. Conversely, let $x = x^2$. Let $e_x \in \mathscr{B}(\mathscr{X})$ be such that $[e_x] = [x]$. From the equalities

$$x = [e_x]\, x = e_x\, x$$

it follows that

$$x(x - e_x) = \mathbf{0}$$

i.e.,

$$[x]\,[x - e_x] = \mathbf{0}.$$

But we have

$$[x - e_x]\,u \leqslant e_x,$$

and this implies that $[x - e_x] \leqslant [x]$; hence $[x - e_x] = \mathbf{0}$ and this shows that $x = e_x$.

The notion of n-root of an element $x \in \mathscr{X}$ is defined as usual by the equation $y^n = x$. If x and y are positive and $n \geqslant 2$, then y is called a positive n-root and it is denoted by $\sqrt[n]{x}$.

For any element $x \geqslant \mathbf{0}$ and any natural number $n \geqslant 2$ the positive n-root of x exists and it is unique.

Indeed, let first x be a (positive) simple element. Then x may be represented by the formula

$$x = \sum_{i=1}^{m} \alpha_i\, e_i,$$

where the elements e_i are unitary and mutually orthogonal, and $\alpha_i \geqslant 0$. In this case we obviously have

$$\sqrt[n]{x} = \sum_{i=1}^{m} \sqrt[n]{\alpha_i}\, e_i.$$

Let x be a (positive) bounded element. It then exists a sequence $\{s_m\}_{m\in\mathbb{N}}$ of positive simple elements, such that $s_m \overset{o}{\to} x$. It is easy to remark that the sequence $\{\sqrt[n]{s_m}\}_{m\in\mathbb{N}}$ is (o)-convergent; by defining

$$y = (o)\text{-}\lim_{m} \sqrt[n]{s_m}$$

we have $y^n = x$.

Let us now show that if y is an arbitrary positive element and if y^n exists, then from $y^n \leqslant x$ it follows that $y \leqslant x + u$.

Indeed, let us put

$$e_1 = [(y - u)_+]u$$

and $e_2 = u - e_1$. Let us denote $y_1 = e_1 y$ and $y_2 = e_2 y$. We obviously have $y_1 + y_2 = y$. From $e_1(y - u) \geqslant \mathbf{0}$ it follows that $y_1 \geqslant e_1$, whereas from $e_2(y - u) \leqslant \mathbf{0}$ it follows that $y_2 \leqslant e_2$.

From $e_1 \leqslant y_1$ we get

$$y_1 = e_1\, y \leqslant y_1^2;$$

by induction we infer that

$$y_1 \leqslant y_1^n \leqslant y^n \leqslant x.$$

By taking into account the fact that

$$y_2 \leqslant e_2 \leqslant u$$

we finally have

$$y = y_1 + y_2 \leqslant x + u.$$

Let x be an arbitrary positive element. Let us define

$$z_m = \sqrt[n]{x \wedge (mu)}$$

for any $m \in \mathbb{N}$. By taking into account what we have already proved, the sequence $\{z_m\}_{m\in\mathbb{N}}$ is bounded and this shows that $y = (o)\text{-}\lim_m z_m$ exists; we obviously have $y^n = x$.

It follows that $\sqrt[n]{x}$ exists for any $x \geqslant 0$. In order to prove the uniqueness, let us suppose that there exist two positive elements y_1 and y_2, such that $y_1^n = y_2^n = x$. Let us suppose that $y_1 < y_2$. By putting $z = y_2 - y_1$, from $\mathbf{0} < z \leqslant y_2$ it follows that z^n exists and $z^n > \mathbf{0}$ (because $[z^n] = [z]$).

If $n_1 + n_2 \leqslant n$, then $y_1^{n_1} y_2^{n_2}$ exists and, consequently $z_1^{n_1} z_2^{n_2}$ exists too. One can then use Newton's formula in order to expand $(y_1 + z)^n$ and this gives

$$y_2^n = (y_1 + z)^n > y_1^n = x.$$

This result contradicts the fact that $y_2^n = x$. It follows that if $y_1 \not\leqslant y_2$, then y_1 and y_2 are not comparable. By putting

$$e = [(y_2 - y_1)_+]u$$

we have $ey_1 < ey_2$ and, at the same time,

$$(ey_1)^n = (ey_2)^n = ex,$$

which is a contradiction. It consequently follows that $y_1 = y_2$ and the positive n-root is, therefore, unique.

4.3.3 Normal σ-complete lattice ordered algebras

By a σ-complete lattice ordered algebra we mean a lattice ordered algebra which is σ-complete as a (vector) lattice.

Definition. A σ-complete lattice ordered algebra is called a *normal* σ-complete lattice ordered algebra if the equality $x_1 x_2 = \mathbf{0}$ is equivalent to $x_1 \perp x_2$.

Lemma 1. If $\mathscr{X}$ is a normal σ-complete lattice ordered algebra, then

$$[xy] \leqslant [x]\,[y], \tag{1}$$

for any $x, y \in \mathscr{X}$.

Proof. Let first x, y be positive elements. Let $z \in \mathscr{X}$ be such that $z \perp x$. Then $|z|x = \mathbf{0}$ and consequently

$$|z|(xy) = (|z|x)y = \mathbf{0},$$

whence the relation $z \perp xy$ follows. Consequently, from $z \perp x$ it follows that $z \perp xy$ and this shows that $[xy] \leqslant [x]$. In a similar manner one can show that $[xy] \leqslant [y]$, and therefore (1) is established for positive elements.

In particular, from (1) one infers that if $x \wedge y = \mathbf{0}$ and $z \geqslant \mathbf{0}$, then

$$xz \wedge y = zx \wedge y = \mathbf{0}.$$

From this we easily obtain that if x and y are arbitrary elements of $\mathscr{X}$, then

$$(x_+y_+ + x_-y_-) \wedge (x_+y_- + x_-y_+) = \mathbf{0}$$

and, therefore,

$$(xy)_+ = x_+y_+ + x_-y_-,$$

$$(xy)_- = x_+y_- + x_-y_+,$$

$$|xy| = |x|\,.\,|y| \tag{2}$$

From (2) it follows that (1) is true for any elements x, y of $\mathscr{X}$.

Lemma 2. *If $\mathscr{X}$ is a normal σ-complete lattice ordered algebra, then*

$$([z]x)y = x([z]y) = [z]\,(xy), \tag{3}$$

for any $x, y, z \in \mathscr{X}$.

Proof. For any elements x, y, z the following equality holds

$$xy = x([z]y) + x([y] - [z]\,[y])y. \tag{4}$$

From inequality (1), in which we substitute $[z]y$ for y, and by using the properties of the principal projectors, we get

$$[x([z]y)] \leqslant [x]\,[[z]y] = [x]\,[z]\,[y] \leqslant [z].$$

We also have

$$[x([y] - [z]\,[y])y] < [x]\,[([y] - [z]\,[y])y] \leqslant$$

$$= [x]([y] - [z][y])\,[y] \leqslant [y] - [z]\,[y].$$

By applying the projector $[z]$ to both members of equality (4) we get

$$[z](xy) = [z](x([z]y) = x([z]y).$$

The equality

$$[z](xy) = ([z]x)y$$

may be established in a similar manner.

Proposition 1. *A σ-complete lattice ordered algebra is a normal σ-complete lattice ordered algebra iff*

$$[xy] = [x][y], \tag{5}$$

for any $x, y \in \mathscr{X}$.

Proof. The condition is obviously sufficient. Let us show that it is also necessary. Let x be an arbitrary element of $\mathscr{X}$. We have (see Section 3. 3.4, proposition 5)

$$[x] - [x^2] = [[x]([x] - [x^2])x].$$

From inequality (1) we get $[x^2] \leqslant [x]$ and, therefore,

$$[x]x^2 = [x^2]x^2.$$

By using lemma 2 we get

$$(([x] - [x^2])x)^2 = ([x] - [x^2])x^2 =$$

$$= [x]x^2 - [x^2]x^2 = \mathbf{0}$$

whence

$$([x] - [x^2])x = \mathbf{0}.$$

Hence we immediately infer that $[x^2] = [x]$, i.e., equality (5) holds for $x = y$.

Let now x and y be any elements. From the relations

$$(|x| \wedge |y|)^2 \leqslant |x||y| = |xy|$$

it follows that

$$[(|x| \wedge |y|)^2] \leqslant [xy]$$

and, therefore, by taking into account what we have already proved

$$[|x| \wedge |y|] \leqslant [xy]. \tag{6}$$

On the other hand it is known that

$$|x| \wedge |y|] = [x][y]$$

(see Section 3.3.4, proposition 6) and, taking into account (1) and (6) we get equality (5).

Proposition 2. *A σ-complete lattice ordered algebra $\mathfrak{X}$ is a normal σ-complete lattice ordered algebra iff the following two conditions are satisfied*:

(i) if $x \perp y$ *then* $xy = \mathbf{0}$;

(ii) if $x \neq \mathbf{0}$, *then* $x^2 > \mathbf{0}$.

Proof. We have to prove only the sufficiency. Let x and y be positive elements such that $xy = \mathbf{0}$. We obviously have

$$(x \wedge y)^2 = \mathbf{0}$$

and from (ii) it follows that $x \wedge y = \mathbf{0}$. Hence we get that for any elements x, y equality (2) holds (see the proof of lemma 1). It follows that if $xy = \mathbf{0}$, then

$$|x|\,|y| = |xy| = \mathbf{0}$$

and this implies that $x \perp y$.

Proposition 3. *In a normal σ-complete lattice ordered algebra from $x_n \xrightarrow{o} x$ and $y_n \xrightarrow{o} y$ it follows that $x_n y_n \xrightarrow{o} xy$.*

Proof. Let first $x_n \downarrow \mathbf{0}$ and $y > \mathbf{0}$.

Let us consider any number $\varepsilon \in (0,1)$ and put $z_n = (x_n - \varepsilon x_1)_+$. The equalit

$$x_n y = (([x_1]) - [z_n])x_n)\, y + ([z_n]x_n)y$$

holds, whereas lemma 2 gives

$$([z_n]\, x_1)y = [z_n]\,(x_1\, y).$$

It follows that

$$x_n y \leqslant (([z_1] - [z_n])\, x_n)\, y + [z_n]\,(x_1 y). \tag{7}$$

Let us now observe that

$$([z_1] - [z_n])\, x_n \leqslant \varepsilon x_1 \tag{8}$$

because we have

$$([z_1] - [z_n])\,(x_n - \varepsilon x_1) = (x_n - \varepsilon x_1) - (x_n - \varepsilon x_1)_+$$

$$= -\,(x - \varepsilon x_1)_- \leqslant \mathbf{0}.$$

On the other hand, from the fact that

$$[z_n]\,(x_n - \varepsilon x_1) \geqslant \mathbf{0}$$

it follows that

$$\mathbf{0} \leqslant [z_n]\, x_1 \leqslant \frac{1}{\varepsilon}\, x_n.$$

Since $x_n \xrightarrow{o} \mathbf{0}$ it follows that $[z_n]\, x_1 \to \mathbf{0}$. But $[z_n]\downarrow$ and $[z_n] \leqslant [x_1]$, whence $[z_n] \downarrow \mathbf{0}$ in the Boolean lattice of projections.

By taking into account inequality (8) in relation (7) from the fact that $[z_n] \downarrow 0$ we get

$$\overline{\lim}\, x_n y_n \leqslant \varepsilon x_1 y;$$

since ε has been chosen arbitrarily in $(0,1)$, we finally get that $x_n y \xrightarrow{o} 0$.

Let $x_n \xrightarrow{o} x$ and y be arbitrary. Let

$$|x_n - x| \leqslant r_n \downarrow 0.$$

From what we have already shown it follows that

$$|x_n y - xy| = |x_n - x||y| \leqslant r_n |y| \xrightarrow{o} 0$$

and this shows that $x_n y \xrightarrow{o} xy$.

One can analogously show that $y x_n \xrightarrow{o} yx$.

Let $x_n \xrightarrow{o} x$ and $y_n \xrightarrow{o} y$. Let $z \in \mathfrak{X}$ be such that $|x_n| \leqslant z$. We have

$$|x_n y_n - xy| \leqslant |x_n y_n - x_n y| + |x_n y - xy|$$

$$= |x_n| \cdot |y_n - y| + |x_n - x||y| \leqslant z\,|y_n - y| + |x_n - x||y|$$

and this implies that $x_n y_n \xrightarrow{o} xy$.

Proposition 4. *If $\mathfrak{X}$ is a σ-complete lattice ordered algebra which has a neutral element $u > 0$ and if u is total, then the product of any elements in $\mathfrak{X}$ coincides with the product of the same elements, which may be defined in $\mathfrak{X}$ when u is considered to be the unit element* [8]).

Proof. Let us first observe that if u is considered as unit element in $\mathfrak{X}$, then for any unitary elements e_1, e_2 the following equalities

$$e_1 e_2 = e_2 e_1 = e_1 \wedge e_2 \tag{9}$$

hold.

Indeed, from $0 \leqslant e_2 \leqslant u$ it follows that

$$0 \leqslant e_1(u - e_2) \leqslant u - e_2$$

and, therefore,

$$e_2 \perp e_1 (u - e_2).$$

If x is an element such that $x \perp e_2$, then from

$$0 \leqslant e_1 e_2 \leqslant u e_2 = e_2$$

---

[8]) i.e. the product defined in Section 4.3.2.

it follows that $x \perp e_1e_2$. Since

$$e_1 = e_1e_2 + e_1(u - e_2)$$

it follows that

$$e_1e_2 = [e_2]e_1 = e_1 \wedge e_2.$$

At the same time one infers that $e_2e_1 = e_2 \wedge e_1$ and this shows that equalities (9) hold.

Let us now remark that if $x_n \xrightarrow{o} x$ and if e is a unitary element, then $ex_n \xrightarrow{o} ex$ and $x_ne \xrightarrow{o} xe$. Indeed, if

$$|x_n - x| \leqslant r_n \downarrow \mathbf{0},$$

then

$$|ex_n - ex| \leqslant e|x_n - x| \leqslant r_n \downarrow \mathbf{0}$$

and this shows that $ex_n \xrightarrow{o} ex$. One can similarly show that $x_ne \xrightarrow{o} xe$.

If x is any bounded element of $\mathscr{X}$ then there exists a sequence $\{s_n\}_{n\in\mathbb{N}}$ of simple elements, such that $s_n \xrightarrow{o} x$.

From (9) we get

$$e\,s_n = s_n\,e = [e]s_n$$

for any unitary element e. By tending to the limit, we get

$$ex = xe = [e]x. \tag{10}$$

Since any element x of the space $\mathscr{X}$ may be represented as the (o)-limit of a sequence of bounded elements, equality (10) holds for any x too.

Let us now show that if $x_n \downarrow \mathbf{0}$ and $a > \mathbf{0}$, then $ax_n \xrightarrow{o} \mathbf{0}$. In order to do this, let us first remark that there exists a $z \geqslant \mathbf{0}$, such that $ax_n \downarrow z$. For any real number λ let us define

$$e_\lambda = [(\lambda u - \;_{+}$$

From $ax_n \xrightarrow{o} z$ it follows that

$$(o)\text{-}\lim_n\, [e_\lambda]\,(ax_n) = [e_\lambda]z \tag{11}$$

for any λ.

On the other hand, from (10) we get

$$[e_\lambda]\,(ax_n) = e_\lambda\,(ax_n) = (e_\lambda\,a)x_n = ([e_\lambda]a)\,x_n$$

and from the definition of e_λ we get

$$[e_\lambda]\;a \leqslant \lambda u.$$

It follows that

$$[e_\lambda](a\,x_n) \leqslant \lambda x_n$$

and since $x_n \downarrow \mathbf{0}$, it follows that $[e_\lambda](a\,x_n) \xrightarrow{o} \mathbf{0}$. From (11) it then follows that $[e_\lambda]z = \mathbf{0}$, for any $\lambda \in \mathbb{R}$. Since $\bigvee_{\lambda \in \mathbb{R}} e_\lambda = u$, it follows that $z = \mathbf{0}$.

We have thus shown that $ax_n \downarrow \mathbf{0}$. One can similarly show that $x_n a \downarrow \mathbf{0}$. Hence it follows that if $x_n \xrightarrow{o} x$ and $y_n \xrightarrow{o} y$, then $x_n y_n \xrightarrow{o} xy$ (see the proof of proposition 3).

By taking into account (9) and the definition of the product of two elements in a space with unit, the proof of the proposition is now at hand.

Proposition 5. *A σ-complete lattice ordered algebra $\mathscr{X}$, which has a neutral element $u > \mathbf{0}$, is a normal σ-complete lattice ordered algebra iff the element u is total.*

Proof. If $\mathscr{X}$ is a normal σ-complete lattice ordered algebra and if u is a neutral element, then for any element $x \in \mathscr{X}$ the equalities

$$[x] = [xu] = [x]\,[u]$$

hold. If $x \perp u$, then

$$[x]\,[u] = [|x| \wedge u] = \mathbf{0},$$

whence $x = \mathbf{0}$. Hence the element u is total.

Conversely, if u is total, then, with proposition 4, the product of any elements of $\mathscr{X}$ coincides with their product as defined in Section 4.3.2, when u is considered to be the unit element of $\mathscr{X}$. In this case it is obvious that $\mathscr{X}$ is a normal σ-complete lattice ordered algebra.

4.4 Special spaces

4.4.1 Discrete spaces

In a vector lattice an element e is called *discrete* (Section 3.2.3) if there exist no orthogonal elements a, b, such that $\mathbf{0} < a \leqslant |e|$, $\mathbf{0} < b \leqslant |e|$.

If $\mathscr{X}$ is a complete vector lattice with unit, and if the unit element u may be represented as $u = \bigvee_{j \in J} e_j$, where $\{e_j\}_{j \in J}$ is a system of unitary, discrete mutually orthogonal elements, then $\mathscr{X}$ is called a *discrete* space.

The spaces (s), (l^p), (m), (Section 3.4.2) are discrete spaces.

Theorem. *Any discrete space is isomorphic to a normal subspace of the space $F(J)$ (for a suitable set J)*[9])

Proof. We shall show that if $\mathscr{X}$ is a discrete space, then any element x of the space may be represented in the form

$$x = \underset{j \in J}{\mathbf{S}}\, \xi_j\, e_j\,, \tag{1}$$

[9]) The space $F(T)$ has been defined in Section 3.4.3.

in a unique manner, where $\{e_j\}_{j\in J}$ has the meaning explained in the definition of the discrete space, whereas ξ_j are real numbers. In particular, from $u = \bigvee_{j\in J} e_j$ it follows that the system $\{e_j\}_{j\in J}$ is total.

If we denote by Q_j the component generated by e_j, then the system $\{Q_j\}_{j\in J}$ is a decomposition of the space $\mathscr{X}$.

Consequently, (see Section 3.3.5), any element $x \in \mathscr{X}$ may be represented as follows

$$x = \mathop{\mathbf{S}}_{j\in J} [e_j]x. \tag{2}$$

We shall now show that any component Q, generated by a discrete unitary element e, is a one-dimensional space; more precisely, for any $a \in Q$ there exists an $\alpha \in \mathbb{R}$, such that $a = \alpha e$.

For this, let us remark that the basis of the space Q, in which the element e is considered as a unit element, consists of the elements e and $\mathbf{0}$. It results that there exists a real number α, such that the spectral function of a (see Section 4.1.3) has the form

$$g^a(\lambda) = \begin{cases} \mathbf{0} & \text{if} \quad \lambda \leqslant \alpha \\ e & \text{if} \quad \lambda > \alpha. \end{cases}$$

From the integral representation of the element a (Section 4.1.3) it follows that $a = \alpha e$.

For any $j \in J$ there exists then a $\xi_j \in \mathbb{R}$, such that $[e_j]x = \xi_j e_j$. From (2), equality (1) now follows.

Let us denote by φ_x the function defined on J by the formula

$$\varphi_x(j) = \xi_j, \ j \in J.$$

The mapping $x \mapsto \varphi_x$ is an isomorphism of the space $\mathscr{X}$ onto a subspace $F_0(J)$ of $F(J)$, as one can easily prove.

Let us now prove that $F_0(J)$ is a normal subspace of $F(J)$. Let, therefore, $|\psi| \leqslant |\varphi|$, where $\psi \in F(J)$, and $\varphi \in F_0(J)$. There exists the element

$$x = \mathop{\mathbf{S}}_{j\in J} \varphi(j)\ e_j$$

and it belongs to $\mathscr{X}$. We infer that

$$|\ \psi(j)\ e_j\ | \leqslant |x|, \quad j \in J$$

and this shows that the element

$$y = \mathop{\mathbf{S}}_{j\in J} \psi(j) e_j$$

exists in $\mathscr{X}$ too. This implies that $\psi \in F_0(J)$. The proof is complete.

4.4.2 Spaces of continuous real functions

If T is any Hausdorff space, then the set $C(T)$ of all continuous real functions, defined on T, is a vector sublattice of the vector lattice $F(T)$.

We shall be particularly interested in the case in which the space T is compact.

A compact space T is called an *extremal* space (or extremally disconnected) if the closure of any open set in T is open (and, obviously, closed). Any extremal space is totally disconnected.

Lemma 1. *In order that a compact, totally disconnected space T be extremal, it is necessary and sufficient that the Boolean lattice $\mathscr{A}$ of all the clopen sets of T be complete.*

Proof. Let us assume that T is extremal. Let $\{A_j\}_{j\in J}$ be a system of sets in $\mathscr{A}$ and let us put $A = \bigcup_{j\in J} A_j$. By hypothesis, its closure $\bar{A} \in \mathscr{A}$. If $B \in \mathscr{A}$ and $A \subset B$, then $\bar{A} \subset B$. This shows that $\bar{A}$ is the least upper bound of the system $\{A_j\}_{j\in J}$ in the Boolean lattice $\mathscr{A}$.

Conversely, let us assume that $\mathscr{A}$ is complete. If A is an open set in T, then A may be represented as a union $A = \bigcup_{j\in J} A_j$, where $\{A_j\}_{j\in J}$ is a system of sets belonging to $\mathscr{A}$. Let $B = \sup_{j\in J} A_j$ in the Boolean lattice $\mathscr{A}$. From $A \subset B$ it then results that $\bar{A} \subset B$. Let $E = B \cap \complement \bar{A}$. If $E \neq \varnothing$, then there exists a clopen non-void set $G \in \mathscr{A}$, such that $G \subset E$. It follows that $A \subset B \cap \complement G$, and this contradicts the definition of B. It follows that $E = \varnothing$ and this implies that $B \subset \bar{A}$. Hence $\bar{A} = B$ and, therefore $\bar{A} \in \mathscr{A}$. The proof is complete.

Lemma 2. *If T is a compact space and if $C(T)$ is a complete vector lattice, then T is totally disconnected.*

Proof [10]. Let t_0 be an arbitrary point of T and W be an open neighbourhood of t_0.

Let φ be a continuous real function, defined on T, such that $-1 \leqslant \varphi(t) \leqslant 1$, or any $t \in T$, and such that $\varphi(t_0) = 1$ and $\varphi(t) = -1$ for any $t \in \complement W$.

Then there exists a clopen set $E \subset T$, such that $t_0 \in E \subset W$.

Indeed, let us show that for any real continuous function f, defined on T, there exist clopen sets E_1, E_2, E_3, which are mutually disjoint, and such that

$$T = E_1 \cup E_2 \cup E_3$$

and $f(t) \geqslant 0$, for $t \in E_1$; $f(t) \leqslant 0$, for $t \in E_2$; $f(t) = 0$ for $t \in E_3$.

In order to do this, let us choose the function $u(t) \equiv 1$ as a unit element in the space $C(T)$ and let us remark that in this case the base of the space consists of all the characteristic functions of the clopen subsets of T.

Let us consider the elements f_+, f_- of $C(T)$ and let us denote $e_1 = [f_+]u$, $e_2 = [f_-]u$. Since the elements e_1, e_2 are unitary, they represent the characteristic

[10]) We shall make use of the following theorem of Urysohn. If T is a compact space and if A and B are two disjoint closed subsets of T, then for any real numbers α and β, such that $\alpha \leqslant \beta$, there exists a continuous function φ, defined on T, such that: (i) $\varphi(t) = \alpha$, $t \in A$, (ii) $\varphi(t) = \beta$, $t \in B$, (iii) $\alpha \leqslant \varphi(t) \leqslant \beta$, for any $t \in T$.

functions of some clopen sets E_1, E_2, respectively. From $e_1 \perp e_2$ it follows that $E_1 \cap E_2 = \varnothing$ and it is easy to remark that

$$\{t; f(t) > 0\} \subset E_1, \quad \{t; f(t) < 0\} \subset E_2.$$

We have only to define $E_3 = \complement(E_1 \cup E_2)$ and we get the required decomposition of T.

From what we have shown, the existence of the above-mentioned set E immediately follows and this implies that T is totally disconnected.

Theorem 1. *If T is a compact space, then $C(T)$ is a complete vector lattice iff T is an extremal space.*

Proof. Let us suppose that $C(T)$ is a complete vector lattice. With lemma 2, the space T is totally disconnected. By choosing as unit element in $C(T)$ the function $u(t) \equiv 1$, the base of the space $C(T)$ is formed of all the characteristic functions of the clopen subsets of T. Since the base of the space $C(T)$ is a complete Boolean lattice, the set of all the clopen subsets of T is a complete Boolean lattice, too. By virtue of lemma 1, the space T is extremal.

Conversely, let us assume that T is an extremal (hence totally disconnected) space. Let A be a majorized subset of $C(T)$. Let $y \in C(T)$ be such that $x \leqslant y$ for any $x \in A$. The function $x_0(t) = \sup\limits_{x \in A} x(t)$ is lower semi-continuous and $x_0 \leqslant y$. For any $t \in T$ let us denote by $W(t)$ an arbitrary neighbourhood of t. If we define

$$a(t) = \inf_{W(t)} \sup_{t' \in W(t)} x_0(t'),$$

we get a continuous function, defined on T, and we have

$$x(t) \leqslant a(t) \leqslant \inf_{W(t)} \sup_{t' \in W(t)} y(t') = y(t);$$

hence $a = \sup A$. The proof is complete.

Theorem 2. *If $\mathscr{X}$ is a complete vector lattice with strong unit, then there exists an extremal compact space T, such that $\mathscr{X}$ be isomorphic to the space $C(T)$.*

Proof. Let u be the unit element of $\mathscr{X}$ and let $\mathscr{B}(\mathscr{X})$ be the base of the space $\mathscr{X}$. As we have already proved (see Section 1.2.4), there exists a totally disconnected, compact topological space T, such that the set $\mathscr{A}$ of all the clopen subsets of T be isomorphic, as a Boolean lattice, to $\mathscr{B}(\mathscr{X})$.

Since $\mathscr{A}$ is complete as a Boolean lattice, the space T is extremal (see lemma 1).

Let x be a simple element of $\mathscr{X}$, given by

$$x = \sum_{i=1}^{n} \alpha_i e_i,$$

where e_i are mutually orthogonal unitary elements. Let us define

$$f_x(t) = \sum_{i=1}^{n} \alpha_i \varphi_{E_i}(t),$$

where, for any index i, E_i is the set from $\mathscr{A}$ which corresponds to e_i by the isomorphism between $\mathscr{B}(\mathscr{X})$ and $\mathscr{A}$, whereas φ_{E_i} is the characteristic function of the set E_i. The mapping $x \mapsto f_x$ of the set $\mathscr{S}(\mathscr{X})$ of the simple elements of $\mathscr{X}$, on the set of simple functions from $C(T)$ is one-to-one and preserves the linear operations. We have $f_u(t) \equiv 1$ and if $x \in \mathscr{S}(\mathscr{X})$, then the inequality $|x| \leqslant \alpha u$ is equivalent to $|f_x(t)| \leqslant \alpha$, for any $t \in T$.

If $x \in \mathscr{X}$, then x may be represented as a limit $x = (\rho)\text{-}\lim\limits_n x_n$, where $x_n \in \mathscr{S}(\mathscr{X})$, and u is the convergence regulator. For any number $\varepsilon > 0$ there exists an $n_\varepsilon \in \mathbb{N}$, such that

$$|x_n - x_m| \leqslant \varepsilon u, \quad (m, n \geqslant n_\varepsilon).$$

Hence we infer that

$$|f_{x_n}(t) - f_{x_m}(t)| \leqslant \varepsilon (m, n \geqslant n_\varepsilon),$$

for any $t \in T$. The sequence of functions $\{f_{x_n}\}_{n \in \mathbb{N}}$ converges, therefore, uniformly to a continuous function f. It is easy to verify that the function f does not depend on the sequence $\{x_n\}_{n \in \mathbb{N}}$. We shall define $f = f_x$. We obtain in this manner a mapping $x \mapsto f_x$ of $\mathscr{X}$ into $C(T)$. The image of $\mathscr{X}$ by this mapping is the entire space $C(T)$.

Indeed, let $g \in C(T)$. Let $\alpha, \beta \in \mathbb{R}$ be such that $\alpha \leqslant g(t) \leqslant \beta$ and let us consider a partition of the interval $[\alpha, \beta]$ given by the points λ_i, i.e.,

$$\alpha = \lambda_0 < \lambda_1 < \ldots < \lambda_n = \beta.$$

If we put

$$A_i = \{t; g(t) < \lambda_i\}, \quad i = 1, 2, \ldots, n,$$

then $\overline{A}_i \in \mathscr{A}$. Let $B_1 = \overline{A}_1$ and $B_i = \overline{A}_i \cap \complement \overline{A}_{i-1}$, for $i = 2, \ldots, n$; we then have $B_i \in \mathscr{A}$, $i = 1, 2, \ldots n$ and the sets B_i are mutually disjoint.

Obviously

$$T = \bigcup_{i=1}^{n} B_i$$

and $\lambda_{i-1} \leqslant f(t) \leqslant \lambda_i$, for $t \in B_i$. If $(\lambda_i - \lambda_{i-1}) \leqslant \varepsilon$, for any i, then

$$\left| g(t) - \sum_{i=1}^{n} \lambda_i \varphi_{B_i}(t) \right| \leqslant \varepsilon,$$

for any $t \in T$. We hence infer that there exists a sequence $\{g_n\}_{n \in \mathbb{N}}$ of simple functions from $C(T)$, which converges uniformly to g on T. For any $n \in \mathbb{N}$ there exists an $x_n \in \mathscr{S}(\mathscr{X})$ such that $g_n = f_{x_n}$(with the notations introduced at the beginning of the proof).

Hence we easily infer that the sequence $\{x_n\}_{n \in \mathbb{N}}$ is (ρ)-convergent (with the regulator u) to an element x, and we have $f_x = g$.

It is also easy to check that the mapping $x \mapsto f_x$ of $\mathscr{X}$ onto $C(T)$ is one-to-one and it preserves linear operations; also, $x \geqslant 0$ iff $f_x \geqslant 0$.

If follows that the spaces $\mathscr{X}$ and $C(T)$ are isomorphic (as ordered linear spaces).

Remark. As it is well known, one calls *set of type* F_σ, in a topological space, any set which may be represented as a union of a countable system of closed sets. A compact space T is called *quasi-extremal* iff the closure of any open set of type F_σ, contained in T, is open (and, obviously, closed, too).

For quasi-extremal spaces we have an analogue of lemma 1:

A compact, totally disconnected space T is quasi-extremal iff the Boolean lattice $\mathscr{A}$, of all the clopen subsets of T, is σ-complete.

With the aid of this lemma one can prove the analogue of theorem 2:

If $\mathscr{X}$ is a σ-complete vector lattice with strong unit, then there exists a compact, quasi-extremal space T, such that $\mathscr{X}$ be isomorphic to $C(T)$.

4.4.3 The concrete representation of arbitrary complete vector lattices[11])

In this section we shall consider functions which take values in the extended real line $[-\infty, +\infty]$.

Let T be an extremal compact space. We shall denote by $C_\infty(T)$ the set of all the continuous functions, defined on T, taking values in $[-\infty, +\infty]$, and such that the values $+\infty$ and $-\infty$ are taken only on rare subsets of T.

If $x, y \in C_\infty(T)$ and if A is the set of the points t of T in which $x(t) \neq \pm\infty$, $y(t) \neq \pm\infty$, then there exists a function $z \in C_\infty(T)$, such that $z(t) = x(t) + y(t)$, for any $t \in A$. We shall then define $z = x + y$ in $C_\infty(T)$. The operation αx for $\alpha \in \mathbb{R}$ and $x \in C_\infty(T)$ may be defined in a natural manner, by taking into account the usual rules regarding the symbols $\pm\infty$. We shall define $x \leqslant y$ iff $x(t) \leqslant y(t)$ for any $t \in T$.

With the operations and the order relation just defined, the set $C_\infty(T)$ becomes a complete vector lattice.

Let now $\mathscr{X}$ be a complete vector lattice with a unit element u, and let $\mathscr{M}(\mathscr{X})$ be the subspace of the bouded elements.

By virtue of theorem 2 from the preceding section, there exists an extremal compact space T, such that $\mathscr{M}(\mathscr{X})$ be isomorphic to the space $C(T)$.

Let x be an arbitrary positive element of $\mathscr{X}$. By putting $x_n = x \wedge nu$, we have $x_n \uparrow x$ and (with the notations from the proof of Theorem 2 from the preceding section) by putting $g(t) = \sup_{n \in \mathbb{N}} f_{x_n}(t)$, we have $g \in C_\infty(T)$.

We shall denote $f_x = g$ in this case. If the element x is not positive, we shall put $f_x = f_{x^+} - f_{x^-}$.

Let us now put $\mathscr{Y} = \{f_x; x \in \mathscr{X}\}$. The set $\mathscr{Y}$ is a normal and total subspace of $C_\infty(T)$ and the mapping $x \mapsto f_x$ of $\mathscr{X}$ onto $\mathscr{Y}$ is an isomorphism of ordered linear spaces.

Let us now consider an arbitrary complete vector lattice $\mathscr{X}$. The space $\mathscr{X}$ may be considered as a normal and total subspace of a complete vector lattice $\mathscr{Y}$, with a unit element (see Section 4.1.2). Let T be an extremal, compact space, such that $\mathscr{Y}$ be isomorphic to $C_\infty(T)$.

It follows that $\mathscr{X}$ is isomorphic to a normal and total subspace of the space $C_\infty(T)$.

Bibliographical comments

The notion of space with unit belongs to H. Freudenthal [1].

A. G. Pinsker [5] showed that any complete vector lattice may be decomposed into components with unit.

---

[11]) Details will be omitted here.

The theorem on the integral representation of elements in σ-complete vector lattices with unit belongs to H. Freudenthal [1]. For another integral representation theorem, see H. Nakano [2].

The spaces with M-unit have been introduced by R. Cristescu [8], [30].

L. V. Kantorovitch [2], [3] considers complete vector lattices which satisfy the condition: if $\{A_n\}_{n\in\mathbb{N}}$ is a sequence of majorized subsets of the space, and if $x = (o)\text{-}\lim_n (\sup A_n)$ exists, then for any n there exists a finite subset $A'_n \subset A_n$, such that $x = (o)\text{-}\lim_n (\sup A'_n)$. One can show that this condition is equivalent to the condition that the space be σ-regular and of (o)-countable type, at the same time (according to the definitions given in this chapter).

By using the continuum hypothesis, one can show that any complete vector lattice, which is σ-regular, is of countable type too. The spaces of countable type have been introduced by A. G. Pinsker. Several propositions concerning the regular spaces, the σ-regular spaces and the spaces of (o)-countable type, have been proved by A. G. Pinsker (see L. V. Kantorovitch, B. Z. Vulich, A. G. Pinsker [1]).

Proposition 1 from Section 4.2.2 has been proved by V. A. Soloviev [1]. The spaces of bounded type have been introduced by L. V. Kantorovitch [17].

The notion of maximal enlargement has been introduced by A. G. Pinsker [2], who proves the existence by giving its construction (proposition 2 from Section 4.2.3). A. G. Pinsker [3] studies the spaces, which he calls enlarged, and which are defined as being isomorphic to their maximal enlargement. The enlarged spaces coincide with the maximal spaces defined in this chapter.

L. V. Kantorovitch gives a method of constructing a maximal space by starting from an arbitrary complete Boolean lattice. (See L. V. Kantorovitch, B. Z. Vulich, A. G. Pinsker [1]). In particular, one can construct the maximal enlargement of a complete vector lattice with unit, by starting with the base of the space. In connection with this methcd, see also J. Olmsted [1] and M.F. Shirokov [2].

The complex vector lattices have been studied by H. Nakano [1], T. Ogasawara [2], R. Cristescu [7].

A. G. Pinsker introduces the discrete or continuous complete vector lattices.

More generally than in Section 4.4, a complete vector lattice $\mathfrak{X}$ is called discrete if any element $x \in \mathfrak{X}$ may be represented as $x = \underset{j\in J}{S}\, x_j$, where $\{x_j\}_{j\in J}$ is a family of mutually orthogonal discrete elements (see L. V. Kantorovitch, B. Z. Vulich, A. G. Pinsker [1]). A complete vector lattice, which does not contain non-zero discrete elements, is said to be continuous.

Any complete vector lattice may be decomposed into a discrete and a continuous component.

The lattice ordered algebras have been studied by S. W. P. Steen, H. Nakano, B. Z. Vulich, V. I. Sobolev, R. Cristescu, etc. S. W. P. Steen [1] considers vector lattices in which the multiplication of the elements is postulated. In a particular case, F. Riesz [3] shows that the multiplication may be introduced without enlarging the system of axioms of the considered vector lattice. More generally B. Z. Vulich [2], [3] first shows that for some pairs of elements of a complete vector lattice with unit one can define the product, such that the space become a commutative lattice ordered quasialgebra with neutral element. By using the concrete representation of some complete vector lattices, B. Z. Vulich [13] gives another method of introducing the product.

L. V. Kantorovitch introduces the product of the elements in a maximal space, by using the construction of the space over a complete Boolean lattice (see L. V. Kantorovitch, B. Z. Vulich, A. G. Pinsker [1]; see also, M. F. Shirokhov [1], [2]).

The multiplication introduced by Vulich has been generalized under the name of V-product by R. Cristescu [6], [7], for the case when the factors belong to different spaces.

The normal σ-complete lattice ordered algebras have been introduced through the condition $[xy] = [x]\,[y]$ and have been studied by H. Nakano [8]. The relationship between the normal algebras and the algebras endowed with Vulich's multiplication has been remarked by R. Cristescu [15]. H. Nakano [1] also studies the complex lattice ordered algebras.

A certain class of ordered algebras has been studied by Al. Ghika [1], [2].

In connection with the lattice ordered algebras, see also B. Brainerd [1]–[3] and S. Leader [1].

Vector lattices of real continuous functions, defined on a compact space, have been studied by M. G. Krein, S. G. Krein, S. Kakutani, B. Z. Vulich, etc.

In connection with the representation of the spaces with strong unit, see M. G. Krein and S. G. Krein [1], [2] and S. Kakutani [2]. Lemma 1 from Section 4.4.2 has been proved by T. Ogasawara. The equivalence between the condition that a space $C(T)$ (with a compact T) be a complete vector lattice and the condition that T be extremal has been shown by B. Z. Vulich [10].

A representation of the spaces with M-unit has been given by G. Grigore [1].

The problem of the concrete representation of the arbitrary complete (or σ-complete) vector lattices, with the aid of the space $C_{\infty}(T)$, has been studied by B. Z. Vulich, to whom the results mentioned in Section 4.4.3 belong (see B. Z. Vulich, [10], [12], [16]).

5

Regular operators and (o)-continuous operators

The notion of a positive operator (i.e., an operator which takes positive values at positive arguments) plays a fundamental role in the theory of operators between ordered vector spaces. A regular operator is, by definition, a difference of two additive, positive operators. The aim of this chapter is to study general regular operators or such particular ones, which satisfy some additional conditions, like continuity with respect to order, etc.

5.1 Regular linear operators

5.1.1 The notion of (o)-bounded operator

Definition. An operator U, which maps an ordered vector space $\mathscr{X}$ into an ordered vector space $\mathscr{Y}$ is called an (o)-*bounded operator* if for any bounded set $A \subset \mathscr{X}$ the set $U(A)$ is bounded [1]).

In particular, one can speak about an (o)-bounded functional.

It is obvious that if U_1 and U_2 are (o)-bounded operators, then $U_1 + U_2$ is an (o)-bounded operator; if U is an (o)-bounded operator then λU is an (o)-bounded operator for any real number $\lambda \in \mathbb{R}$.

It is also easy to observe that in order that an operator U, defined on an ordered vector space $\mathscr{X}$ and taking values in a space $\mathscr{Y}$ of bounded type, be an (o)-bounded operator, it is sufficient that for any bounded sequence $\{x_n\}_{n\in\mathbb{N}}$ of elements of $\mathscr{X}$, the sequence $\{U(x_n)\}_{n\in\mathbb{N}}$ be bounded.

We shall denote by $\mathscr{X}^{\#}$ the set of all (o)-bounded linear functionals, defined on an ordered vector space $\mathscr{X}$; by θ we shall denote the topology of (o)-boundedness in $\mathscr{X}$ (see Section 3.1.4).

Proposition 1. *For any ordered vector space $\mathscr{X}$ the following equality holds*

$$\mathscr{X}^*_\theta = \mathscr{X}^{\#}.$$

---

[1]) The zero elements of the spaces $\mathscr{X}$ and $\mathscr{Y}$, as well as those of other vector spaces which will be considered, will be denoted by the same symbol $\mathbf{0}$.

Proof. Let $\mathcal{M}_o$ be the set of all (o)-bounded subsets of $\mathcal{X}$, and $\mathcal{M}_\theta$ the set of all (θ)-bounded subsets. If $f \in \mathcal{X}_\theta^*$, then f is bounded on any (o)-bounded set, because $\mathcal{M}_o \subset \mathcal{M}_\theta$. It follows that $f \in \mathcal{X}^\#$. Conversely, let $f \in \mathcal{X}^\#$ and let us define

$$W = \{x;\ |f(x)| \leqslant 1\}.$$

The set W is obviously equilibrated and convex. If $A \in \mathcal{M}_o$, then there exists an $\alpha > 0$, such that $|f(x)| \leqslant \alpha$, for any $x \in A$; consequently $A \subset \alpha W$. It follows that W is a neighbourhood of the origin for the topology θ, and this implies that f is (θ)-continuous, i.e., $f \in \mathcal{X}_\theta^*$.

Remark. From proposition 1 it follows that if the topology θ is separated, then on the space $\mathcal{X}$ there exists a sufficient system of (o)-bounded linear functionals [2]).

Conversely, if on $\mathcal{X}$ there exists a sufficient system of (o)-bounded linear functionals, then the topology θ is separated and we have

$$\theta = \mu(\mathcal{X}, \mathcal{X}^\#),$$

where μ is Mackey's topology.

Proposition 2. *In order that a linear functional φ, defined on a vector lattice $\mathcal{X}$, be (o)-bounded, it is necessary and sufficient that for any decreasing sequence $\{x_n\}_{n \in \mathbb{N}}$ of positive elements of $\mathcal{X}$, the sequence $\left\{\frac{1}{n}\varphi(x_n)\right\}_{n\in\mathbb{N}}$ be bounded.*

Proof. If $\mathbf{0} \leqslant x_n \downarrow$, then $x_n \in [\mathbf{0}, x_1]$ for any n; consequently, if φ is (o)-bounded, then $\{\varphi(x_n)\}_{n\in\mathbb{N}}$ is bounded.

Conversely, let us assume that the condition in the statement of the proposition is satisfied.

Since for any bounded set $A \subset \mathcal{X}$, there exists an element $a \in \mathcal{X}$, such that $x_+, x_- \in [\mathbf{0}, a]$, for any $x \in A$, it is sufficient to prove that for any element $x > \mathbf{0}$ of $\mathcal{X}$, the set $\varphi([\mathbf{0}, x])$ is bounded.

Let us assume that an element $x_0 > \mathbf{0}$ exists, such that the set $\varphi([\mathbf{0}, x_0])$ be not bounded. There then exists an $x_1 \in [0, x_0]$, such that $|\varphi(x_1)| \geqslant 1$, and such that the set $\varphi([\mathbf{0}, x_1])$ be not bounded. Indeed, there exists an $x' \in [\mathbf{0}, x_0]$, such that

$$|\varphi(x')| \geqslant |\varphi(x_0)| + 1. \tag{1}$$

In this case, at least one of the sets $\varphi([\mathbf{0}, x'])$, $\varphi([\mathbf{0}, x_0 - x'])$ is not bounded, because if both are bounded, then let $\mathbf{0} \leqslant x \leqslant x_0$. From $x_0 = x' + (x_0 - x')$ it follows that $x = a + b$, where $\mathbf{0} \leqslant a \leqslant x'$ and $\mathbf{0} \leqslant b \leqslant x_0 - x'$ and we have

$$\varphi(x) = \varphi(a) + \varphi(b) \in \varphi([\mathbf{0}, x']) + \varphi([\mathbf{0}, x_0 - x']),$$

i.e., the set $\varphi([\mathbf{0}, x_0])$ is bounded.

---

[2]) This means that for any non-zero element x there exists an (o)-bounded linear functional f, such that $f(x) \neq 0$.

If the set $\varphi([\mathbf{0}, x'])$ is not bounded, we shall put $x_1 = x'$. Otherwise, we shall put $x_1 = x_0 - x'$, and, by virtue of (1), we have $|\varphi(x_0 - x')| \geqslant 1$.

By induction, one can construct a sequence $\{x_n\}_{n\in\mathbb{N}}$, such that $\mathbf{0} \leqslant x_n \downarrow$ and such that $|\varphi(x_n)| \geqslant n^2$. But in this case the sequence $\left\{\frac{1}{n}\varphi(x_n)\right\}_{n\in\mathbb{N}}$ is not bounded, and this is a contradiction. The proposition is proved.

5.1.2 Positive operators

Definition 1. An operator U, which maps an ordered linear space $\mathscr{X}$ into an ordered linear space $\mathscr{Y}$, is said to be a *positive operator* iff $x \geqslant \mathbf{0}$ implies $U(x) \geqslant \mathbf{0}$.

In particular, one can speak about positive functionals. If U is a positive operator, we shall write $U \geqslant \mathbf{0}$.

If $-U \geqslant \mathbf{0}$, then U is said to be a negative operator.

It is easy to check that if U is an additive and positive operator, then $x_1 \leqslant x_2$ implies $U(x_1) \leqslant U(x_2)$.

If $\mathscr{X}$ and $\mathscr{Y}$ are vector lattices and if U is an additive and positive operator, which maps $\mathscr{X}$ into $\mathscr{Y}$, then

$$|U(x)| \leqslant U(|x|),$$

for any $x \in \mathscr{X}$. This fact is easily established.

Proposition 1. *If U is an additive and positive operator, defined on the vector lattice $\mathscr{X}$, and taking values in an Archimedean space $\mathscr{Y}$, then U is homogeneous.*

Proof. If α is an arbitrary real number, then there exist sequences $\{\rho_n\}_{n\in\mathbb{N}}$ and $\{\rho_n'\}_{n\in\mathbb{N}}$ of rational numbers, such that $\rho_n \uparrow \alpha$ and $\rho_n' \downarrow \alpha$. If $x > \mathbf{0}$, $x \in \mathscr{X}$, then

$$\rho_n x \leqslant \alpha x \leqslant \rho_n' x,$$

and this implies

$$\rho_n U(x) \leqslant U(\alpha x) \leqslant \rho_n' U(x).$$

Since the space $\mathscr{Y}$ is Archimedean, from the preceding inequalities one gets

$$\alpha U(x) \leqslant U(\alpha x) \leqslant \alpha U(x)$$

(see Section 3.1.3, proposition 3), and this shows that

$$U(\alpha x) = \alpha U(x).$$

Corollary. *Any additive and positive functional, defined on a vector lattice, is homogeneous.*

In the following lemma we shall denote by $\mathscr{X}_+$ the set of the positive elements of a vector lattice $\mathscr{X}$.

Lemma 1. *Let $\mathscr{X}$ be a vector lattice and $\mathscr{Y}$ a linear space. If U_0 is a mapping of $\mathscr{X}_+$ into $\mathscr{Y}$, which satisfies the condition*

$$U_0(x_1 + x_2) = U_2(x_1) + U_0(x_2),$$

for any $x_1, x_2 \in \mathscr{X}_+$, then U_0 can be extended in a unique manner as an additive operator. defined on $\mathscr{X}$. This extension is given by the formula

$$U(x) = U_0(x_+) - U_0(x_-). \tag{1}$$

Proof. Formula (1) defines an additive operator on the space $\mathscr{X}$. Indeed, if x', x'' are arbitrary elements from $\mathscr{X}$, then there exists an element $a \geqslant \mathbf{0}$, of $\mathscr{X}$, such that

$$(x' + x'')_+ + a = x'_+ + x'_+,$$
$$(x' + x'')_- + a = x'_- + x''_-.$$

Hence we get

$$U_0((x' + x'')_+) + U_0(a) = U_0(x'_+) + U_0(x''_+),$$
$$U_0((x' + x'')_-) + U_0(a) = U_0(x'_-) + U_0(x''_-).$$

By substracting member by member, we get

$$U(x' + x'') = U(x') + U(x'').$$

It is obvious that U is an extension of U_0.

The uniqueness of the extension is obvious.

In the next proposition we shall denote by $\mathscr{X}^l$ the set of all linear functionals defined on a linear space $\mathscr{X}$.

If φ and ψ are two functionals on an ordered vector space $\mathscr{X}$, we shall write $\varphi \leqslant \psi$ if $\varphi - \psi \geqslant \mathbf{0}$ (i.e., $\varphi - \psi$ is a positive functional).

Lemma 2. If φ is a positive linear functional, defined on a vector lattice $\mathscr{X}$, then for any element $x \in \mathscr{X}$ the following equality holds

$$\varphi(x_+) = \sup_{\substack{\mathbf{0} \leqslant \psi \leqslant \varphi \\ \psi \in \mathscr{X}^l}} \psi(x).$$

Proof. Let us consider an arbitrary element $a \in \mathscr{X}$ and let us define

$$\lambda = \sup_{\substack{\mathbf{0} \leqslant \psi \leqslant \varphi \\ \psi \in \mathscr{X}^l}} \psi(a).$$

It is obvious that $\lambda \leqslant \varphi(a_+)$.

Let us now consider the function χ_0 defined on the set $\mathscr{X}_+$ of the positive elements of $\mathscr{X}$ by the formula

$$\chi_0(x) = \sup_{0 \leqslant \alpha \in \mathbb{R}} \varphi(x \wedge \alpha a_+).$$

Let us prove that for any positive elements x_1, x_2, the following equality holds

$$\chi_0(x_1 + x_2) = \chi_0(x_1) + \chi_0(x_2). \tag{2}$$

In order to do this, let us first remark that for any $\alpha \geqslant 0$ we have

$$(x_1 + x_2) \wedge \alpha a_+ \leqslant (x_1 \wedge \alpha a_+) + (x_2 \wedge \alpha a_+).$$

(see Section 3.2.3, lemma); hence we shall infer that

$$\varphi((x_1 + x_2) \wedge \alpha a_+) \leqslant \varphi(x_1 \wedge \alpha a_+) + \varphi(x_2 \wedge \alpha a_+).$$

By taking into account the definition of χ_0, we immediately infer that

$$\chi_0(x_1 + x_2) \leqslant \chi_0(x_1) + \chi_0(x_2).$$

In order to prove the opposite inequality, let us now remark that for any elements b_1, b_2, b_3, b_4 of $\mathscr{X}_+$, the following inequality holds

$$(b_1 \wedge b_2) + (b_3 \wedge b_4) \leqslant (b_1 + b_3) \wedge (b_2 + b_4).$$

It follows that for any numbers $\alpha_1, \alpha_2 \geqslant 0$, we have

$$\varphi(x_1 \wedge \alpha_1 a_+) + \varphi(x_2 \wedge \alpha_2 a_+) \leqslant \varphi((x_1 + x_2) \wedge (\alpha_1 + \alpha_2) a_+),$$

whence

$$\chi_0(x_1) + \chi_0(x_2) \leqslant \chi_0(x_1 + x_2).$$

We have therefore proved equality (2). By taking into account the preceding lemma, we infer that χ_0 can be uniquely extended as an additive functional χ, defined on $\mathscr{X}$, which, moreover, is positive.

Since for any $x \geqslant \mathbf{0}$ we have

$$\chi(x) = \chi_0(x) \leqslant \varphi(x),$$

it follows that $\chi \leqslant \varphi$. It follows that $\mathbf{0} \leqslant \chi \leqslant \varphi$ and $\chi \in \mathscr{X}^l$.

Since $\chi(a_-) = 0$, it follows that

$$\chi(a) = \chi_0(a_+) = \varphi(a_+).$$

This shows that $\lambda = \varphi(a_+)$, and the lemma is proved.

Proposition 2. *Let $\mathscr{X}$ be a vector lattice on which a sufficient set of positive linear functionals exists*[3]). *If $a \in \mathscr{X}$ and if $\varphi(a) \geqslant \mathbf{0}$, for any positive linear functional φ, then $a \geqslant \mathbf{0}$.*

Proof. If φ and ψ are linear functionals and if $\mathbf{0} \leqslant \psi \leqslant \varphi$ then $\psi(a) \leqslant \varphi(a)$. From lemma 2 it follows that $\varphi(a_+) = \varphi(a)$, for any positive linear functional φ.

This implies that $a = a_+$, because on $\mathscr{X}$ there exists a sufficient set of positive linear functionals. The proposition is proved.

Definition 2. A vector subspace $\mathscr{X}_0$, of an ordered vector space $\mathscr{X}$ is said to be a *majorizing vector subspace*, if for any element $x \in \mathscr{X}$, there exist $x_1, x_2 \in \mathscr{X}_0$, such that $x_1 \leqslant x \leqslant x_2$.

Theorem. *Let $\mathscr{X}$ be a vector lattice, $\mathscr{X}_0$ a majorizing vector subspace of $\mathscr{X}$ and $\mathscr{Y}$ a complete vector lattice. If U_0 is a linear positive operator, defined on $\mathscr{X}_0$ and taking values in $\mathscr{Y}$, then there exists a linear and positive operator U, defined on $\mathscr{X}$, and taking values in $\mathscr{Y}$, such that $U(x) = U_0(x)$ for any $x \in \mathscr{X}_0$.*

Proof. Let $v \in \complement\mathscr{X}_0$ and $x_1, x_2 \in \mathscr{X}_0$ be such that $x_1 \leqslant v \leqslant x_2$. Since the operator U_0 is linear and positive, we have

$$U_0(x_1) \leqslant U_0(x_2).$$

Let us define

$$a = \sup_{\substack{x \leqslant v \\ x \in \mathscr{X}_0}} U_0(x), \quad b = \inf_{\substack{x \geqslant v \\ x \in \mathscr{X}_0}} U_0(x)$$

and let us consider an element y_1, such that $a \leqslant y_1 \leqslant b$.

Let $\mathscr{X}_1$ be the linear hull of $\mathscr{X}_0 \cup \{v\}$ and let us define on $\mathscr{X}_1$ an operator U_1, taking values in $\mathscr{Y}$, by the formula

$$U_1(x + \alpha v) = U_0(x) + \alpha y_1,$$

where $x \in \mathscr{X}_0$ and $\alpha \in \mathbb{R}$.

The operator U_1 is positive. Indeed, let $x + \alpha v \geqslant \mathbf{0}$, where $\alpha \neq 0$. If $\alpha > 0$, then

$$v \geqslant -\frac{1}{\alpha} x$$

---

[3]) This means that for any non-zero element x there exists a positive linear functional φ, such that $\varphi(x) \neq 0$.

and this implies

$$y_1 \geqslant U_0\left(-\frac{1}{\alpha}x\right) = -\frac{1}{\alpha}U_0(x),$$

whence we immediately infer that $U_1(x + \alpha v) \geqslant \mathbf{0}$. For $\alpha < 0$ we get the same result.

Let now $\mathscr{U}$ be the set of all linear, positive extensions of U_0, i.e. $U' \in \mathscr{U}$ iff U' is a linear, positive operator, defined on a vector subspace $\mathscr{X}'$ of $\mathscr{X}$, taking values in $\mathscr{Y}$, and such that $\mathscr{X}_0 \subset \mathscr{X}'$; moreover, it is required that $U'(x) = U_0(x)$ for $x \in \mathscr{X}_0$. We shall order the set $\mathscr{U}$ by putting $U' \leqslant U''$ iff U'' is an extension of U'. Endowed with this order relation, the set $\mathscr{U}$ satisfies the conditions of Zorn's lemma. Therefore, there exists a maximal element U, and, by virtue of what we have already established, it follows that U is defined on the whole space $\mathscr{X}$. The theorem is proved.

5.1.3 Regular operators

Definition. An operator U, defined on an ordered vector space $\mathscr{X}$, taking values in an ordered vector space $\mathscr{Y}$, is called a *regular operator* iff it may be written as a difference of two additive positive operators.

We shall denote by $(\mathscr{X}, \mathscr{Y})^r$ the set of all regular operators which map $\mathscr{X}$ into $\mathscr{Y}$.

If U_1 and U_2 are two operators defined on an ordered vector space $\mathscr{X}$, with values in an ordered vector space $\mathscr{Y}$, we shall write $U_1 \leqslant U_2$ iff $U_1(x) \leqslant U_2(x)$, for any $x \geqslant \mathbf{0}$, i.e., iff $U_2 - U_1 \geqslant \mathbf{0}$. We shall say that U_1 is majorized by U_2.

If U is a regular operator, i.e., if U can be written in the form $U = U_1 - U_2$, where U_1, U_2 are additive, positive operators, then U is majorized by U_1.

Conversely, if U is an additive operator and if U is majorized by an additive, positive operator U_1, then

$$U = U_1 - (U_1 - U),$$

where $U_1 - U \geqslant \mathbf{0}$, and this shows that U is regular.

Proposition 1. *Any regular operator is (o)-bounded.*

Proof. Let U be a regular operator and U_1 an additive, positive operator, such that U be majorized by U_1.

If $a \leqslant x \leqslant b$, then $x - a \geqslant \mathbf{0}$ and $b - x \leqslant b - a$ and this shows that

$$U(x) = U(a) + U(x - a) \leqslant U(a) + U_1(x - a) \leqslant U(a) + U_1(b - a).$$

Analogously, one can show that

$$U(b) + U_1(a - b) \leqslant U(x).$$

Proposition 2. *Any regular operator, defined on a vector lattice and taking values in an Archimedean space, is linear.*

This proposition is an immediate consequence of proposition 1 from the preceding section.

Theorem 1. *If $\mathscr{X}$ is a vector lattice and if $\mathscr{Y}$ is a complete vector lattice, then the set $(\mathscr{X}, \mathscr{Y})^r$ of all regular operators which map $\mathscr{X}$ into $\mathscr{Y}$ is a complete vector lattice* [4]).

Proof. The set $(\mathscr{X}, \mathscr{Y})^r$ is obviously an ordered vector space. Since for any regular operator U there exists an additive and positive operator U_1, such that $U \leqslant U_1$, the set $(\mathscr{X}, \mathscr{Y})^r$ is a directed vector space. One has only to show that any bounded from above subset of $(\mathscr{X}, \mathscr{Y})^r$ has a lowest upper bound.

Let $\mathscr{U} = \{U_j\}_{j\in J}$ be a bounded from above system of elements of $(\mathscr{X}, \mathscr{Y})^r$ and let U_0 be an upper bound for this system. For any element $x \geqslant 0$ let us define

$$G(\mathscr{U}; x) = \left\{\sum_{i=1}^{n} U_{j_i}(x_i);\ j_i \in J,\ n \in \mathbb{N},\ \mathbf{0} \leqslant x_i \in \mathscr{X},\ \sum_{i=1}^{n} x_i = x\right\}. \tag{1}$$

The set $G(\mathscr{U}; x)$ is bounded from above, because from $x_i \geqslant \mathbf{0}$ it follows that

$$U_{j_i}(x_i) \leqslant U_0(x_i),$$

and, by taking into account the fact that $\sum x_i = x$, we get

$$\sum_{i=1}^{n} U_{j_i}(x_i) \leqslant U_0(x), \tag{2}$$

Let us put

$$U(x) = \sup G(\mathscr{U}; x). \tag{3}$$

From (2) we get $U(x) \leqslant U_0(x)$ and, at the same time, we obviously have $U_j(x) \leqslant U(x)$, for any $j \in J$. It follows that

$$U_j(x) \leqslant U(x) \leqslant U_0(x),\ \forall\, j \in J, \forall\, x \geqslant \mathbf{0}. \tag{4}$$

Formula (3) defines a mapping of $\mathscr{X}_+$ into $\mathscr{Y}$, which has the property

$$U(x_1 + x_2) = U(x_1) + U(x_2), \tag{5}$$

for any $x_1, x_2 \in \mathscr{X}_+$.

Indeed, if

$$x_1 = \sum_{i=1}^{n} a_i,\quad x_2 = \sum_{k=1}^{m} b_k,$$

---

[4]) In $(\mathscr{X}, \mathscr{Y})^r$ one considers the usual operations $U_1 + U_2$, αU and the order relation $U_1 \leqslant U_2$ given by $U_2 - U_1 \geqslant \mathbf{0}$.

where $a_i, b_k \geqslant 0$, then

$$\sum_{i=1}^{n} U_{j_i'}(a_i) + \sum_{k=1}^{m} U_{j_k''}(b_k) \leqslant U(x_1 + x_2),$$

whence we immediately get

$$U(x_1) + U(x_2) \leqslant U(x_1 + x_2). \tag{6}$$

If

$$x_1 + x_2 = \sum_{i=1}^{n} c_i,$$

where $c_i \geqslant \mathbf{0}$, then there exist elements $a_i \geqslant \mathbf{0}$, $b_i \geqslant \mathbf{0}$, $i = 1, 2, \ldots, n$, such that

$$a_i + b_i = c_i, \quad i = 1, 2, \ldots, n,$$

$$\sum_{i=1}^{n} a_i = x_1, \quad \sum_{i=1}^{n} b_i = x_2$$

(see Section 3.2.1, proposition 3), and we infer

$$\sum_{i=1}^{n} U_{j_i}(c_i) = \sum_{i=1}^{n} U_{j_i}(a_i) + \sum_{i=1}^{n} U_{j_i}(b_i) \leqslant U(x_1) + U(x_2).$$

We get that

$$U(x_1 + x_2) \leqslant U(x_1) + U(x_2). \tag{7}$$

From (6) and (7) we get (5).

As one knows, the formula

$$U(x) = U(x_+) - U(x_-),$$

where x is an arbitrary element of $\mathscr{X}$, then defines an additive operator on the space $\mathscr{X}$. From (4) we infer that $U \leqslant U_0$, and this shows that U is a regular operator. Inequalities (4) also prove that U is an upper bound of the system U (because $U_j \leqslant U$, $\forall j \in J$). Since $U \leqslant U_0$ and since U_0 is an arbitrary upper bound of U, it follows that $U = \sup \mathscr{U}$.

The theorem is proved.

Remark. If $\{U_j\}_{j\in J}$ is a bounded from above system of elements of the space $(\mathscr{X}, \mathscr{Y})^r$, and if this system is increasingly directed with respect to the order relation of $(\mathscr{X}, \mathscr{Y})^r$, then for any element $x \geqslant \mathbf{0}$ we have

$$(\bigvee_{j\in J} U_j)(x) = \bigvee_{J\in j} U_j(x).$$

This fact is easy to prove.

Lemma. *If $\mathscr{X}$ is a vector lattice, $\mathscr{Y}$ a complete vector lattice and if $U \in (\mathscr{X}, \mathscr{Y})^r$, then*

$$|U(x)| \leqslant |U|(|x|), \tag{8}$$

for any $x \in \mathscr{X}$.

Proof. If $x \geqslant 0$, then $U(x) \leqslant |U|(x)$ and $-U(x) \leqslant |U|(x)$, whence (8) immediately follows. If x is arbitrary, then

$$|U(x)| = |U(x_+) - U(x_-)| \leqslant |U(x_+)| + |U(x_-)|$$

$$\leqslant |U|(x_+) + |U|(x_-) = |U|(|x|).$$

Proposition 3. *If $\mathscr{X}$ is a vector lattice and $\mathscr{Y}$ a complete vector lattice, then for any $U \in (\mathscr{X}, \mathscr{Y})^r$ and any element $x \geqslant \mathbf{0}$ of $\mathscr{X}$ the following equalities hold*

$$U_+(x) = \sup_{0\leqslant x'\leqslant x} U(x') \tag{9}$$

$$U_-(x) = -\inf_{0\leqslant x'\leqslant x} U(x') \tag{10}$$

$$|U|(x) = \sup_{|x'|\leqslant x} U(x') = \sup_{|x'|\leqslant x} |U(x')|. \tag{11}$$

Proof. Formula (9) can be immediately obtained if in (1) we take $\mathscr{U} = \{U, \mathbf{0}\}$.
Formula (10) immediately follows from (9). In order to prove (11) let us first observe that if in (1) we take $\mathscr{U} = \{U, -U\}$, we get

$$|U|(x) = \sup_{\substack{x_1+x_2=x\\ x_1,x_2\geqslant 0}} \{U(x_1) - U(x_2)\}. \tag{12}$$

On the other hand, the set $\{U(x');\ |x'| \leqslant x\}$ is bounded from above and we obviously have

$$\sup_{|x'|\leqslant x} U(x') = \sup_{|x'|\leqslant x} \{U(x') \vee (-U(x'))\} = \sup_{|x'|\leqslant x} |U(x)| \leqslant |U|(x).$$

If $x = x_1 + x_2$ and $x_1, x_2 \geqslant 0$, then

$$|x_1 - x_2| \leqslant x$$

and

$$U(x_1) - U(x_2) = U(x_1 - x_2) \leqslant \sup_{|x'| \leqslant x} U(x')$$

and from (12) we get

$$|U|(x) \leqslant \sup_{|x'| \leqslant x} U(x').$$

The proposition is proved.

Proposition 4. *Let $\mathscr{X}$ be a vector lattice and $\mathscr{Y}$ a complete vector lattice. If $U_n \xrightarrow{o} U$ in the space $(\mathscr{X}, \mathscr{Y})^r$, then $U_n(x) \xrightarrow{o} U(x)$, for any $x \in \mathscr{X}$.*

Proof. It is sufficient to consider the case in which the sequence $\{U_n\}_{n \in \mathbb{N}}$ is monotonous. Let $U_n \uparrow U$. For any element $x \geqslant \mathbf{0}$ the sequence $\{U_n(x)\}_{n \in \mathbb{N}}$ is increasing and bounded from above by $U(x)$. There then exists an element $V_0(x) \in \mathscr{Y}$, such that $U_n(x) \uparrow V_0(x)$. In particular, $V_0(x) \leqslant U(x)$. It is obvious that

$$V_0(x_1 + x_2) = V_0(x_1) + V_0(x_2),$$

for any $x_1, x_2 \geqslant 0$. By putting

$$V(x) = V_0(x_+) - V_0(x_-),$$

for any element $x \in \mathscr{X}$, we get an additive operator on $\mathscr{X}$, taking values in $\mathscr{Y}$. Since for any $x \geqslant \mathbf{0}$ we have

$$V(x) = V_0(x) \leqslant U(x),$$

it follows that V is a regular operator. From

$$U_n \leqslant V \leqslant U$$

and from $U_n \uparrow U$ we infer that $V = U$. Finally, from $U_n(x) \uparrow U(x)$ for any $x \geqslant \mathbf{0}$ it obviously follows that $U_n(x) \xrightarrow{o} U(x)$, for any element x. The proof is similar in the case of a decreasing sequence $\{U_n\}_{n \in \mathbb{N}}$.

Corollary. *If $U_n \in (\mathscr{X}, \mathscr{Y})^r$, $U_n \uparrow$ (or $U_n \downarrow$) and if $U_n(x) \xrightarrow{o} U(x)$ for any $x \in \mathscr{X}$, then $U_n \uparrow U$ ($U_n \downarrow U$, respectively).*

Indeed, from the hypothesis it immediately follows that U is a regular operator and if (for example) $U_n \uparrow$, then $U_n \leqslant U$ and it therefore exists $V = (o)\text{-}\lim U_n$. But in this case $V(x) = (o)\text{-}\lim U_n(x) = U(x)$, for any x and this shows that $V = U$.

Proposition 5. *If $\mathscr{X}$ is a vector lattice, $\mathscr{Y}$ is a complete vector lattice, and if $\mathscr{Y}_0$ is a component of the space $\mathscr{Y}$, then $(\mathscr{X}, \mathscr{Y}_0)^r$ is a component of the space $(\mathscr{X}, \mathscr{Y})^r$.*

Proof. Let us denote $\mathscr{Z} = (\mathscr{X}, \mathscr{Y})^r$ and $\mathscr{Z}_0 = (\mathscr{X}, \mathscr{Y}_0)^r$. It is obvious that $\mathscr{Z}_0$ is a vector subspace of $\mathscr{Z}$.

If $U \in \mathscr{Z}_0$, then $|U| \in \mathscr{Z}_0$ (here $|U|$ is the absolute value of U in the space $\mathscr{Z}$). Indeed, for any element $x \geqslant \mathbf{0}$ from $\mathscr{X}$ we have

$$\{U(x');\ |x'| \leqslant x\} \subset \mathscr{Y}_0,$$

and since $\mathscr{Y}_0$ is a component of $\mathscr{Y}$, it follows that $|U|(x) \in \mathscr{Y}_0$. Hence we immediately get that for an arbitrary element x we have $|U|(x) \in \mathscr{Y}_0$ and this implies that $|U| \in \mathscr{Z}_0$.

Let $U \in \mathscr{Z}_0$, $V \in \mathscr{Z}$ and $|V| \leqslant |U|$ in the space $\mathscr{Z}$. For any element $x \geqslant \mathbf{0}$ from $\mathscr{X}$ the following inequalities hold

$$|V(x)| \leqslant |V|(x) \leqslant |U|(x).$$

Since $|U|(x) \in \mathscr{Y}_0$ and since $\mathscr{Y}_0$ is a normal subspace of $\mathscr{Y}$, it follows that $V(x) \in \mathscr{Y}_0$. Hence we immediately infer that for an arbitrary element $x \in \mathscr{X}$ we have $V(x) \in \mathscr{Y}_0$. This shows that $V \in \mathscr{Z}_0$, i.e. $\mathscr{Z}_0$ is a normal subspace of $\mathscr{Z}$.

Let $U = \{U_j\}_{j \in J}$ be a system of elements of $\mathscr{Z}_0$, such that $U = \bigvee U_j$ exists in the space $\mathscr{Z}$. For any element $x \geqslant 0$ from $\mathscr{X}$ we have $U_j(x) \in \mathscr{Y}_0$ and consequently (by using formula (1)) we have $G(\mathscr{U}; x) \subset \mathscr{Y}_0$. Since $\mathscr{Y}_0$ is a component of $\mathscr{Y}$, it follows that $U(x) \in \mathscr{Y}_0$. We hence infer that for any x we also have $U(x) \in \mathscr{Y}$ and, therefore, $U \in \mathscr{Z}_0$.

Since $\mathscr{Z}$ is a complete vector lattice, from what we have already shown (see Section 3.3.3, proposition 5) it follows that $\mathscr{Z}_0$ is a component of $\mathscr{Z}$.

Theorem 2. *An additive operator U, defined on a vector lattice $\mathscr{X}$, taking values in a complete vector lattice $\mathscr{Y}$, is a regular operator iff it is an (o)-bounded operator.*

Proof. From proposition 1 it follows that any regular operator is (o)-bounded. Conversely, let us assume that U is an additive (o)-bounded operator. In particular, for any element $x \geqslant \mathbf{0}$ of $\mathscr{X}$, the set $\{U(x');\ \mathbf{0} \leqslant x' \leqslant x\}$ is bounded and, if $V(x)$ denotes its least upper bound, we get a mapping of $\mathscr{X}_+$ into $\mathscr{Y}$.

For any elements $x_1, x_2 \geqslant \mathbf{0}$ of $\mathscr{X}$ we have [5])

$$V(x_1 + x_2) = V(x_1) + V(x_2) \tag{13}$$

and, therefore, by putting

$$V(x) = V(x_+) - V(x_-)$$

[5]) Equality (13) can be directly checked, by taking into account the fact that if $\mathbf{0} \leqslant a \leqslant x_1 + x_2$, where $x_1, x_2 \geqslant \mathbf{0}$, then there exist $b, c \in \mathscr{X}$, such that $a = b + c$, $\mathbf{0} \leqslant b \leqslant x_1$, $\mathbf{0} \leqslant c \leqslant x_2$. From the proof of theorem 1, by taking into account the fact that $V(x) = \sup G(\mathscr{U}; x)$, where $\mathscr{U} = \{U, \mathbf{0}\}$, we also get (13).

we obtain an additive operator which maps $\mathscr{X}$ into $\mathscr{Y}$. This operator is also positive and we obviously have $U \leqslant V$.

Corollary. *An additive functional on a vector lattice is regular iff it is (o)-bounded.*

5.2 (o)-Continuous linear operators

5.2.1 General facts about (o)-continuous operators

Let $\mathscr{X}$ and $\mathscr{Y}$ be ordered vector spaces.
An operator U, which is defined on $\mathscr{X}$ and takes values in $\mathscr{Y}$, is said to be (o)-*continuous*[6]) at a point x_0 iff for any sequence $\{x_n\}_{n \in \mathbb{N}}$ of elements of $\mathscr{X}$, such that $x_n \xrightarrow{o} x_0$ it follows that $U(x_n) \xrightarrow{o} U(x_0)$.

If an additive operator is (o)-continuous at a point x_0, then it is (o)-continuous at any point.

If U is an additive and positive operator and if from $x_0 \downarrow \mathbf{0}$ it follows that $U(x_n) \downarrow \mathbf{0}$, then U is (o)-continuous.

Indeed, from the properties of U it follows that the condition $x_n \uparrow \mathbf{0}$ implies $U(x_n) \uparrow \mathbf{0}$. But if $x_n \xrightarrow{o} \mathbf{0}$, then there exist sequences $\{a_n\}_{n \in \mathbb{N}}$, $\{b_n\}_{n \in \mathbb{N}}$ such that $a_n \uparrow 0$, $b_n \downarrow \mathbf{0}$ and $a_n \leqslant x_n \leqslant b_n$, for any $n \in \mathbb{N}$. It follows that

$$U(a_n) \leqslant U(x_n) \leqslant U(b_n)$$

(for any $n \in \mathbb{N}$) and $U(a_n) \uparrow \mathbf{0}$, $U(b_n) \downarrow \mathbf{0}$. This shows that $U(x_n) \xrightarrow{o} \mathbf{0}$.

If $\mathscr{X}$ and $\mathscr{Y}$ are Archimedean directed vector spaces, then any additive and (o)-continuous operator mapping $\mathscr{X}$ into $\mathscr{Y}$ is homogeneous.

It is easy to prove this assertion.

It is also easy to prove that if $\mathscr{X}$ and $\mathscr{Y}$ are Archimedean vector lattices and if U is a linear (o)-continuous operator which maps $\mathscr{X}$ into $\mathscr{Y}$, then for any (o)-annihilating set[7]) $E \subset \mathscr{X}$ the set $U(E)$ is (o)-annihilating. Conversely, if $\mathscr{X}$ is moreover, a regular space and U is a linear operator mapping $\mathscr{X}$ into $\mathscr{Y}$ and mapping any (o)-annihilating set $E \subset \mathscr{X}$ onto an (o)-annihilating set $U(E)$, then U is (o)-continuous.

Indeed, if $x_n \xrightarrow{o} 0$, then there exists a sequence $\{\alpha_n\}_{n \in \mathbb{N}}$ of real numbers, such that $\alpha_n \to +\infty$ and $\alpha_n x_n \xrightarrow{o} 0$. It follows that

$$U(x_n) - \frac{1}{\alpha_n} U(\alpha_n x_n) \xrightarrow{o} 0.$$

and this shows that U is (o)-continuous.

---

[6]) One should say sequentially (o)-continuous, but in the theory of ordered vector spaces the operators which are continuous, with respect to the generalized (o)-convergent sequences have other names, which differ according to the author; for example: completely continuous, universally continuous, M. S. (o)-continuous, etc.

[7]) For the definition of an (o)-annihilating set see Section 4.2.2, definition 2.

5.2.2 (o)-Continuous regular operators

Let $\mathscr{X}$ be a vector lattice and $\mathscr{Y}$ a complete vector lattice.

We shall denote by $(\mathscr{X}, \mathscr{Y})_o^r$ the set of all the (o)-continuous regular operators which map $\mathscr{X}$ into $\mathscr{Y}$.

Theorem. *The set $(\mathscr{X}, \mathscr{Y})_o^r$ is a component of the space $(\mathscr{X}; \mathscr{Y})^r$.*

Proof. The set $(\mathscr{X}, \mathscr{Y})_o^r$ is obviously a vector subspace of the space $(\mathscr{X}, \mathscr{Y})^r$.

Let us now show that if $U \in (\mathscr{X}, \mathscr{Y})_o^r$, then $U_+ \in (\mathscr{X}, \mathscr{Y})_o^r$. By making use of a remark from Section 5.2.1, it is sufficient to show that if $x_n \downarrow \mathbf{0}$, then $U_+(x_n) \downarrow \mathbf{0}$.

Let then $x_n \downarrow \mathbf{0}$ and let us write

$$y_0 = \bigwedge_{n \in \mathbb{N}} U_+(x_n).$$

If $\mathbf{0} \leqslant a \leqslant x_1$, then

$$\mathbf{0} \leqslant a - a \wedge x_n = a \vee x_n - x_n \leqslant x_1 - x_n$$

and, by taking into account formula (9) from Section 5.1.3, it follows that

$$U(a - a \wedge x_n) \leqslant U_+(x_1 - x_n)$$

and this implies that

$$U_+(x_n) - U(a \wedge x_n) \leqslant U_+(x_1) - U(a). \tag{1}$$

Since $a \wedge x_n \downarrow \mathbf{0}$ and since U is (o)-continuous, from (1) it follows that

$$U(a) \leqslant U_+(x_1) - y_0.$$

Since this inequality holds for any element $a \in [\mathbf{0}, x_1]$ we further get

$$U_+(x_1) \leqslant U_+(x_1) - y_0.$$

and, therefore, $y_0 = \mathbf{0}$.

We have thus shown that $U_+ \in (\mathscr{X}, \mathscr{Y})_0^r$. Hence we infer that $(\mathscr{X}, \mathscr{Y})_o^r$ is a vector sublattice of the space $(\mathscr{X}, \mathscr{Y})^r$, and even a normal subspace, as one can easily prove.

Let $\{U_j\}_{j \in J}$ be a system of elements of $(\mathscr{X}, \mathscr{Y})_o^r$ and let $U = \bigvee_{j \in J} U_j$ in $(\mathscr{X}, \mathscr{Y})^r$. Let us show that $U \in (\mathscr{X}, \mathscr{Y})_o^r$.

Let us consider the case in which the system $\{U_j\}_{j \in J}$ is directed to the right and is formed of positive elements. In this case, for any element $x \geqslant \mathbf{0}$ of $\mathscr{X}$, the following equality holds

$$U(x) = \bigvee_{j \in J} U_j(x). \tag{2}$$

Let $x_n \downarrow 0$. From $x_1 - x_n \geqslant 0$ it follows that

$$U_j(x_1 - x_n) \leqslant U(x_1 - x_n), \quad (\forall j \in J; \forall n \in \mathbb{N}),$$

i.e.,

$$U(x_n) - U_j(x_n) \leqslant U(x_1) - U_j(x_1), \quad (\forall j \in J, \forall n \in \mathbb{N}).$$

By tending to the limit with respect to n, one obtains

$$\bigwedge_{n \in \mathbb{N}} U(x_n) \leqslant U(x_1) - U_j(x_1), \quad (\forall j \in J),$$

whence we immediately infer, by taking into account also (2), that the left-hand member is equal to zero. Consequently, $U \in (\mathscr{X}, \mathscr{Y})^r_o$.

If the system $\{U_j\}_{j \in J}$ is not directed, we consider the system of the least upper bounds of all its finite sub-systems. If the system $\{U_j\}_{j \in J}$ is directed, but its elements are not positive, we shall consider the system of the elements $U_j - U_{j_0}$ (with a fixed $j_0 \in J$), such that $U_j \geqslant U_{j_0}$. Consequently, $(\mathscr{X}, \mathscr{Y})^r_o$ is a component, by virtue of proposition 5, from Section 3.3.3.

Corollary. *Any regular operator U, defined on a vector lattice $\mathscr{X}$, taking values in a complete vector lattice $\mathscr{Y}$, can be uniquely decomposed in the form $U = U_1 + U_2$, where U_1 is an (o)-continuous regular operator and U_2 is orthogonal to any (o)-continuous regular operator.*

Remarks. (1) If $\mathscr{X}$ is a regular space, then any regular operator (which maps $\mathscr{X}$ into $\mathscr{Y}$) is (o)-continuous. In particular

$$(\mathscr{X}, \mathscr{Y})^r_o = (\mathscr{X}, \mathscr{Y})^r.$$

(2) If $\mathscr{X}$ is an Archimedean vector lattice and if $\mathscr{Y}$ is a complete vector lattice of bounded type, then any additive and (o)-continuous operator (which maps $\mathscr{X}$ into $\mathscr{Y}$) is regular.

These assertions are easy to prove, by taking into account the remarks made at the end of section 1.

5.2.3 Other types of continuity

Let $\mathscr{X}$ and $\mathscr{Y}$ be vector lattices.

An operator U which maps $\mathscr{X}$ into $\mathscr{Y}$ is said to be (o^*)-continuous[8]) iff $x_n \xrightarrow{o^*} x$ implies $U(x_n) \xrightarrow{o^*} U(x)$.

Any (o^*)-continuous, positive and additive operator is (o)-continuous.

[8]) For the definition of the (o^*)-convergent sequence see Section 1.3.1, definition 2.

Indeed, we have only to take into account the fact that for the monotonous sequences, the (o^*)-convergence is equivalent to the (o)-convergence.

If, moreover, $\mathscr{X}$ and $\mathscr{Y}$ are Archimedean spaces, then an operator U (which maps $\mathscr{X}$ into $\mathscr{Y}$) is said to be (ρ)-continuous[9]) if $x_n \xrightarrow{\rho} x$ implies $U(x_n) \xrightarrow{\rho} U(x)$.

If $\mathscr{X}$ is an Archimedean vector lattice and if $\mathscr{Y}$ is a complete vector lattice, then any regular operator U which maps $\mathscr{X}$ into $\mathscr{Y}$ is (ρ)-continuous.

Indeed, if

$$|x_n| \leqslant \varepsilon_n r$$

where, $0 < \varepsilon_n \to 0$, then

$$|U(x_n)| \leqslant |U|(|x_n|) \leqslant \varepsilon_n |U|(r),$$

and this shows that $U(x_n) \xrightarrow{\rho} \mathbf{0}$.

One can also consider (o^*o)-continuous operators (U is (o^*o)-continuous iff $x_n \xrightarrow{o^*} x$ implies $U(x_n) \xrightarrow{o} U(x)$), or (ρo)-continuous operators (whose definition is obvious).

Any additive and (ρo)-continuous operator U, defined on an Archimedean vector lattice $\mathscr{X}$, and taking values in a complete vector lattice of bounded type $\mathscr{Y}$, is a regular operator.

Indeed, let E be a bounded subset of $\mathscr{X}$ and let $a \in \mathscr{X}$ be such that $|x| \leqslant a$ for any $x \in E$. Let $y_n = U(x_n)$, where $x_n \in E$. If $\{\lambda_n\}_{n \in \mathbb{N}}$ is a sequence of real numbers which converges to $\mathbf{0}$, then $\lambda_n x_n \xrightarrow{\rho} \mathbf{0}$ and, therefore $\lambda_n y_n \xrightarrow{o} \mathbf{0}$.

The set $U(E)$ is thus (o)-annihilating and hence, it is bounded. It follows that the operator U is regular.

In particular, if $\mathscr{X}$ is an Archimedean vector lattice, and if $\mathscr{Y}$ is a complete vector lattice of bounded type, then an additive operator U which maps $\mathscr{X}$ into $\mathscr{Y}$ is (ρ)-continuous iff it is (ρo)-continuous.

One can also consider operators which are continuous with respect to the convergence of generalized sequences.

Let $\mathscr{X}$ be a vector lattice and $\mathscr{Y}$ a complete vector lattice.

A linear operator U, which maps $\mathscr{X}$ into $\mathscr{Y}$, is said to be (ω)*-continuous* iff $x_\delta \xrightarrow[\delta \in \Delta]{\omega} x$ (where $\{x_\delta\}_{\delta \in \Delta}$ is a generalized sequence of elements of $\mathscr{X}$) implies $U(x_\delta) \xrightarrow[\delta \in \Delta]{\omega} U(x)$. Any (ω)-continuous operator is (o)-continuous. If U is linear and positive and if $x_\delta \underset{\delta \in \Delta}{\downarrow} \mathbf{0}$ implies $U(x_\delta) \underset{\delta \in \Delta}{\downarrow} \mathbf{0}$, then U is (ω)-continuous. This fact is easy to check.

If $\mathscr{X}$ is a complete vector lattice of countable type (and if $\mathscr{Y}$ is any complete vector lattice), then any (o)-continuous regular operator U, which maps $\mathscr{X}$ into $\mathscr{Y}$, is (ω)-continuous.

Indeed, if U is regular and (o)-continuous, then U_+ and U_- are (o)-continuous. It is, therefore, sufficient to consider the case in which U is positive. If $x_\delta \underset{\delta \in \Delta}{\downarrow} \mathbf{0}$, then

---

[9]) For the definition of the (ρ)-convergent sequence see Section 3.2.4.

there exists an increasing sequence of indices $\{\delta_n\}_{n\in\mathbb{N}}$, such that $x_{\delta_n}\downarrow \mathbf{0}$ (see Section 4.2.2, proposition 2, corollary). It follows that $U(x_{\delta_n})\downarrow \mathbf{0}$, whence $U(x_\delta)\underset{\delta\in\Delta}{\downarrow}\mathbf{0}$, i.e., U is (ω)-continuous.

If $\mathscr{X}$ is any vector lattice, whereas $\mathscr{Y}$ is a complete vector lattice, then the set $(\mathscr{X},\mathscr{Y})^r_\omega$ of all (ω)-continuous regular operators, which map $\mathscr{X}$ into $\mathscr{Y}$, is a component of the space $(\mathscr{X},\mathscr{Y})^r$.

The proof is similar to that given for the theorem concerning the set of all (o)-continuous regular operators.

5.2.4 The extension of some continuous, positive, linear operators

If A is a subset of a vector lattice $\mathscr{X}$, we shall call the *Borelian hull* of A the smallest set $B\subset\mathscr{X}$ which contains A and has the property: if $\{x_n\}_{n\in\mathbb{N}}$ is a sequence of elements of B and $x_n\xrightarrow{o}x$, then $x\in B$.

We shall denote $B = o(A)$.

Let $\mathscr{X}$ be a complete vector lattice, $\mathscr{X}_0$ a majorizing vector sublattice (Section 5.1.2, definition 2) and $\mathscr{Y}$ a complete vector lattice of (o)-countable type, which we shall also assume to be σ-regular. Let U_0 be a linear and positive operator, defined on $\mathscr{X}_0$, taking values in $\mathscr{Y}$, such that it is (o)-continuous on $\mathscr{X}_0$ with respect to the (o)-convergence of $\mathscr{X}$ (i.e., if $\{x_n\}_{n\in\mathbb{N}}$ is a sequence of elements of $\mathscr{X}_0$, and x an element of $\mathscr{X}_0$, and if $x_n\xrightarrow{o}x$ in $\mathscr{X}$, then $U_0(x_n)\xrightarrow{o}U_0(x)$). Assuming these conditions, we shall prove that U_0 can be extended to a linear and positive operator U, defined on a vector sublattice $\mathscr{X}_1$, and (o)-continuous on $\mathscr{X}_1$ (with respect to the (o)-convergence in $\mathscr{X}$), such that:

(i) $o(\mathscr{X}_0)\subset\mathscr{X}_1$;

(ii) $o(\mathscr{X}_1)=\mathscr{X}_1$;

(iii) if there exists a vector subspace $\mathscr{X}_2$ of $\mathscr{X}$ such that $\mathscr{X}_0\subset\mathscr{X}$ and $o(\mathscr{X}_2)=\mathscr{X}_2$ and if V is an (o)-continuous, linear operator on $\mathscr{X}_2$ (the continuity being considered with respect to the (o)-convergence in $\mathscr{X}$), such that $V(x) = U_0(x)$, for $x\in\mathscr{X}_0$, then $V(x) = U(x)$, for any $x\in\mathscr{X}_1\cap\mathscr{X}_2$.

For the proof, let us denote by G the set of all elements $g\in\mathscr{X}$ such that there exists a sequence $\{x_n\}_{n\in\mathbb{N}}$ of elements of $\mathscr{X}_0$, such that $x_n\uparrow g$. It is easy to check the following properties of the set G.

(1) if $g', g''\in G$, then the elements $g'+g''$, $g'\wedge g''$, $g'\vee g''$ belong to G;

(2) if $g\in G$, and $\alpha>0$, then $\alpha g\in G$;

(3) if $x_n\uparrow g$ and $x'_n\uparrow g$ (where x_n, $x'_n\in\mathscr{X}_0$), then the sequences $\{U_0(x_n)\}_{n\in\mathbb{N}}$, $\{U_0(x'_n)\}_{n\in\mathbb{N}}$ are (o)-convergent and

$$(o)\text{-lim } U_0(x_n) = (o)\text{-lim } U_0(x'_n).$$

If we denote $H=\{-g; g\in G\}$, then it is obvious that

$$\mathscr{X}_0\subset G\cap H.$$

Now let $g \in G$ and $x_n \uparrow g$, where $x_n \in \mathscr{X}_0$. We shall define

$$U_1(g) = (o)\text{-}\lim_n U_0(x_n).$$

It is easy to check that

$$U_1(g' + g'') = U_1(g') + U_1(g''),$$

$$U_1(\alpha g) = \alpha U_1(g),$$

for any elements $g, g', g'' \in G$ and any number $\alpha > 0$; also, it is easy to see that $g' \leqslant g''$ implies $U_1(g') \leqslant U_1(g'')$.

For any $h \in H$ we shall define

$$U_1(h) = -U_1(-h).$$

If $h \leqslant g (h \in H, g \in G)$, then $U_1(h) \leqslant U_1(g)$.

Indeed, if $x_n \uparrow g$ and $x'_n \downarrow h$ $(x_n, x'_n \in \mathscr{X}_0)$, then

$$x'_n \leqslant x_n \vee x'_n = x_n + (x'_n - x_n)_+$$

and this implies

$$U_0(x'_n) \leqslant U_0(x_n) + U_0((x'_n - x_n)_+).$$

Since

$$(x'_n - x_n)_+ \xrightarrow{o} (h - g)_+ = \mathbf{0},$$

it follows that

$$U_1(h) \leqslant U_1(g).$$

Let us also observe that if $g_m \uparrow v (g_m \in G)$, then $v \in G$, because from $x_{mn} \underset{n \in \mathbb{N}}{\uparrow} g_m$ $(x_{mn} \in \mathscr{X}_0)$, by putting

$$x_n = \bigvee_{i=1}^{n} x_{in}$$

we get $x_n \in \mathscr{X}_0$ and $x_n \uparrow g$. Moreover

$$U_1(v) = (o)\text{-}\lim_m U_1(g_m),$$

as one can easily prove.

For any element $x \in \mathscr{X}$ let us put

$$\underline{U}(x) = \bigvee_{\substack{h \leqslant x \\ h \in H}} U_1(h),$$

$$\overline{U}(x) = \bigwedge_{\substack{g \geqslant x \\ g \in G}} U_1(g).$$

It is obvious that $\underline{U}(x) \leqslant \overline{U}(x)$ for any x. If for an x we have $\underline{U}(x) = \overline{U}(x)$, we shall denote $U(x) = \underline{U}(x) = \overline{U}(x)$.

Let us consider the set

$$\mathscr{X}_1 = \{x;\ \underline{U}(x) = \overline{U}(x) = U(x)\}.$$

In order that $x \in \mathscr{X}_1$, it is necessary and sufficient that there exists an element $y_0 \in \mathscr{Y}$ such that for any number $\varepsilon > 0$ there exists a $g \in G$ and an $h \in H$, such that $h \leqslant x \leqslant g$ and

$$U_1(g) - U_1(h) \leqslant \varepsilon y_0. \tag{1}$$

If there exists a y such that

$$U_1(h) \leqslant y \leqslant U_1(g), \tag{2}$$

for any $\varepsilon > 0$ and for the corresponding elements h, g, then $y = U(x)$.

Indeed, let $x \in \mathscr{X}_1$. Since the space $\mathscr{Y}$ is of (o)-countable type, let

$$h_n \leqslant x \leqslant g_n \ (n \in \mathbb{N}), \quad h_n \in H, \quad g_n \in G,$$

be such that

$$\bigwedge_{n \in \mathbb{N}} U_1(g_n) = \bigvee_{n \in \mathbb{N}} U_1(h_n) = U(x).$$

By putting

$$h'_n = \bigvee_{i=1}^{n} h_i, \qquad g'_n = \bigwedge_{i=1}^{n} g_i,$$

we have

$$(o)\text{-}\lim_n U_1(g'_n) = (o)\text{-}\lim_n U_1(h'_n) = U(x).$$

If y_0 is a common convergence regulator of the two sequences, then there exists an $n_\varepsilon \in N$, such that

$$U_1(g'_{n_\varepsilon}) - U_1(h'_{n_\varepsilon}) \leqslant \varepsilon y_0,$$

i.e., (1) is satisfied.

Conversely, if condition (1) is satisfied (under the stated assumptions), then

$$\overline{U}(x) - \underline{U}(x) \leqslant \varepsilon y_0,$$

whence

$$\overline{U}(x) = \underline{U}(x),$$

and this implies that $x \in \mathscr{X}_1$.

Finally, let us observe that from (1) and (2) it follows that

$$|U(x) - y| \leqslant \varepsilon y_0$$

and, since ε is arbitrary, it follows that $y = U(x)$.

Let us now establish the stated properties of the set $\mathscr{X}_1$ and of the operator U.

The facts that $\mathscr{X}_1$ is a vector subspace and that U is a positive, linear operator, defined on $\mathscr{X}_1$, are easy to prove. Let us show that $\mathscr{X}_1$ is a vector sublattice of $\mathscr{X}$.

Let $x \in \mathscr{X}_1$. Let $\varepsilon > 0$, $y_0 \in \mathscr{Y}$, $g \in G$ and $h \in H$ be such that $h \leqslant x \leqslant g$ and such that (1) hold. We have

$$h_+ \leqslant x_+ \leqslant g_+,$$

where $h_+ \in H$ and $g_+ \in G$. Since

$$\begin{aligned} U_1(g_+) - U_1(h_+) &= U_1(g_+ - h_+) \leqslant U_1(g - h) \\ &= U_1(g) - U_1(h) \leqslant \varepsilon y_0 \end{aligned}$$

it follows that $x_+ \in \mathscr{X}_1$; but this implies that $\mathscr{X}_1$ is a vector sublattice of $\mathscr{X}$.

Now let $\{x_n\}_{n \in \mathbb{N}}$ be a sequence of elements of $\mathscr{X}_1$ such that $x_n \uparrow x$ in the space $\mathscr{X}$. There then exists the element $y = (o)\text{-}\lim\limits_n U(x_n)$. Let

$$h_n \leqslant x_n \leqslant g_n; \quad U_1(g_n) - U_1(h_n) \leqslant \frac{\varepsilon}{2^n} y_0$$

(where ε, y_0, h_n, g_n have obvious meaning). If we put

$$g_n' = \bigvee_{i=1}^{n} g_i$$

one can show by induction that

$$U_1(g_n') \leqslant U(x_n) + \left(1 - \frac{1}{2^n}\right) \varepsilon\, y_0. \tag{3}$$

If we now put

$$g = \bigvee_{n \in \mathbb{N}} g'_n$$

then $g \geqslant x$ and, by taking into account the fact that

$$U(x_n) \leqslant U_1(g'_n),$$

from (3) we get, by tending to the limit,

$$y \leqslant U_1(g) \leqslant y + \varepsilon y_0.$$

One can also show that for a certain $h \leqslant x$ we have

$$y \geqslant U_1(h) \geqslant y - \varepsilon y_0.$$

It follows that $h \leqslant x \leqslant g$ and

$$U_1(h) \leqslant y \leqslant U_1(g); \; U_1(g) - U_1(h) \leqslant 2\varepsilon y_0.$$

It follows that $x \in \mathscr{X}_1$ and

$$U(x) = y = \lim_n U(x_n).$$

The same result can be obtained in the case in which $x_n \downarrow x$.

Let $\{x_n\}_{n \in \mathbb{N}}$ be a sequence of positive elements of $\mathscr{X}_1$ and let $x_n \overset{o}{\to} \mathbf{0}$ in $\mathscr{X}$. Let us put

$$f_{nm} = \bigvee_{i=n}^{m} x_i$$

$$y_n = (o)\text{-}\lim_m U(f_{nm})$$

$$y = (o)\text{-}\lim y_n.$$

Let r be a common convergence regulator of the sequences $\{f_{nm}\}_{m \in \mathbb{N}}$ and $\{j_n\}_{n \in \mathbb{N}}$ an increasing sequence of natural numbers such that

$$U(f_{nj_n}) \geqslant y_n - \frac{\varepsilon}{2^n} r. \tag{4}$$

Let

$$v_m = \bigwedge_{i=1}^{m} f_{ij_i}.$$

Then $v_m \downarrow \mathbf{0}$, hence $U(v_m) \downarrow \mathbf{0}$. But [10])

$$v_m + \sum_{i=1}^{m-1} (f_{ij_m} - f_{ij_i}) \geqslant f_{mj_m}$$

and from the inequality

$$y_i \geqslant U(f_{ij_m})$$

it follows that $U(f_{ij_i}) \geqslant y_i - \frac{\varepsilon}{2^i} r \geqslant U(f_{ij_m}) - \frac{\varepsilon}{2^i} r$.

Consequently

$$U(v_m) \geqslant U(f_{mj_m}) - \sum_{i=1}^{m-1} \frac{\varepsilon}{2^i} r.$$

From this inequality and from (4) it follows that

$$U(v_m) \geqslant y_m - \frac{2^m - 1}{2^m} \varepsilon r;$$

by tending to the limit, we get

$$0 \leqslant y \leqslant \varepsilon r$$

and this implies that $y = \mathbf{0}$.

From

$$\mathbf{0} \leqslant U(x_n) \leqslant y_n$$

it follows that $U(x_n) \xrightarrow{o} \mathbf{0}$.

If the elements of the sequence $\{x_n\}_{n \in \mathbb{N}}$ are not positive, then from

$$|U(x_n)| \leqslant U(|x_n|)$$

it also follows that $U(x_n) \xrightarrow{o} \mathbf{0}$.

---

[10]) We have

$$(f_{1j_1} \wedge \ldots \wedge f_{mj_m}) + (f_{1j_m} - f_{1j_1}) + \ldots + (f_{m-1 j_m} - f_{m-1, j_{m-1}})$$
$$\geqslant \bigwedge_{i=1}^{m} f_{ij_m} = f_{mj_m}.$$

Consequently, the operator U is (o)-continuous on $\mathscr{X}_1$ with respect to the (o)-convergence in $\mathscr{X}$.

The fact that $o(\mathscr{X}_1) = \mathscr{X}_1$ is an immediate consequence of what we have already shown: if $\{x_n\}_{n\in\mathbb{N}}$ is a sequence of elements of $\mathscr{X}_1$ and if $x_n \xrightarrow{o} x$ in $\mathscr{X}$, then by putting

$$x'_{nm} = \bigvee_{i=n}^{m} x_i, \qquad x''_n = \bigvee_{i=n}^{\infty} x_i$$

we have $x'_{nm} \underset{m\in\mathbb{N}}{\uparrow} x''_n$ and $x''_n \downarrow x$, whence we immediately get $x \in \mathscr{X}_1$.

From $\mathscr{X}_0 \subset \mathscr{X}_1$ we immediately get that $o(\mathscr{X}_0) \subset \mathscr{X}_1$.

We have only to prove the uniqueness of the extension of U (in the sense explained in (iii)).

Let then V and $\mathscr{X}_2$ have the properties stated in (iii) and let us show that $V(x) = U(x)$ if $x \in \mathscr{X}_1 \cap \mathscr{X}_2$.

If $g \in G$ and $x_n \uparrow g$ (where $x_n \in \mathscr{X}_0$), then $g \in \mathscr{X}_2$ and

$$V(g) = (o)\text{-}\lim_n V(x_n) = (o)\text{-}\lim_n U_0(x_n) = U_1(g).$$

One can similarly show that $H \subset \mathscr{X}_2$ and $V(h) = U_1(h)$, for any $h \in H$.

Let now $x \in \mathscr{X}_1 \cap \mathscr{X}_2$ and let $h_n \in H$, $g_n \in G$ be such that:

$$g_n \leqslant x_n \leqslant g_n, \qquad n \in \mathbb{N},$$

and such that

$$(o)\text{-}\lim_n U_1(h_n) = (o)\text{-}\lim_n U_1(g_n) = U(x).$$

We have

$$U_1(h_n) = V(h_n) \leqslant V(x) \leqslant V(g_n) = U_1(g_n),$$

whence $V(x) = U(x)$.

Remarks and applications. (1) We have previously proved the following property of the operator U: if $\{x_n\}_{n\in\mathbb{N}}$ is a sequence of elements of $\mathscr{X}_1$ and if $x_n \xrightarrow{o} \mathbf{0}$ in $\mathscr{X}$, then $U(x_n) \xrightarrow{o} \mathbf{0}$. More generally, the following proposition is true:

Let $\mathscr{X}$ be a vector lattice, $\mathscr{X}_1$ a majorizing vector sublattice of $\mathscr{X}$ and $\mathscr{Y}$ a σ-regular space. Let U be an additive and positive operator which maps $\mathscr{X}_1$ into $\mathscr{Y}$, which moreover, has the following property: if $\{x_n\}_{n\in\mathbb{N}}$ is a sequence of elements of $\mathscr{X}_1$ and if $x_n \downarrow \mathbf{0}$ in $\mathscr{X}$, then $U(x_n) \downarrow \mathbf{0}$. It follows that if $\{x_n\}_{n\in\mathbb{N}}$ is a sequence of elements of $\mathscr{X}_1$ and if $x_n \xrightarrow{o} x$ in $\mathscr{X}$, and if $x \in \mathscr{X}_1$, then $U(x_n) \xrightarrow{o} U(x)$.

(2) The results obtained in this section can be applied, in particular, to the construction of the Lebesgue or the Radon integrals.

Let $\mathscr{X} = B(T)$ be the space of all bounded real functions defined on a closed domain T of $\mathbb{R}^n$ (with the usual order relation and algebraic operations). Let

$\mathscr{X}_0 = C(T)$ be the space of all continuous real functions defined on T. By taking as a functional φ on $C(T)$ the Riemann integral

$$\varphi(x) = \int_T x(t)\,\mathrm{d}t, \quad x \in C(T),$$

and, by extending the functional φ by the method described in this section, one obtains the Lebesgue integral for measurable, bounded, functions.

The Radon integral on an interval T of the real axis may be obtained by extending the functional ψ, defined on $C(T)$ by the Stieltjes integral

$$\psi(x) = \int_T x(t)\,\mathrm{d}k(t), \quad x \in C(T),$$

where k is an increasing function defined on T.

5.2.5 Normal functionals

Definition. A regular functional g, defined on a vector lattice $\mathscr{X}$, is said to be *abnormal* if g vanishes on a normal and total subspace of $\mathscr{X}$. A regular functional f defined on $\mathscr{X}$ is said to be *normal* if $f \perp g$ for any abnormal functional g.

In what follows we shall assume that $\mathscr{X}$ is an Archimedean vector lattice. We shall denote by $\widetilde{\mathscr{X}}$ the Dedekind extension of $\mathscr{X}$ and $\mathscr{X}^r = (\mathscr{X}, \mathbb{R})^r$.

Lemma 1. *Let f and g be two positive linear functionals on $\mathscr{X}$, such that $f \wedge g = \mathbf{0}$. If $\tilde{f}$ and $\tilde{g}$ are positive linear functionals on $\widetilde{\mathscr{X}}$ such that $\tilde{f}(x) = f(x)$ and $\tilde{g}(x) = g(x)$, for $x \in \mathscr{X}$, then $\tilde{f} \wedge \tilde{g} = \mathbf{0}$.*

Proof. Let $h = \tilde{f} \wedge \tilde{g}$. For any positive element $x \in \mathscr{X}$ we have

$$h(x) = \inf_{\substack{y_1+y_2=x \\ 0\leqslant y_1,y_2\in\widetilde{\mathscr{X}}}} \{\tilde{f}(y_1) + \tilde{g}(y_2)\} \leqslant \inf_{\substack{x_1+x_2=x \\ 0\leqslant x_1,x_2\in\mathscr{X}}} \{f(x_1) + g(x_2)\} = (f \wedge g)(x) = 0,$$

hence $h(x) = 0$.

If $0 \leqslant y \in \widetilde{\mathscr{X}}$, then there exists an $x \in \mathscr{X}$ such that $y \leqslant x$ and, therefore, $h(y) = 0$. It follows that $h(y) = 0$ for any $y \in \widetilde{\mathscr{X}}$, i.e., $\tilde{f} \wedge \tilde{g} = \mathbf{0}$.

Lemma 2. *Let $f \in \mathscr{X}^r$ and $\tilde{f} \in \widetilde{\mathscr{X}}^r$ be such that $\tilde{f}(x) = f(x)$, for any $x \in \mathscr{X}$. If $\tilde{f}$ is abnormal, then f is abnormal too.*

Proof. There exists a normal and total subspace E of $\widetilde{\mathscr{X}}$, such that $\tilde{f}(y) = 0$, for any $y \in E$. By putting $G = E \cap \mathscr{X}$, the set G is a normal and total subspace of $\mathscr{X}$ and for any $x \in G$ we have $f(x) = \tilde{f}(x) = 0$.

Lemma 3. *Let f be a positive normal functional on $\mathscr{X}$. If $\tilde{f}$ is a positive linear functional on $\widetilde{\mathscr{X}}$, such that $f(x) = \tilde{f}(x)$ for $x \in \mathscr{X}$, then $\tilde{f}$ is normal.*

Proof. Let $\tilde{g}$ be an abnormal positive functional on $\tilde{\mathscr{X}}$. By taking into account lemma 2, the restriction g of $\tilde{g}$ to $\mathscr{X}$ is abnormal too. We then have $f \wedge g = \mathbf{0}$, and from lemma 1 we infer that $\tilde{f} \wedge \tilde{g} = \mathbf{0}$. One can then immediately infer that if $\tilde{g}$ is any abnormal functional, then $\tilde{f} \perp \tilde{g}$ and, therefore, $\tilde{f}$ is normal.

Proposition 1. *For any positive normal functional f on $\mathscr{X}$ there exists a positive normal functional $\tilde{f}$ on $\tilde{\mathscr{X}}$, which is an extension of f.*

Proof. Since $\mathscr{X}$ is a majorizing vector subspace of $\tilde{\mathscr{X}}$, there exists a positive linear functional $\tilde{f}$ on $\tilde{\mathscr{X}}$, which is an extension of f (Section 5.1.2). From lemma 3 one infers that $\tilde{f}$ is normal.

Proposition 2. *In order that a regular functional f, defined on the Archimedean vector lattice $\mathscr{X}$ be normal, it is necessary and sufficient that it be (ω)-continuous.*

Proof. Let f be a positive normal functional on $\mathscr{X}$. From proposition *1*, we deduce that there exists a positive normal functional $\tilde{f}$ on $\tilde{\mathscr{X}}$, which is an extension of f to $\tilde{\mathscr{X}}$. The functional $\tilde{f}$ is (ω)-continuous. Indeed, if this be not true, then there should exist a generalized sequence $\{y_\delta\}_{\delta \in \Delta}$ of elements of $\tilde{\mathscr{X}}$ and a number $\lambda > 0$ such that $y_\delta \downarrow \mathbf{0}$ and $\tilde{f}(y_\delta) \geqslant \lambda$, for any $\delta \in \Delta$. One can also show that the sequence $\{y_\delta\}_{\delta \in \Delta}$ can be required to satisfy the condition $[y_\delta]\, y_{\delta_0} \downarrow 0$, for a certain index [11] δ_0. Let us define

$$\tilde{g}(y) = \lim_{\delta \in \Delta} \tilde{f}([y_\delta]\, y), \quad y \in \tilde{\mathscr{X}}.$$

The functional $\tilde{g}$ is obviously additive and since $\mathbf{0} \leqslant \tilde{g} \leqslant \tilde{f}$ it follows that $\tilde{g}$ is normal.

From $[y_\delta]\, y_{\delta_0} \geqslant y_\delta$ for $\delta \geqslant \delta_0$ it follows that

$$\tilde{g}(y_{\delta_0}) = \inf_{\delta \geqslant \delta_0} \tilde{f}([y_\delta]\, y_{\delta_0}) \geqslant \inf_{\delta \geqslant \delta_0} \tilde{f}(y_\delta) \geqslant \lambda > 0,$$

and this implies that $\tilde{g} > \mathbf{0}$.

Let us denote by G the set of those elements $y \in \tilde{\mathscr{X}}$ for which a y_δ exists, such that $y \perp y_\delta$. One can show that G is a total and normal subspace [12]) and, since $\tilde{g}(y) = 0$, for any $y \in G$, it follows that $\tilde{g}$ is abnormal.

---

[11]) Let $z_\delta \downarrow \mathbf{0}$ in $\tilde{\mathscr{X}}$. Let us fix an index $\delta_0 \in \Delta$. Consider a number $\varepsilon > 0$ and let us define $a_\delta = [(\varepsilon z_{\delta_0} - z_\delta)_+]\, z_{\delta_0}$ and $b_\delta = z_{\delta_0} - a_\delta$. Let $\delta \geqslant \delta_0$. In the component H generated by z_{δ_0}, the element z_{δ_0} can be taken to be a unit (and, moreover, $z_\delta \in H$ for $\delta \geqslant \delta_0$). We have, therefore, $z_\delta = [a_\delta]\, z_\delta + [b_\delta]\, z_\delta$, $[a_\delta]\, z_\delta \leqslant \varepsilon a_\delta$ and $[b_\delta]\, z_\delta \geqslant \varepsilon b_\delta$. It follows that $b_\delta \downarrow \mathbf{0}$, $[b_\delta]\, b_{\delta_0} \downarrow \mathbf{0}$ and $\tilde{f}(z_\delta) \leqslant \varepsilon \tilde{f}(z_{\delta_0}) + \tilde{f}(b_\delta)$.

[12]) If $z \perp G$ and $z \neq \mathbf{0}$, then $z \in y_\delta^{\perp\perp} (\forall \delta \in \Delta)$, hence $z \wedge y_\delta \neq \mathbf{0}$. There follows $[y_\delta]\, y_{\delta_0} \geqslant [z] y_{\delta_0} \geqslant z \wedge y_{\delta_0} > \mathbf{0}$, which contradicts $[y_\delta]\, y_{\delta_0} \downarrow \mathbf{0}$.

Since $\tilde{g}$ is, at the same time, normal and abnormal, it follows that $\tilde{g} = \mathbf{0}$, and this is a contradiction.

It follows that $\tilde{f}$ is (ω)-continuous.

If $x_\delta \downarrow \mathbf{0}$ in $\mathscr{X}$, then $x_\delta \downarrow \mathbf{0}$ in $\tilde{\mathscr{X}}$ too and since $\tilde{f}(x_\delta) = f(x_\delta)$, it follows that $f(x_\delta) \to 0$. Consequently, f is (ω)-continuous.

Conversely, let f be a positive linear functional which, moreover, is (ω)-continuous on $\mathscr{X}$. Let us assume that f is not normal. Then there exists an abnormal functional $g > \mathbf{0}$, such that $h = f \wedge g \neq \mathbf{0}$. From $\mathbf{0} < h \leqslant g$ it follows that h is abnormal too; hence, there exists a normal and total subspace $E \subset \mathscr{X}$, such that $h(x) = 0$ for any $x \in E$. From $\mathbf{0} < h \leqslant f$ it follows that h is (ω)-continuous. Since for any element $x \geqslant \mathbf{0}$ of $\mathscr{X}$ there exists a generalized sequence $\{x_\delta\}_{\delta \in \Delta}$ of elements of E, such that $\mathbf{0} \leqslant x_\delta \uparrow x$ (Section 3.3.2, proposition 4, corollary), it follows that $h(x) = 0$ for any $x \in \mathscr{X}$, and this is a contradiction. The proposition is proved.

5.2.6 Almost integral operators

Let $\mathscr{X}$ be a vector lattice and $\mathscr{Y}$ a complete vector lattice. As in the preceding section, we shall denote $\mathscr{X}^r = (\mathscr{X}, \mathbb{R})^r$. We shall denote by $\mathscr{Y}^*_\omega$ the space of all (ω)-continuous linear functionals, defined on $\mathscr{Y}$. If $f \in \mathscr{X}^r$ and $y \in \mathscr{Y}$, we shall denote by $V_{f,y}$ the operator which maps $\mathscr{X}$ into $\mathscr{Y}$ and is given by the formula

$$V_{f,y}(x) = f(x)\, y, \quad x \in \mathscr{X}.$$

Since $(\mathscr{X}, \mathscr{Y})^r$ is a complete vector lattice, for any subset $\mathscr{A}$ of this space one can consider the component generated by $\mathscr{A}$, i.e., the intersection of all components which contain $\mathscr{A}$ (see Section 3.3.3, proposition 5).

Definition. An operator $U \in (\mathscr{X}, \mathscr{Y})^r$ is said to be *almost integral* if it belongs to the component generated by the set $\{V_{f,y}; f \in \mathscr{X}^r, y \in \mathscr{Y}\}$.

Proposition 1. *If $\mathscr{Y}$ is a discrete space, then any operator $U \in (\mathscr{X}, \mathscr{Y})^r$ is almost integral.*

Proof. Let $\{e_j\}_{j \in J}$ be a total system of discrete positive, mutually orthogonal elements of the space $\mathscr{Y}$. We shall denote by $\Phi(J)$ the set of all finite subsets of J, ordered by inclusion.

Let $U \in (\mathscr{X}, \mathscr{Y})^r$. For any $j \in J$ we shall put

$$U_j(x) = [e_j]\, |U|\, (x),$$

and for any $H \in \Phi(J)$ we shall denote

$$U_H = \sum_{j \in H} U_j.$$

For any $j \in J$ we shall consider the functional g_j given by

$$g_j(y)\, e_j = [e_j]\, y, \quad y \in \mathscr{Y},$$

(see Section 4.4.1) and we shall denote $h_j = g_j \circ |U|$. Since

$$[e_j]\,|U|\,(x) = g_j(|U|\,(x))\, e_j = h_j(x)\, e_j = V_{h_j, e_j}(x),$$

it follows that $U_j = V_{h_j, e_j}$. From

$$\mathbf{0} \leqslant U_H \underset{H \in \Phi(J)}{\uparrow} |U|$$

one gets that $|U|$, and therefore U too, are almost integral operators.

Lemma. If $\mathscr{Y}$ is a space with unit in which there are no non-zero discrete elements, then for any (ω)-continuous linear functional $f > 0$, defined on $\mathscr{Y}$, for any unitary element $e > \mathbf{0}$ of $\mathscr{Y}$ and any number $\varepsilon > 0$ there exists a unitary element e_0, such that $\mathbf{0} < e_0 < e$ and $f(e_0) < \varepsilon$.

Proof. Let $\mathscr{E}$ be the set of all subsets A of $\mathscr{Y}$, consisting of unitary elements and having the following properties: $\mathbf{0} \notin A$; A is totally ordered with respect to the order relation induced by that of $\mathscr{Y}$; $e \in A$ and it is the greatest element of A. By ordering it by inclusion, the set $\mathscr{E}$ satisfies the hypothesis of Zorn's lemma. Consequently, there exists a maximal E in $\mathscr{E}$. Let $e_1 = \inf E$.

The element e_1 is unitary (Section 4.1.1, proposition 1) and, since E is maximal in $\mathscr{Y}$, whereas in $\mathscr{Y}$ there are no non-zero discrete elements, it follows that $e_1 = \mathbf{0}$.

Since f is an (ω)-continuous linear functional, it follows that $\inf_{y \in E} f(y) = 0$, and the lemma is proved.

Proposition 2. *If in the space $\mathscr{Y}$ there exist no non-zero discrete elements, then the identity operator, which maps $\mathscr{Y}$ onto $\mathscr{Y}$, is an element of $(\mathscr{Y}, \mathscr{Y})^r$, which is orthogonal to the component of all almost integral operators.*

Proof. We shall denote by I the identity operator which maps $\mathscr{Y}$ onto $\mathscr{Y}$. The proposition will be proved if we show that $I \perp V_{f,y}$ for any positive f and y. We shall first suppose that $\mathscr{Y}$ is a space with unit u. In this case it is enough to show that $I \perp V_{f,u}$ for any $f > 0$.

Let $\mathbf{0} < f \in \mathscr{Y}^r$ and $f_0 = [\mathscr{Y}^*_\omega]\, f$. We shall show that

$$(I \wedge V_{f_0,u})\,(u) = \mathbf{0}. \tag{1}$$

Let us denote by y_0 the left-hand member in (1) and let us assume that $y_0 > 0$. In this case, let $e > 0$ be a unitary element and let $\alpha > 0$ be a number such that $y_0 \geqslant \alpha e$ (Section 4.1.2, proposition 3). Let $\varepsilon > 0$ be an arbitrary positive number. By virtue of the preceding lemma, there exists a unitary element e_0, such that $\mathbf{0} < e_0 < e$ and $f_0\,(e_0) < \varepsilon$. From the equality

$$(I \wedge V_{f_0, u})\,(u) = \inf \{\, y_1 + f_0(y_2)\, u;\;\; y_1 + y_2 = u,\; y_1, y_2 \geqslant \mathbf{0}\},$$

it follows that

$$\alpha e \leqslant y_0 \leqslant (u - e_0) + f_0(e_0)\, u$$

and, by applying the projector $[e_0]$, we get

$$\alpha e_0 \leqslant f_0(e_0)\, e_0,$$

whence $\alpha e_0 < \varepsilon e_0$. It follows that $\alpha < \varepsilon$ and this is a contradiction, since ε is arbitrary. It follows that $y_0 = \mathbf{0}$, i.e., (1) is true.

Let us now observe that, by denoting $f_1 = f - f_0$, from

$$\mathbf{0} \leqslant I \wedge V_{f_1, u} \leqslant I$$

it follows that $I \wedge V_{f_1, u}$ is (ω)-continuous. Let us denote by $(\mathscr{Y}, \mathscr{Y})^r_\omega$ the space of all (ω)-continuous regular operators which map $\mathscr{Y}$ into $\mathscr{Y}$. From $f_1 \perp \mathscr{Y}^*_\omega$ it follows that $V_{f_1, u} \perp (\mathscr{Y}, \mathscr{Y})^r_\omega$ and, therefore, from

$$\mathbf{0} \leqslant I \wedge V_{f_1, u} \leqslant V_{f_1, u}$$

it follows that

$$I \wedge V_{f_1, u} \perp (\mathscr{Y}, \mathscr{Y})^r_\omega.$$

This relation implies that $I \wedge V_{f_1, u} = \mathbf{0}$. Consequently,

$$I \wedge V_{f, u} = I \wedge V_{f_0, u} + I \wedge V_{f_1, r} = \mathbf{0}.$$

If $\mathscr{Y}$ has no total elements, we shall consider a decomposition of $\mathscr{Y}$ into spaces with unit. If $\{\mathscr{Y}_j\}_{j \in J}$ is such a decomposition, whereas u_j is the unit of $\mathscr{Y}_j$, then the identity operator I_j on $\mathscr{Y}_j$ is orthogonal to any operator V_{f, u_j}, whence

$$(I \wedge V_{f, u_j})\,(y) = \mathbf{0}, \quad \forall j \in J, \quad \mathbf{0} \leqslant y \in \mathscr{Y}_j.$$

We infer, for this case too, that $I \perp V_{f,y}$ for any positive f and y.

5.2.7 Operators on spaces of vector sequences

Let $\mathscr{X}$ be a vector lattice.

We shall denote by $s(\mathscr{X})$ the set of all sequences of elements of $\mathscr{X}$. If $X = \{x_n\}_{n \in \mathbb{N}} \in s(\mathscr{X})$, we shall write $P_n(X) = x_n$.

We shall consider in $s(\mathscr{X})$ the operations and the order relations defined as follows.

If X, $Y \in s(\mathscr{X})$, and $\alpha \in \mathbb{R}$, then

$$X + Y = \{P_n(X) + P_n(Y)\}_{n \in \mathbb{N}},$$

$$\alpha X = \{\alpha P_n(X)\}_{n \in \mathbb{N}},$$

$$X \leqslant Y \Leftrightarrow P_n(Y) \leqslant P_n(Y), \qquad \forall n \in \mathbb{N}.$$

The set $s(\mathscr{X})$ becomes, in this manner, a vector lattice. We shall denote by $s_0(\mathscr{X})$ the subset of $s(\mathscr{X})$ defined by the condition; $X \in s_0(\mathscr{X})$ if there exists a $k \in \mathbb{N}$, such that $P_n(X) = \mathbf{0}$, for $n \geqslant k$.

For any $m \in \mathbb{N}$ we shall consider the operator I_m, which maps $\mathscr{X}$ into $s_0(\mathscr{X})$ and is given by the formula

$$I_m(x) = \{\delta_{mn} x\}_{n \in \mathbb{N}}, \ (x \in \mathscr{X}),$$

where δ_{mn} is Kronecker's symbol. If $X \in s(\mathscr{X})$, and $n \in \mathbb{N}$, then we shall put

$$S_n(X) = \sum_{j=1}^{n} I_j(P_j(X)).$$

We shall denote by $\mathscr{X}_o^r$ the complete vector lattice of all regular (o)-continuous functionals, defined on $\mathscr{X}$.

Proposition 1. *Let $v(\mathscr{X})$ be a vector sublattice of $s(\mathscr{X})$ such that $s_0(\mathscr{X}) \subset v(\mathscr{X})$. The general form of the (o)-continuous regular functionals on $v(\mathscr{X})$ is given by the formula*

$$\varphi(X) = \sum_{n=1}^{\infty} f_n(x_n), \tag{1}$$

where $X = \{x_n\}_{n \in \mathbb{N}} \in v(\mathscr{X})$, $F = \{f_n\}_{n \in \mathbb{N}} \in s(\mathscr{X}_o^r)$, whereas the series

$$\sum_{n=1}^{\infty} |f_n|(|x_n|) \tag{2}$$

is converging for any $X = \{x_n\}_{n \in \mathbb{N}} \in v(\mathscr{X})$.

Proof. Let φ be a regular (o)-continuous functional, defined on $v(\mathscr{X})$. For any $m \in \mathbb{N}$ we shall define

$$f_m(x) = \varphi(I_m(x)), \quad x \in \mathscr{X}.$$

Since the operator $I_m : \mathscr{X} \to v(\mathscr{X})$ is positive and (o)-continuous, f_m is a regular, (o)-continuous functional defined on $\mathscr{X}$. On the other hand, we have $S_n(X) \xrightarrow{o} X$ in $v(\mathscr{X})$, for any $X \in v(\mathscr{X})$ and, therefore,

$$\varphi(X) = \lim_n \varphi(S_n(X)).$$

Hence, by taking into consideration the equalities

$$\varphi(S_n(X)) = \sum_{j=1}^{n} \varphi(I_j(P_j(X)) = \sum_{j=1}^{n} f_j(x_j),$$

equality (1) follows immediately.

It is easy to prove that

$$|\varphi|(I_m(x)) = |f_m|(x)$$

and, from the preceding result, we get

$$|\varphi|(|X|) = \sum_{n=1}^{\infty} |f_n|(|x_n|),$$

where the series is convergent for any $X = \{x_n\}_{n\in\mathbb{N}} \in v(\mathscr{X})$.

Conversely, let φ be a functional defined on $v(\mathscr{X})$, having the form (1) and such that the series (2) is converging for any $X = \{x_n\}_{n\in\mathbb{N}} \in v(\mathscr{X})$. The functional φ is obviously regular.

If we put

$$\psi(X) = \sum_{n=1}^{\infty} |f_n|(x_n),$$

then ψ is a positive linear functional, defined on $v(\mathscr{X})$, and

$$|\varphi(X)| \leqslant \psi(|X|).$$

In order to prove the (o)-continuity of φ it is, therefore, enough to prove the (o)-continuity of ψ.

Let $X_m \downarrow \mathbf{0}$ and assume that $X_m = \{x_{mn}\}_{n\in\mathbb{N}}$. Since $m \geqslant m_0$ implies $x_{mn} \leqslant x_{m_0 n}$, for any $n \in \mathbb{N}$, for any number $\varepsilon > \mathbf{0}$ there exists a number $n_0 = n_0(\varepsilon, m_0)$, such that

$$\sum_{n=n_0+1}^{\infty} |f_n|(x_{mn}) < \frac{\varepsilon}{2} \quad (m \geqslant m_0). \tag{3}$$

On the other hand, from $X_m \downarrow \mathbf{0}$ it follows that $(o)\text{-}\lim_m x_{mn} = \mathbf{0} (\forall n \in \mathbb{N})$, whence $\lim_m |f_n|(x_{mn}) = 0$. Then there exists an $m_1 \geqslant m_0$, such that

$$\sum_{n=1}^{n_0} |f_n|(x_{mn}) < \frac{\varepsilon}{2} \quad (m \geqslant m_1). \tag{4}$$

From (4) and (5) it follows that $\psi(X_n) < \varepsilon$, for any $m \geqslant m_1$. The proposition is proved.

Remark. If we put

$$\tilde{v}(\mathscr{X}) = \left\{ F \in s(\mathscr{X}_o^r);\ \sum_{n=1}^{\infty} |f_n|(|x_n|) < +\infty,\ \forall X \in v(\mathscr{X}) \right\}.$$

then the space of all regular and (o)-continuous functionals, defined on $v(\mathscr{X})$, is isomorphic to the space $\tilde{v}(\mathscr{X})$ by the correspondence $\varphi \mapsto F$ given by formula (1).

In the next proposition we shall denote by $m(\mathscr{X})$ the space of all bounded sequences of elements of $\mathscr{X}$ (which is a normal subspace of $s(\mathscr{X})$).

If $\mathscr{Z}$ is a σ-complete vector lattice, we shall denote by $l(\mathscr{Z})$ the space of all sequences $Z = \{z_n\}_{n \in \mathbb{N}}$ of elements of $\mathscr{Z}$, such that the series $\Sigma\, |z_n|$ is (o)-converging. The space $l(\mathscr{Z})$ is a normal subspace of the space $s(\mathscr{Z})$.

We remark that if $\mathscr{Z}$ is either σ-complete or complete, the space $s(\mathscr{Z})$ has the same property.

Proposition 2. *Let $\mathscr{X}$ be a vector lattice and $\mathscr{Y}$ a complete vector lattice. An operator U, which maps $m(\mathscr{X})$ into $\mathscr{Y}$, is regular and (o)-continuous iff it can be written as*

$$U(X) = (o)\text{-}\sum_{n=1}^{\infty} U_n(x_n), \tag{5}$$

where $X = \{x_n\}_{n \in \mathbb{N}}$ and

$$\{U_n\}_{n \in \mathbb{N}} \in l((\mathscr{X}, \mathscr{Y})_o^r). \tag{6}$$

Proof. Let $U \in (m(\mathscr{X}), \mathscr{Y})_o^r$ and define $U_n = U \circ I_n$. Obviously, $U_n \in (\mathscr{X}, \mathscr{Y})_o^r$. It is enough to establish (5) and (6) for $U \geqslant \mathbf{0}$. In this case $U_n \geqslant \mathbf{0}$ ($\forall_n \in \mathbb{N}$).

For any $x \in \mathscr{X}$ we shall denote $\langle x \rangle = \{x, x, x, \ldots\}$ Obviously, we have $\langle x \rangle \in m(\mathscr{X})$, and

$$\left(\sum_{j=1}^{n} U_j \right)(x) \leqslant U(\langle x \rangle),$$

for any $n \in \mathbb{N}$ and $x \geqslant \mathbf{0}$. By putting $V(x) = U(\langle x \rangle)$, $x \in \mathscr{X}$, we have $V \in (\mathscr{X}, \mathscr{Y})_o^r$ and

$$\sum_{j=1}^{n} U_j \leqslant V,\ \forall\, n \in \mathbb{N}.$$

Hence we infer that $\{U_n\}_{n \in \mathbb{N}} \in l((\mathscr{X}, \mathscr{Y})_o^r)$.

From

$$U(S_n(X)) = \sum_{j=1}^{n} U_j(x_j)$$

and from the fact $S_n(X) \xrightarrow{o} X$ in $m(\mathscr{X})$, equality (5) follows.

Conversely, let $U: m(\mathscr{X}) \to \mathscr{Y}$ be such that (5) and (6) are satisfied.

From (6) it follows that the series from formula (5) is absolutely (o)-converging, for any $X \in m(\mathscr{X})$, and U is a linear operator.

Let us now put $U_n \geqslant \mathbf{0}$, $n \in \mathbb{N}$. In this case $U \geqslant \mathbf{0}$. If (for $k \in \mathbb{N}$) we put

$$T_k(X) = \sum_{j=1}^{k} U_j(x_j), \; X \in m(\mathscr{X}),$$

then T_k is a positive, (o)-continuous, linear operators. It is obvious that $T_k \leqslant U$, for any $k \in \mathbb{N}$. If $T \in (m(\mathscr{X}), \mathscr{Y})^r$ and if $T_k \leqslant T (k \in \mathbb{N})$ then

$$\sum_{j=1}^{k} U_j(x_j) \leqslant T(X), \quad \forall X \geqslant \mathbf{0}.$$

It follows that $U(X) \leqslant T(X)$, $\forall X \geqslant \mathbf{0}$, and hence $U \leqslant T$.

Consequently, $U = \bigvee_{k \in \mathbb{N}} T_k$ in the space $(m(\mathscr{X}), \mathscr{Y})^r$. Since the set $(m(\mathscr{X}), \mathscr{Y})^r_o$ is a component of the space $(m(\mathscr{X}), \mathscr{Y})^r$, and since $T_k \in (m(\mathscr{X}), \mathscr{Y})^r_o$, for any $k \in \mathbb{N}$, it follows that $U \in (m(\mathscr{X}, \mathscr{Y})^r_o$.

We assumed that $U_n \geqslant \mathbf{0}$. $n \in \mathbb{N}$. If this condition is not satisfied, we shall take into account the fact that $U_n = (U_n)_+ - (U_n)_-$, whereas $(U_n)_+$ and $(U_n)_-$ are (o)-continuous, positive linear operators.

5.3 Disjunctive operators and multiplicative operators

5.3.1 Disjunctive operators

Let $\mathscr{X}$ and $\mathscr{Y}$ be vector lattices.

Definition. An operator U which maps $\mathscr{X}$ into $\mathscr{Y}$, is said to be *disjunctive* if $x_1 \perp x_2$ ($x_1, x_2 \in \mathscr{X}$) implies $U(x_1) \perp U(x_2)$.

Proposition 1. *If U is a disjunctive, positive and additive operator, then*

$$U(x_1 \vee x_2) = U(x_1) \vee U(x_2), \tag{1}$$

$$U(x_1 \wedge x_2) = U(x_1) \wedge U(x_3), \tag{2}$$

for any elements x_1, x_2.

Proof. For any elements x_1, x_2 the following inequality holds

$$(x_1 - (x_1 \wedge x_2)) \wedge (x_2 - (x_1 \wedge x_2)) = \mathbf{0}$$

whence

$$(U(x_1) - U(x_1 \wedge x_2)) \wedge (U(x_2) - U(x_1 \wedge x_2)) = \mathbf{0}$$

and this equality is equivalent to (2).

Equality (1) is an obvious consequence of (2).

Proposition 2. *If U is a regular and disjunctive operator, which takes values in a complete vector lattice, then the operators U_+, U_-, $|U|$ are disjunctive.*

Proof. If $x_1 \perp x_2$ and $|x'| \leqslant |x_1|$, $|x''| \leqslant |x_2|$, then $U(x') \perp U(x'')$, whence (see Section 5.1.3, proposition 3)

$$|U|(|x_1|) \perp |U|(|x_2|)$$

and therefore

$$|U|(x_1) \perp |U|(x_2).$$

This shows that the operator $|U|$ is disjunctive. From $\mathbf{0} \leqslant U_+ \leqslant |U|$ and $\mathbf{0} \leqslant U_- \leqslant |U|$ it obviously follows that the operators U_+ and U_- are disjunctive.

Proposition 3. *If φ is a regular and disjunctive functional, then φ is either a positive, or a negative, functional.*

Proof. Let us assume that the functionals φ_+, φ_- are non-zero. Then there exist positive elements a_1, a_2, such that $\varphi_+(a_1) > 0$ and $\varphi_-(a_2) > 0$. If x_0 is an element such that $x_0 \geqslant a_1, a_2$, then $\varphi_+(x_0) > 0$ and $\varphi_-(x_0) > 0$. Let us define

$$\varepsilon = \frac{1}{2} \min(\varphi_+(x_0), \varphi_-(x_0)).$$

Since $\varphi_+ \wedge \varphi_- = \mathbf{0}$, there exist positive elements x_1, x_2 such that $x_1 + x_2 = x_0$ and

$$\varphi_+(x_1) + \varphi_-(x_2) < \varepsilon.$$

Let us denote $x_3 = x_1 \wedge x_2$, $a = x_1 - x_3$, $b = x_2 - x_3$. It is obvious that $a \wedge b = \mathbf{0}$ and

$$\varphi_+(x_3) \leqslant \varphi_+(x_1) < \varepsilon,$$

$$\varphi_-(x_3) \leqslant \varphi_-(x_2) < \varepsilon.$$

It follows that

$$\varphi_+(b) = \varphi_+(x_0) - \varphi_+(x_1) - \varphi_+(x_3) > \varphi_+(x_0) - 2\varepsilon \geqslant 0$$

$$\varphi_-(a) = \varphi_-(x_0) - \varphi_-(x_2) - \varphi_-(x_3) > \varphi_-(x_0) - 2\varepsilon \geqslant 0;$$

this implies that there exist elements a_1, b_1, such that $\mathbf{0} \leqslant a_1 \leqslant a$, $\mathbf{0} \leqslant b_1 \leqslant b$ and $\varphi(a_1) < 0$, $\varphi(b_1) > 0$. But in this case we have $\min\ (|\varphi(a_1)|, |\varphi(b_1)|) > 0$, although $a_1 \wedge b_1 = \mathbf{0}$, and this contradicts the fact that the functional φ is disjunctive.

The proposition is proved.

In the next propositions we shall denote (as in Section 4.1) by $\mathscr{B}(\mathscr{X})$ the basis of a space $\mathscr{X}$ with unit.

Proposition 4. *Let $\mathscr{X}$ and $\mathscr{Y}$ be σ-complete vector lattice with unit and U a linear (o)-continuous operator which maps $\mathscr{X}$ into $\mathscr{Y}$. If $U(\mathscr{B}(\mathscr{X})) \subset \mathscr{B}(\mathscr{Y})$, then U is a disjunctive operator.*

Proof. If $e_1, e_2 \in \mathscr{B}(\mathscr{X})$ and $e_1 \perp e_2$, then $e_1 + e_2 \in \mathscr{B}(\mathscr{X})$ and therefore $U(e_1 + e_2) \in \mathscr{B}(\mathscr{Y})$. Since $U(e_1)$ and $U(e_2)$ are unitary elements and

$$U(e_1) + U(e_2) = U(e_1 + e_2),$$

it follows that $U(e_1) \perp U(e_2)$ (see Section 4.1.1, proposition 1).

Let x_1 and x_2 be two simple element of $\mathscr{X}$. They can be represented with respect to the same unitary elements, i.e.,

$$x_j = \sum_{i=1}^{m} \alpha_{ij} e_i, \quad j = 1, 2,$$

where $e_i \in \mathscr{B}(\mathscr{X})$ and $e_i \perp e_j$ for $i \neq j$. If $x_1 \perp x_2$, then $\alpha_{i1}\alpha_{i2} = 0$ for any i and since

$$U(x_j) = \sum_{i=1}^{m} \alpha_{ij} U(e_i), \quad j = 1, 2,$$

it follows that $U(x_1) \perp U(x_2)$.

If x_1, x_2 are bounded elements, then there exist sequences $\{x_{nj}\}_{n\in\mathbb{N}}$, $j = 1, 2$, of simple elements such that $x_{nj} \xrightarrow{o} x_j$, $j = 1, 2$. If $x_1 \perp x_2$, then one can assume that $x_{n1} \perp x_{n2}$ for any n (because if this be not true, then one could replace x_{nj} by $[x_j]\, x_{nj}$). From $U(x_{n1}) \perp U(x_{n2})$ and $U(x_{nj}) \xrightarrow{o} U(x_j)$ one infers that $U(x_1) \perp U(x_2)$.

If x_1, x_2 are arbitrary positive elements, then $x_j = (o)\text{-}\lim_n (x_j \wedge (nu))$ (where u is the unit element in $\mathscr{X}$, whereas $j = 1, 2$). If $x_1 \perp x_2$, then $x_1 \wedge (nu) \perp x_2 \wedge (nu)$ and one can deduce, as before, that $U(x_1) \perp U(x_2)$.

If x_1, x_2 are quite arbitrary, such that $x_1 \perp x_2$, then the relation $U(x_1) \perp U(x_2)$ can be easily established.

Proposition 5. *Let $\mathscr{X}$ be a σ-complete vector lattice with unit, $\mathscr{Y}$ a σ-complete vector lattice in which total elements exist, and let U be a linear, positive and disjunctive operator, which maps $\mathscr{X}$ into $\mathscr{Y}$. Under these assumption one can choose a unit element in $\mathscr{Y}$ such that $U(\mathscr{B}(\mathscr{X})) \subset \mathscr{B}(\mathscr{Y})$.*

Proof. We shall denote by u the unit element of $\mathscr{X}$. Let $y_0 > \mathbf{0}$ be a total element in $\mathscr{Y}$, and let us define

$$v = U(u) + y_0 - [U(u)]y_0.$$

From

$$[U(u)]\,(y_0 - [U(u)]\,y_0) = \mathbf{0}$$

it follows that

$$U(u) \perp y_0 - [U(u)]\,y_0$$

and, therefore,

$$[v] = [U\,(u)] + [y_0 - [U\,(u)]\,y_0]. \tag{3}$$

(see Section 3.3.4, proposition 7).
Since

$$[U\,(u)] = [U\,(u)]\,[y_0] = [[U\,(u)]\,y_0]$$

(see Section 3.3.4, proposition 5), from (3) one infers that $[v] = [y_0]$, i.e., v is a total element in $\mathscr{Y}$. We shall choose v to be the unit element in $\mathscr{Y}$.

If $e \in \mathscr{B}\,(\mathscr{X})$, then from the equality

$$v = U\,(e) + U\,(u - e) + y_0 - [U\,(u)]\,y_0$$

and from the relation

$$U\,(e) \perp U\,(u - e) + y_0 - [U\,(u)]\,y_0$$

it follows that $U\,(e) \in \mathscr{B}\,(\mathscr{Y})$. The proposition is proved.

5.3.2 Multiplicative operators

Let $\mathscr{X}$ and $\mathscr{Y}$ be σ-complete vector lattices with unit. For each of these spaces we shall consider the multiplication introduced in Section 4.3.2.

Definition. An operator U, which maps $\mathscr{X}$ into $\mathscr{Y}$, is said to be a *multiplicative operator* if the existence of the product x_1x_2 ($x_1, x_2 \in \mathscr{X}$) implies the existence of the product $U\,(x_1)\,U\,(x_2)$ and the equality

$$U\,(x_1\,x_2) = U\,(x_1)\,U\,(x_2).$$

Proposition 1. If U is a multiplicative operator, then U is positive and $U\,(\mathscr{B}\,(\mathscr{X})) \subset \mathscr{B}\,(\mathscr{Y})$.

Proof. If $x \geqslant \mathbf{0}$ ($x \in \mathscr{X}$), then $\sqrt{x}$ exists (see Section 4.3.2) and

$$U\,(x) = U\,(\sqrt{x}\cdot\sqrt{x}) = U\,(\sqrt{x})\,U\,(\sqrt{x}) = (U\,(\sqrt{x}))^2,$$

which implies that $U(x) \geqslant \mathbf{0}$. If $e \in \mathscr{B}(\mathscr{X})$, then $e = e^2$, whence

$$U(e) = U(e^2) = (U(e))^2,$$

and this shows that $U(e) \in \mathscr{B}(\mathscr{Y})$ (see the cited section).

Proposition 2. *If U is a multiplicative operator and if* $U(\mathbf{0}) = \mathbf{0}$, *then U is disjunctive.*

The proof is immediate.

Remark. In any σ-complete vector lattice $\mathscr{X}$ with unit, the projector is a multiplicative operator, Indeed, any projector in $\mathscr{X}$ is determined by a unitary element e and $[e]\,x = ex$, for any $x \in \mathscr{X}$. Consequently, we have

$$[e]\,(xy) = e\,(xy) = (ex)\,(ey) = ([e]\,x)\,([e]\,y),$$

i.e., the projection $[e]$ is multiplicative.

One also knows that a projector is a linear (o)-continuous operator and $x \geqslant 0$ implies $[e]\,(x) \leqslant x$.

One can prove the following proposition.

Proposition 3. *Let* $\mathscr{X}$ *be a complete vector lattice with a unit element and U an* (o)*-continuous, multiplicative, linear operator, which maps* $\mathscr{X}$ *into* $\mathscr{X}$. *In order that U be a projector it is necessary and sufficient that* $x \geqslant \mathbf{0}$ *imply* $U(x) \leqslant x$.

Proof. According to the previous remark, one has only to show that the condition is sufficient. From a proposition which was established in Section 3.3.3 (proposition 6) we have only to show that $U(U(x)) = U(x)$, for any $x \in \mathscr{X}$.

Let first $e \in \mathscr{B}(\mathscr{X})$ and denote $e_1 = U(e)$. From proposition 1 we have $e_1 \in \mathscr{B}(\mathscr{X})$ and $e_1 \leqslant e$, by hypothesis. We have

$$U(e) = U(e_1) + U(e - e_1) = U(U(e)) + U(e - e_1),$$

and

$$\mathbf{0} \leqslant U(e - e_1) \leqslant e - e_1,$$

Since $e - e_1 \perp e_1$, it follows that

$$U(e - e_1) \perp U(e).$$

On the other hand,

$$U(e - e_1) \leqslant U(e)$$

and this implies that $U(e - e_1) = \mathbf{0}$, i.e.,

$$U(e) = U(U(e)).$$

The equality

$$U(x) = U(U(x)) \tag{1}$$

is now an immediate consequence, if x is a simple element.

With the help of the (o)-continuity, equality (1) also obtains for an arbitrary bounded element x and, finally, for a quite arbitrary element x. The proposition is proved.

5.3.3 Translators

Let $\mathscr{X}$ be a σ-complete vector lattice.

Definition. A *translator* is any linear operator U, which maps $\mathscr{X}$ into $\mathscr{X}$ and which satisfies the following conditions.

(i) the inequality $[x_1] \leqslant [x_2]$ is equivalent to the inequality $[U(x_1)] \leqslant [U(x_2)]$;
(ii) for any element $y \in \mathscr{X}$ there exists an element $x \in \mathscr{X}$, such that $[y] = [U(x)]$.

Proposition 1. *Any translator is an invertible and disjunctive operator.*

Proof. If U is a translator, then from $U(x) = \mathbf{0}$ it follows that $[x] = 0$, by taking into account condition (i) of the definition; consequently, $x = \mathbf{0}$.

Let $x_1 \perp x_2$. Let x_0 be such that

$$[U(x_0)] = [[U(x_1)]\ U(x_2)] = [U(x_1)]\ [U(x_2)].$$

From

$$[U(x_0)] \leqslant [U(x_1)]$$

it follows that $[x_0] \leqslant [x_1]$. Similarly, we get the inequality $[x_0] \leqslant [x_2]$ and, consequently,

$$[x_0] \leqslant [x_1]\,[x_2].$$

But $[x_1]\,[x_2] = 0$, and this implies that $x_0 = \mathbf{0}$. It follows that

$$[U(x_1)]\ U(x_2)] = \mathbf{0}$$

and this shows that $U(x_1) \perp U(x_2)$.

Lemma. *If U is a translator, then*

$$U([v]\,x) = [U(v)]\ U(x),$$

for any elements v and x.

Proof. For any elements $v, x \in \mathscr{X}$ we have the following equality

$$U(x) = U([v]\,x) + U(([x] - [x]\,[v])\,x). \tag{1}$$

From the inequality

$$[[v]\,x] \leqslant [v]$$

and from property (i) of a translator, it follows that

$$[U([v]\,x)] \leqslant [U(v)]. \tag{2}$$

By taking into account the fact that U is disjunctive, from

$$[v]\,[([x] - [x]\,[v])\,x] = \mathbf{0},$$

we get

$$[U(v)]\,[U(([x] - [x]\,[v])\,x)] = \mathbf{0}.$$

Hence, by taking into account (1) and (2), it follows that

$$[U(v)]\,U(x) = [U(v)]\;U([v]\,x) = U([v]\,x).$$

Proposition 2. *Any positive* [13]) *translator is an* (*o*)*-continuous operator.*

Proof. Let $x_n \downarrow \mathbf{0}$. Let us consider an arbitrary number $\varepsilon > 0$ and for each $n \in N$ let us define

$$v_n = (x_n - \varepsilon x_1)_+.$$

We have

$$[v_n]\,(x_n - \varepsilon x_1) = (x_n - \varepsilon x_1)_+ \geqslant \mathbf{0}$$

and, therefore

$$\varepsilon\,[v_n]\,x_1 \leqslant [v_n]\,x_n \leqslant x_n$$

whence we get that $[v_n]\,x_1 \downarrow \mathbf{0}$. On the other hand, we have

$$([x_1] - [v_n])\,(x_n - \varepsilon x_1) = -\,([x_1] - [v_n])\,(x_n - \varepsilon x_1)_- \leqslant \mathbf{0}$$

and, consequently,

$$([x_1] - [v_n])\,x_n \leqslant \varepsilon\,([x_1] - [v_n])\,x_1 \leqslant \varepsilon x_1,$$

whence

$$x_n = [v_n]\,x_n + ([x_1] - [v_n])\,x_n \leqslant [v_n]\,x_1 + \varepsilon x_1.$$

---

[13]) A positive translator is, by definition, a translator, which is a positive operator, at the same time.

From the preceding lemma we get

$$U(x_n) \leqslant [U(v_n)]\, U(x_1) + \varepsilon\, U(x_1). \tag{3}$$

By putting

$$[y] = \bigwedge_{n \in \mathbb{N}} [U(v_n)]$$

we infer that there exists an $a \in \mathscr{X}$, such that

$$[y] = [U(a)].$$

From

$$[U(a)] \leqslant [U(v_n)]$$

it follows that $[a] \leqslant [v_n]$. From here, and from the fact that $[v_n]\, x_1 \downarrow 0$, it follows that $[a]\, x_1 = \mathbf{0}$. Consequently,

$$[U(a)]\, U(x_1) = \mathbf{0}$$

whence

$$[U(v_n)]\, U(x_1) \downarrow \mathbf{0}.$$

From this relation and from (3), by taking into account the fact that $\{U(x_n)\}_{n \in \mathbb{N}}$ is decreasing, it follows that

$$(o)\text{-}\lim_{n} U(x_n) \leqslant \varepsilon U(x_1).$$

Since ε has been arbitrary, we get that $U(x_n) \xrightarrow{o} \mathbf{0}$ and thus the proposition is proved.

★

Let U be a positive translator on the space $\mathscr{X}$. For any element $a \in \mathscr{X}$ and any natural number n, let us put

$$a^n = \frac{1}{n} \sum_{i=0}^{n} U^i(a)$$

where $U^0(a) = a$ and $U^i(a) = U(U^{i-1}(a))$, for $i \in \mathbb{N}$.

Any element $x \in \mathscr{X}$, such that $U(x) = x$, is said to be an *invariant element* (with respect to U).

Any functional f on $\mathscr{X}$, such that $f(U(x)) = f(x)$, for any $x \in \mathscr{X}$, is said to be *an invariant functional* (with respect to U).

By definition, the space $\mathscr{X}$ has property $\mathscr{E}$ with respect to U, iff for any invariant element $x > \mathbf{0}$, there exists an (o)-continuous, positive, linear functional f, invariant with respect to U, such that $f(x) > \mathbf{0}$.

One can prove the following theorems [14]).

(i) If $\mathscr{X}$ has property $\mathscr{E}$ with respect to U, and if the sequence $\{a^n\}_{n\in\mathbb{N}}$ is bounded, then this sequence is (o)-convergent.

(ii) If $\mathscr{X}$ has property $\mathscr{E}$ with respect to U, then for any element v, invariant with respect to U, there exists a sequence of elements $\{v_m\}_{m\in\mathbb{N}}$, such that $[v_m] \uparrow [v]$ and such that (for any $m \in \mathbb{N}$) the sequence $\{[v_m]a^n\}_{n\in\mathbb{N}}$ should converge individually [15]).

5.4 Regular bilinear operators and (o)-continuous bilinear operators

5.4.1 Regular bilinear operators

Let $\mathscr{X}$ and $\mathscr{Y}$ be vector lattices and $\mathscr{Z}$ a complete vector lattice.

Lemma 1. *If U_0 is a mapping of $\mathscr{X}_+ \times \mathscr{Y}_+$ into $\mathscr{Z}$, which satisfies the conditions*

$$U_0(x_1 + x_2, y) = U_0(x_1, y) + U_0(x_2, y),$$

$$U_0(x, y_1 + y_2) = U_0(x, y_1) + U_0(x, y_2),$$

for any $x_1, x_2 \in \mathscr{X}_+$ and any $y_1, y_2 \in \mathscr{Y}_+$, then U_0 can be extended to a biadditive operator U defined on $\mathscr{X} \times \mathscr{Y}$, and this extension is unique. This extension is given by the formula

$$U(x, y) = U_0(x_+, y_+) - U_0(x_+, y_-) - U_0(x_-, y_+) + U_0(x_-, y_-). \tag{1}$$

Proof. We shall first define a mapping U_1 of $\mathscr{X} \times \mathscr{Y}_+$ into $\mathscr{Z}$ by the formula

$$U_1(x, y) = U_0(x_+, y) - U_0(x_-, y).$$

The operator U_1 is additive with respect to the first argument (Section 5.1.2 lemma 1). Now, by putting

$$U(x, y) = U_1(x, y_+) - U_1(x, y_-),$$

[14]) See H. Nakano [9].

[15]) The definition of the individual convergence has been given in Section 3.3.4.

we get an operator defined on $\mathscr{X}\times\mathscr{Y}$, taking values in $\mathscr{Z}$, which is additive with respect to the first argument, since it is the difference of two additive operators.

Analogously, one can show that the operator U, defined by the formula (1) is additive with respect to the second argument.

The operator U obviously is an extension of U_0. The uniqueness of the extension is obvious.

Definition 1. An operator U, which maps $\mathscr{X}\times\mathscr{Y}$ into $\mathscr{Z}$, is said to be a *positive operator* if $x \geqslant \mathbf{0}$ and $y \geqslant \mathbf{0}$ implies $U(x, y) \geqslant \mathbf{0}$.

If U is a positive operator, we shall write $U \geqslant \mathbf{0}$. One can immediately show that if U is biadditive and positive, then: $x_1 \leqslant x_2$ ($x_1, x_2 \in \mathscr{X}$) and $y \geqslant \mathbf{0}$ ($y \in \mathscr{Y}$) implies $U(x_1, y) \leqslant U(x_2, y)$; $U(x, y_1) \leqslant U(x, y_2)$, for any elements $x \geqslant \mathbf{0}$ and $y_1 \leqslant y_2$ ($x \in \mathscr{X}$ and $y_1, y_2 \in \mathscr{Y}$); one has

$$|U(x, y)| \leqslant U(|x|, |y|),$$

for any elements $x \in \mathscr{X}$, $y \in \mathscr{Y}$.

Any biadditive and positive operator which maps $\mathscr{X}\times\mathscr{Y}$ into $\mathscr{Z}$ is a bilinear operator.

Definition 2. An operator U, which is defined on $\mathscr{X}\times\mathscr{Y}$ and takes values in $\mathscr{Z}$, is said to be a *regular operator* if it may be represented as the difference of two positive biadditive [16]) operators.

In particular, any regular operator U, which maps $\mathscr{X}\times\mathscr{Y}$ into $\mathscr{Z}$, is a bilinear operator.

If U_1 and U_2 are biadditive operators, defined on $\mathscr{X}\times\mathscr{Y}$ and taking values in $\mathscr{Z}$, we shall write $U_1 \leqslant U_2$ if $U_2 - U_1 \geqslant 0$. We shall say that U_1 is majorized by U_2.

A biadditive operator U is regular iff U is majorized by a positive biadditive operator.

If U is a regular biadditive operator, then there exists a positive biadditive operator U_0, such that

$$|U(x, y)| \leqslant U_0(|x|, |y|), \tag{2}$$

for any $x \in \mathscr{X}$, $y \in \mathscr{Y}$.

Indeed, if $U = U_1 - U_2$, where U_1 and U_2 are positive biadditive operators, then one can take $U_0 = U_1 + U_2$ and (2) is satisfied.

Remark. The notions of positive operator and of regular operator make sense in the case when $\mathscr{Z}$ is an arbitrary vector lattice. If $\mathscr{Z}$ is an Archimedean vector lattice and if U is a biadditive and positive (or a regular) operator, defined on $\mathscr{X}\times\mathscr{Y}$ and taking values in $\mathscr{Z}$, then U is bilinear. The other remarks we have previously made are in force for the case of an arbitrary vector lattice $\mathscr{Z}$.

---

[16]) We consider the usual operations $U_1 + U_2$ and αU on biadditive operators.

Theorem 1. *A biadditive operator U, which maps $\mathscr{X} \times \mathscr{Y}$ into $\mathscr{Z}$, is a regular operator iff for any positive elements $x \in \mathscr{X}$, $y \in \mathscr{Y}$ the set*

$$G(U; x, y) = \left\{ \sum_{i,j} |U(x_i, y_j)|;\ \mathbf{0} \leqslant x_i \in \mathscr{X}, \sum_{i=1}^{m} x_i = x,\ \mathbf{0} \leqslant y_j \in \mathscr{Y}, \sum_{j=1}^{n} y_j = y \right\}$$

is bounded in $\mathscr{Z}$.

Proof. If U is a regular operator, then there exists a positive biadditive operator U_0, such that (2) is satisfied. For any element of the set $G(U; x, y)$ we, therefore, have

$$\sum_{i,j} |U(x_i, y_j)| \leqslant \sum_{i,j} U_0(x_i, y_j) = U_0(x, y)$$

and this shows that the set $G(U; x, y)$ is bounded.

Conversely, if for any positive elements $x \in \mathscr{X}$, $y \in \mathscr{Y}$ the set $G(U; x, y)$ is bounded, then by defining

$$V(x, y) = \sup G(U; x, y),$$

we get a mapping of $\mathscr{X}_+ \times \mathscr{Y}_+$ into $\mathscr{Z}$.

Let us show that

$$V(a + b, y) = V(a, y) + V(b, y), \tag{3}$$

for any $a, b \geqslant \mathbf{0}$ and $y \geqslant \mathbf{0}$.

Let

$$a + b = \sum_{i=1}^{m} x_i, \quad y = \sum_{j=1}^{n} y_j,$$

where $x_i \geqslant \mathbf{0}$, $y_j \geqslant \mathbf{0}$. Then there exist positive elements a_i, b_i $(i = 1,2, \ldots, m)$ of $\mathscr{X}$, such that

$$a_i + b_i = x_i, \ i = 1,2, \ldots, m,$$

$$\sum_{i=1}^{m} a_i = a, \ \sum_{i=1}^{m} b_i = b$$

and we have

$$\sum_{i,j} |U(x_i, y_j)| = \sum_{i,j} |U(a_i + b_i, y_j)|$$

$$= \sum_{i,j} |U(a_i, y_j) + U(b_i, y_j)| \leqslant \sum_{i,j} |U(a_i, y_j)| + \sum_{i,j} |U(b_i, y_j)|$$

$$\leqslant V(a, y) + V(b, y).$$

It follows that

$$V(a+b, y) \leqslant V(a, y) + V(b, y).$$

In order to prove the opposite inequality, let

$$a = \sum_{i=1}^{m} a_i,\ b = \sum_{j=1}^{n} b_j,$$

$$y = \sum_{k=1}^{n'} y'_k, = \sum_{l=1}^{n''} y''_l,$$

where $a_i, b_j \geqslant \mathbf{0}$ and $y'_k, y''_l \geqslant \mathbf{0}$. Then there exist positive elements y_{kl} $(k = 1, 2, \ldots, n'; l = 1, 2, \ldots, n'')$, of $\mathscr{Y}$ such that

$$\sum_{l=1}^{n''} y_{kl} = y'_k;\quad \sum_{k=1}^{n'} y_{kl} = y''_l\,.$$

Let us put

$$x_n = \begin{cases} a_h & \text{if}\quad 1 \leqslant h \leqslant m \\ b_{h-m} & \text{if}\quad m \leqslant h \leqslant m+n. \end{cases}$$

Then the following relations hold

$$\sum_{i,k} |\,U(a_i, y'_k)\,| + \sum_{j,l} |\,U(b_j, y''_l)\,|$$

$$\leqslant \sum_{i,k} \sum_{l} |\,U(a_i, y_{kl})\,| + \sum_{j,l} \sum_{k} |\,U(b_j, y_{kl})\,|$$

$$= \sum_{h,k,l} |\,U(x_h, y_{kl})\,| \leqslant V(a+b, y),$$

whence

$$V(a, y) + V(b, y) \leqslant V(a+b, y).$$

We have thus proved relation (3).

One can analogously prove that if $x \geqslant \mathbf{0}$ $(x \in \mathscr{X})$ and $c, d \geqslant \mathbf{0}$ $(c, d \in \mathscr{Y})$, then

$$V(x, c+d) = V(x, c) + V(x, d).$$

By taking into account lemma 1, one infers that the mapping V can be uniquely extended as biadditive operator on $\mathscr{X} \times \mathscr{Y}$.

We shall denote also by V the extended operator, which is obviously positive, and one can easily see that $U \leqslant V$.

Consequently, U is a regular operator.

The theorem is proved.

If U is an operator which maps $\mathscr{X} \times \mathscr{Y}$ into $\mathscr{Z}$, then for any element $x \in \mathscr{X}$ we shall denote by U_x the operator (which maps $\mathscr{Y}$ into $\mathscr{Z}$) defined by the formula

$$U_x(y) = U(x, y), \; y \in \mathscr{Y}.$$

We shall also consider the operator F_U, which maps $\mathscr{X}$ into the set of the operators defined on $\mathscr{Y}$ and taking values in $\mathscr{Z}$, and which is given by the formula

$$F_U(x) = U_x, \; x \in \mathscr{X}.$$

Theorem 2. *The set $(\mathscr{X} \times \mathscr{Y}, \mathscr{Z})^r$ of all regular, bilinear operators (defined on $\mathscr{X} \times \mathscr{Y}$ and taking values in $\mathscr{Z}$) is a complete vector lattice; it is isomorphic to the space $(\mathscr{X}, (\mathscr{Y}, \mathscr{Z})^r)^r$, by the mapping $U \mapsto F_U$.*

Proof. The set $(\mathscr{X} \times \mathscr{Y}, \mathscr{Z})^r$, endowed with the usual operations $U_1 + U_2$ and αU and with the order relation already defined (i.e., $U_1 \leqslant U_2$, iff $x \geqslant \mathbf{0}$ and $y \geqslant \mathbf{0}$ implies $U_1(x, y) \leqslant U_2(x, y)$) obviously is a directed vector space.

One can also directly show that $(\mathscr{X} \times \mathscr{Y}, \mathscr{Z})^r$ is a complete vector lattice. Indeed, one can observe that for any $U \in (\mathscr{X} \times \mathscr{Y}, \mathscr{Z})^r$, the operator V, which has been considered in the proof of theorem 1, exists and it is the absolute value of U. Hence we infer that $(\mathscr{X} \times \mathscr{Y}, \mathscr{Z})^r$ is a vector lattice. If $\{U_j\}_{j \in J}$ is a bounded from above and directed system of positive elements of the space $(\mathscr{X} \times \mathscr{Y}, \mathscr{Z})^r$, then there exists its least upper bound, and it is given by the operator U, which, for positive elements, is given by the formula

$$U(x, y) = \bigvee_{j \in J} U_j(x, y),$$

(whereas for arbitrary elements, it is given by the method described in lemma 1).

Let us prove that the mapping $U \mapsto F_U$ is an isomorphism between the ordered vector spaces $(\mathscr{X} \times \mathscr{Y}, \mathscr{Z})^r$ and $(\mathscr{X}, (\mathscr{Y}, \mathscr{Z})^r)^r$.

If $U = U' - U''$, where U' and U'' are positive, bilinear operators, then

$$U_x(y) = U'_x(y) - U''_x(y), \; y \in \mathscr{Y}.$$

If $x \geqslant \mathbf{0}$ then U'_x and U''_x are obviously positive linear operators, hence $U_x \in (\mathscr{Y}, \mathscr{Z})^r$.

If x is an arbitrary element of $\mathscr{X}$, then

$$U_x = U_{x_+} - U_{x_-},$$

and, therefore, $U_x \in (\mathscr{Y}, \mathscr{Z})^r$, for any $x \in \mathscr{X}$.

Consequently, F_U maps the space $\mathscr{X}$ into the space $(\mathscr{Y}, \mathscr{Z})^r$.
For any elements $x_1, x_2 \in \mathscr{X}$ the following equality holds

$$U_{x_1+x_2} = U_{x_1} + U_{x_2}$$

and, therefore, the operator F_U is additive. From the equalities

$$F_U(x) = U_x = U'_x - U''_x = F_{U'}(x) - F_{U''}(x),$$

which hold for any $x \in \mathscr{X}$, it follows that

$$F_U = F_{U'} - F_{U''}.$$

But the operators $F_{U'}$, $F_{U''}$ are obviously additive and positive and, consequently, $F_U \in (\mathscr{X}, (\mathscr{Y}, \mathscr{Z})^r)^r$.

If $F \in (\mathscr{X}, (\mathscr{Y}, \mathscr{Z})^r)^r$, then, by putting

$$U(x, y) = (F(x))(y),$$

one obtains an operator $U \in (\mathscr{X} \times \mathscr{Y}, \mathscr{Z})^r$ and $F(x) = U_x$, whence $F = F_U$.

The mapping $U \mapsto F_U$ is obviously one-to-one. It is easily verified that

$$F_{\alpha U_1 + \beta U_2} = \alpha F_{U_1} + \beta F_{U_2}$$

for any $\alpha, \beta \in \mathbb{R}$ and any $U_1, U_2 \in (\mathscr{X} \times \mathscr{Y}, \mathscr{Z})^r$.

If $U \in (\mathscr{X} \times \mathscr{Y}, \mathscr{Z})^r$ and $U \geqslant \mathbf{0}$, then $U_x \geqslant \mathbf{0}$, for any $x \geqslant \mathbf{0}$ $(x \in \mathscr{X})$. It follows that $F_U \geqslant \mathbf{0}$.

Conversely, if $F_U \geqslant \mathbf{0}$, then for any element $x \geqslant \mathbf{0}$ $(x \in \mathscr{X})$ we have

$$F_U(x) = U_x \geqslant \mathbf{0}.$$

Hence it follows that for $\mathbf{0} \leqslant x \in \mathscr{X}$ and $\mathbf{0} \leqslant y \in \mathscr{Y}$ we have

$$U(x, y) = U_x(y) \geqslant \mathbf{0}$$

i.e., $U \geqslant \mathbf{0}$. The theorem is proved.

Remarks. (1) As we have remarked in the proof of theorem 2, if $U \in (\mathscr{X} \times \mathscr{Y}, \mathscr{Z})^r$ and $\mathbf{0} \leqslant x \in \mathscr{X}$, $\mathbf{0} \leqslant y \in \mathscr{Y}$, then

$$|U|(x, y) = \sup\left\{\sum_{i,j} |U(x_i, y_j)|;\ \mathbf{0} \leqslant x_i \in \mathscr{X},\ \sum_{i=1}^{m} x_i = x,\ \ \mathbf{0} \leqslant y_j \in \mathscr{Y},\ \sum_{j=1}^{n} y_j = y\right\}.$$

For U_+ and U_- one obtains similar formulae by replacing $\sum_{i,j} |U(x_i, y_j)|$ by $\sum_{i,j} (U(x_i, y_j))_+$ and $\sum_{i,j}(U(x_i, y_j))_-$, respectively.

(2) For any element $x \geqslant \mathbf{0}$ $(x \in \mathscr{X})$, we have

$$|U_x| = \bigvee_{|x'| \leqslant x} U_{x'}, \tag{4}$$

whereas if x is an arbitrary element of $\mathscr{X}$, then

$$|U_x| \leqslant |U|_{|x|}.$$

These are immediate consequences of theorem 2.

5.4.2 (*o*)-Continuous bilinear operators

Let $\mathscr{X}, \mathscr{Y}$ and $\mathscr{Z}$ be as in the preceding section. An operator U, defined on $\mathscr{X} \times \mathscr{Y}$ and taking values in $\mathscr{Z}$, is said to be (*o*)-*continuous* if for any (*o*)-convergent sequence $\{x_n\}_{n \in \mathbb{N}}$ of elements of $\mathscr{X}$ and for any (*o*)-convergent sequence $\{y_n\}_{n \in \mathbb{N}}$ of elements of $\mathscr{Y}$, from $x_n \xrightarrow{o} x$ and $y_n \xrightarrow{o} y$ it follows that $U(x_n, y_n) \xrightarrow{o} U(x, y)$.

Any biadditive and (*o*)-continuous operator, which maps $\mathscr{X} \times \mathscr{Y}$ into $\mathscr{Z}$, is bilinear.

Proposition. *Let U be a regular operator, defined on $\mathscr{X} \times \mathscr{Y}$ and taking values in $\mathscr{Z}$. If for any sequence $\{x_n\}_{n \in \mathbb{N}}$ of elements of $\mathscr{X}$, which is (o)-convergent to* $\mathbf{0}$, *and any sequence $\{y_n\}_{n \in \mathbb{N}}$ of elements of $\mathscr{Y}$, which is (o)-convergent to* $\mathbf{0}$, *it follows that $U(x_n, y) \xrightarrow{o} \mathbf{0}$ and $U(x, y_n) \xrightarrow{o} \mathbf{0}$, for any $x \in \mathscr{X}$ and $y \in \mathscr{Y}$, then U is (o)-continuous.*

Proof. Under the assumptions of the proposition, the operators U_x and U_y (where the notations are those of the preceding section) are regular and (*o*)-continuous, for any $x \in \mathscr{X}$ and $x \in \mathscr{Y}$. From formula (4) in the preceding section, and from the theorem of Section 5.2.2, it follows that $|U|_x \in (\mathscr{Y}, \mathscr{Z})_0^r$. Analogously, one has $|U|^y \in (\mathscr{X}, \mathscr{Z})_0^r$.

Let $x_n \xrightarrow{o} x$, $y_n \xrightarrow{o} y$ and let $|x_n| \leqslant a (n \in \mathbb{N})$. From the inequalities

$$|U(x_n, y_n) - U(x, y)| \leqslant |U(x_n, y_n) - U(x_n, y)|$$

$$+ |U(x_n, y) - U(x, y)| \leqslant |U|(a, |y_n - y|)$$

$$+ |U|(|x_n - x|, |y|),$$

it immediately follows that

$$U(x_n, y_n) \xrightarrow{o} U(x, y)$$

and the proposition is proved.

Theorem. *The set* $(\mathscr{X} \times \mathscr{Y}, \mathscr{Z})_o^r$ *of all regular* (*o*)*-continuous operators, which map* $\mathscr{X} \times \mathscr{Y}$ *into* $\mathscr{Z}$, *is a component of the space* $(\mathscr{X} \times \mathscr{Y}, \mathscr{Z})^r$, *and it is isomorphic to the space* $(\mathscr{X}, (\mathscr{Y}, \mathscr{Z})_o^r)_o^r$.

Proof. Let $U \in (\mathscr{X}, \times \mathscr{Y}, \mathscr{Z})_o^r$ and $\{r_n\}_{n\in\mathbb{N}}$ be a sequence of elements of $\mathscr{X}$, such that $r_n \downarrow 0$. For any $y \in \mathscr{Y}$, the operator $|U|^y$ is (*o*)-continuous; consequently, $|U|(r_n, y) \xrightarrow{o} 0$. On the other hand, the sequence $\{|U|_{r_n}\}_{n\in\mathbb{N}}$ is decreasing, hence $|U|_{r_n} \xrightarrow{o} \mathbf{0}$ in the space $(\mathscr{Y}, \mathscr{Z})_o^r$ (see 5.1.3, corollary of proposition 4). If $x_n \xrightarrow{o} x$ in the space $\mathscr{X}$, then there exists a sequence $\{r_n\}_{n\in\mathbb{N}}$ of elements of $\mathscr{X}$, such that $r_n \downarrow \mathbf{0}$ and $|x_n - x| \leqslant r_n$. From the inequalities

$$|U_{x_n} - U_x| \leqslant |U|_{|x_n - x|} \leqslant |U|_{r_n}$$

it follows that $U_{x_n} \xrightarrow{o} U_x$.

We have thus shown that if $U \in (\mathscr{X} \times \mathscr{Y}, \mathscr{Z})_0^r$, then $F_U \in (\mathscr{X}, (\mathscr{Y}, \mathscr{Z}_o^r)_o^r$ (where F_U is defined as in the preceding section: $F_U(x) = U_x$).

Let $U \in (\mathscr{X} \times \mathscr{Y}, \mathscr{Z})^r$ and let us assume that $F_U \in (\mathscr{X}, (\mathscr{Y}, \mathscr{Z})_o^r)_o^r$.

If $x_n \xrightarrow{o} x$ in the space $\mathscr{X}$, then $F_U(x_n) \xrightarrow{o} F_U(x)$, i.e., $U_{x_n} \xrightarrow{o} U$ in the space $(\mathscr{Y}, \mathscr{Z})_o^r$, whence it follows that $U_{x_n}(y) \xrightarrow{o} U_x(y)$ in $\mathscr{Z}$, for any $y \in \mathscr{Y}$. On the other hand, $U_x \in (\mathscr{Y}, \mathscr{Z})_o^r$, for any $x \in \mathscr{X}$; this shows that $y_n \xrightarrow{o} y$ in the space $\mathscr{Y}$ implies that $U_x(y_n) \xrightarrow{o} U_x(y)$ in $\mathscr{Z}$, for any $x \in \mathscr{X}$. From the preceding proposition it follows that $U \in (\mathscr{X} \times \mathscr{Y}, \mathscr{Z})_o^r$.

The spaces $(\mathscr{X} \times \mathscr{Y}, \mathscr{Z})_o^r$ and $(\mathscr{X}, (\mathscr{Y}, \mathscr{Z})_o^r)_o^r$ are therefore isomorphic (as ordered vector spaces). Since $(\mathscr{Y}, \mathscr{Z})_o^r$ is a component in $(\mathscr{Y}, \mathscr{Z})^r$ (Section 5.2.2, theorem), the set $(\mathscr{X}, (\mathscr{Y}, \mathscr{Z})_o^r)^r$ is a component in $(\mathscr{X}, (\mathscr{Y}, \mathscr{Z})^r)^r$ (Section 5.1.3, proposition 5). Since $(\mathscr{X}, (\mathscr{Y}, \mathscr{Z})_o^r)_o^r$ is a component in $(\mathscr{X}, (\mathscr{Y}, \mathscr{Z})_o^r)^r$, it follows that $(\mathscr{X}, (\mathscr{Y}, \mathscr{Z})_o^r)_o^r$ is a component in $(\mathscr{X}, (\mathscr{Y}, \mathscr{Z})^r)^r$. The theorem is proved.

5.4.3 Operators of type (V)

Let $\mathscr{X}, \mathscr{Y}, \mathscr{Z}$ be vector lattices.

If $\mathscr{D} \subset \mathscr{X} \times \mathscr{Y}$, then for any $y \in \mathscr{Y}$ and $x \in \mathscr{X}$ we shall define

$$\mathscr{D}^y = \{x \in \mathscr{X};\ (x, y) \in \mathscr{D}\},$$

$$\mathscr{D}_x = \{y \in \mathscr{Y};\ (x, y) \in \mathscr{D}\}.$$

A subset $\mathscr{D}$ of $\mathscr{X} \times \mathscr{Y}$ will be called a *normal domain* iff [17] $\mathscr{D}^0 = \mathscr{X}, \mathscr{D}_0 = \mathscr{Y}$, whereas for any $y \in \mathscr{Y}$ and $x \in \mathscr{X}$, the sets $\mathscr{D}^y$ and $\mathscr{D}_x$ are normal subspaces of $\mathscr{X}$, of $\mathscr{Y}$ respectively. If U is an operator defined on a subset $\mathscr{D}$ of $\mathscr{X} \times \mathscr{Y}$, taking values in $\mathscr{Z}$, and if y is an arbitrary element of $\mathscr{Y}$, we shall denote by U^y the operator defined

---

[17]) As in the preceding sections, the zero elements of the spaces $\mathscr{X}, \mathscr{Y}, \mathscr{Z}$ will be denoted by the same symbol $\mathbf{0}$.

on $\mathscr{D}^y$ by the formula

$$U^y(x) = U(x, y).$$

We shall also denote by U_x the operator defined on $\mathscr{D}_x$ by the formula $U_x(y)=U(x,y)$.

Let U be an operator defined on a normal domain $\mathscr{D} \subset \mathscr{X} \times \mathscr{Y}$, taking values in $\mathscr{Z}$. If U is biadditive on $\mathscr{D}$ (i.e., if U^y and U_x are additive operators on $\mathscr{D}^y$, and $\mathscr{D}_x$ respectively, for any $x \in \mathscr{X}$ and $y \in \mathscr{Y}$) and if U is positive (i.e., $U(x, y) \geqslant 0$, if $(x, y) \in \mathscr{D}$ and $x \geqslant 0$, $y \geqslant 0$) then U has the following properties.

(i) $U(x, \mathbf{0}) = U(\mathbf{0}, y) = \mathbf{0}$, for any $x \in \mathscr{X}$, $y \in \mathscr{Y}$;

(ii) if $(x_1, y_1) \in \mathscr{D}$ and $\mathbf{0} \leqslant x \leqslant y_1$, $\mathbf{0} \leqslant y \leqslant y_1$, then $(x, y) \in \mathscr{D}$, and

$$U(x, y) \leqslant U(x_1, y_1);$$

(iii) if $(x, y) \in \mathscr{D}$, then $(|x|, |y|) \in \mathscr{D}$ and

$$|U(x, y)| \leqslant U(|x|, |y|);$$

(iv) if $(x, y) \in \mathscr{D}$, then $(\alpha x, y) \in \mathscr{D}$ and $(x, \alpha y) \in \mathscr{D}$, for any real number α, and if $\mathscr{Z}$ is an Archimedean space, then

$$U(\alpha x, y) = U(x, \alpha y) = \alpha U(x, y).$$

Definition 1. An operator U, defined on a subset $\mathscr{D}$ of $\mathscr{X} \times \mathscr{Y}$, taking values in $\mathscr{Z}$, is said to be *bidisjunctive* if it satisfies the following condition: if $(x_1, y_1) \in \mathscr{D}$, $(x_2, y_2) \in \mathscr{D}$ and if $x_1 \perp x_2$ or $y_1 \perp y_2$, then $U(x_1, y_1) \perp U(x_2, y_2)$.

Proposition. *If U is a biadditive operator, which, moreover, is positive and bidisjunctive and if it is defined on a normal domain $\mathscr{D} \subset \mathscr{X} \times \mathscr{Y}$, taking values in $\mathscr{Z}$, then for any $(x, y) \in \mathscr{D}$ the following equalities hold*

$$(U(x, y))_+ = U(x_+, x_+) + U(x_-, y_-), \tag{1}$$

$$(U(x, y))_- = U(x_+, y_-) + U(x_-, y_+), \tag{2}$$

$$|U(x, y)| = U(|x|, |y|). \tag{3}$$

If $(x_1, y) \in \mathscr{D}$, $(x_2, y) \in \mathscr{D}$ and $y \geqslant \mathbf{0}$, then $(x_1 \vee x_2, y) \in \mathscr{D}$, $(x_1 \wedge x_2, y) \in \mathscr{D}$, and the following equalities hold:

$$U(x_1 \vee x_2, y) = U(x_1, y) \vee U(x_2, y), \tag{4}$$

$$U(x_1 \wedge x_2, y) = U(x_1, y) \wedge U(x_2, y). \tag{5}$$

Proof. Indeed, equalities (1) and (2) follow from the equality

$$U(x, y) = U(x_+, y_+) + U(x_-, y_-) - (U(x_+, y_-) + U(x_-, y_+)),$$

and from the fact that

$$U(x_+, y_+) + U(x_-, y_-) \perp U(x_+, y_-) + U(x_-, y_+).$$

From (1) and (2) formula (3) follows immediately.

If $(x, y) \in \mathscr{D}$ and $y \geqslant \mathbf{0}$, then from (1) it follows that

$$(U(x, y))_+ = U(x_+, y).$$

On the other hand, if $(x_1, y) \in \mathscr{D}$ and $(x_2, y) \in \mathscr{D}$, then from

$$|x_1 \vee x_2| \leqslant |x_1| + |x_2|$$

one immediately infers that $(x_1 \vee x_2, y) \in \mathscr{D}$. From the equality

$$x_1 \vee x_2 = x_1 + (x_2 - x_1)_+,$$

and from the fact that y is positive it follows that

$$U(x_1 \vee x_2, y) = U(x_1, y) + U((x_2 - x_1)_+, y)$$

$$= U(x_1, y) + (U(x_2 - x_1, y))_+ = U(x_1, y) + (U(x_2, y) - U(x_1, y))_+$$

$$= U(x_1, y) \vee U(x_2, y),$$

and, consequently, equality (4) holds.

The fact that $(x_1 \wedge x_2, y) \in \mathscr{D}$, as well as equality (5), is a consequence of the equality

$$(x_1 \wedge x_2) = -((-x_1) \vee (-x_2)),$$

and of what we have already proved.

Definition 2. We shall call *operator of type* (V) any bidisjunctive positive and bilinear operator U, defined on a normal domain $\mathscr{D} \subset \mathscr{X} \times \mathscr{Y}$, such that the following condition be satisfied: if $(x_n, y_n) \in \mathscr{D}$ $(n \in \mathbb{N})$, $x_n \xrightarrow{o} x$, $y_n \xrightarrow{o} y$ and if the sequence $\{U(x_n, y_n)\}_{n \in \mathbb{N}}$ is bounded, then $(x, y) \in \mathscr{D}$ and $U(x_n, y_n) \xrightarrow{o} U(x, y)$.

In the next theorem we shall assume that $\mathscr{X}$ and $\mathscr{Y}$ are spaces with unit and we shall denote by u', u'' the unit elements of $\mathscr{X}$, $\mathscr{Y}$, respectively. We shall denote by $\mathscr{B}(\mathscr{V})$ the base of a space $\mathscr{V}$ with unit.

Theorem. *Let $\mathscr{X}, \mathscr{Y}, \mathscr{Z}$ be σ-complete vector lattices with unit, and $\mathscr{D}$ a normal domain in $\mathscr{X} \times \mathscr{Y}$, such that $\mathscr{D}^{u''} = \mathscr{X}$, $\mathscr{D}_{u'} = \mathscr{Y}$. If U is an operator of type (V), defined on $\mathscr{D}$ and taking values in $\mathscr{Z}$, then, with a suitable selection of the unit element of $\mathscr{Z}$, there exist (o)-continuous, disjunctive, positive linear operators, F_1, F_2, defined on $\mathscr{X}, \mathscr{Y}$, respectively and taking values in $\mathscr{Z}$, such that* [18])

$$U(x, y) = F_1(x)F_2(y), \tag{6}$$

for any $(x, y) \in \mathscr{D}$.

Conversely, any operator of the form (6), defined on the set of all elements $(x, y) \in \in \mathscr{X} \times \mathscr{Y}$, such that $F_1(x)F_2(y)$ exists (where F_1, F_2 are (o)-continuous disjunctive, positive, linear operators), is an operator of type (V).

Proof. For any $x \in \mathscr{X}$, $y \in \mathscr{Y}$, let us put

$$F_1(x) = U(x, u''), \; F_2(y) = U(u', y).$$

The operators, F_1, F_2 (defined on $\mathscr{X}, \mathscr{Y}$ respectively, and taking values in $\mathscr{Z}$) are linear, positive, disjunctive and (o)-continuous. These facts are easy to prove. One can then choose the unit element u of $\mathscr{Z}$, such that $e \in \mathscr{B}(\mathscr{X})$ imply $F_1(e) \in \mathscr{B}(\mathscr{Z})$ (Section 5.3.1, proposition 5). In particular, it follows that $U(u', u'') \in \mathscr{B}(\mathscr{Z})$. One can also infer that if $f \in \mathscr{B}(\mathscr{Y})$, then $F_2(f) \in \mathscr{B}(\mathscr{Z})$.

We shall now show that if $e \in \mathscr{B}(\mathscr{X})$ and $f \in \mathscr{B}(\mathscr{Y})$, then

$$U(e, f) = U(e, u'') \wedge U(u', f). \tag{7}$$

We have

$$U(e, f) \wedge U(e, u'' - f) = 0$$

and, therefore,

$$U(e, u'') = U(e, f) + U(e, u'' - f) = U(e, f) \vee U(e, u'' - f).$$

On the other hand

$$U(e, u'' - f) \wedge U(u', f) = 0$$

and this implies that

$$U(e, u'') \wedge U(u', f) = (U(e, f) \vee U(e, u'' - f)) \wedge U(u', f)$$

$$= U(e, f) \wedge U(u', f) = U(e, f),$$

i.e., (7) is proved.

Since $U(e, u'')$ and $U(u', f)$ are unitary elements, equality (7) can also be written as

$$U(e, f) = F_1(e)F_2(f).$$

---

[18]) In the right hand member of formula (6) the multiplication in a space with unit element, as defined in Section 4.3.2, is considered.

Hence one can easily infer that if x and y are simple elements (of $\mathscr{X}, \mathscr{Y}$ respectively), then

$$U(x, y) = F_1(x)F_2(y).$$

If x and y are bounded elements (of $\mathscr{X}, \mathscr{Y}$, respectively), then there exist sequences of simple elements $\{x_n\}_{n\in\mathbb{N}}$, $\{y_n\}_{n\in\mathbb{N}}$, as well as real numbers ξ, η, such that $x_n \xrightarrow{o} x$ $y_n \xrightarrow{o} y$ and $|x_n| \leqslant \xi u'$, $|y_n| \leqslant \eta u''$, for any $n \in \mathbb{N}$. Since

$$U(x_n, y_n) = F_1(x_n)F_2(y_n), \tag{8}$$

and since the sequence $\{U(x_n, y_n)\}_{n\in\mathbb{N}}$ is bounded, it follows that $F_1(x)\ F_2(y)$ exists and that equality (6) holds.

If $(x, y) \in \mathscr{D}$ and x, y are arbitrary positive elements, then by putting $x_n = x \wedge (nu')$, $y_n = y \wedge (nu'')$, we have $(x_n, y_n) \in \mathscr{D}$ and the sequence $\{U(x_n, y_n)\}_{n\in\mathbb{N}}$ is bounded. Equality (6) obtains in this case too.

Finally, if $(x, y) \in \mathscr{D}$ and if x, y are arbitrary elements, then, by taking into account the fact that the elements (x_+, y_+), (x_+, y_-), (x_-, y_+), (x_-, y_-) belong to $\mathscr{D}$ and, by using the representation of U for positive elements, equality (6) obtains again.

Conversely, if F_1, F_2 are (*o*)-continuous, disjunctive, positive, linear operators (defined on $\mathscr{X}, \mathscr{Y}$ respectively and taking values in $\mathscr{Z}$), and if $\mathscr{D}$ is the set of those elements $(x, y) \in \mathscr{X} \times \mathscr{Y}$, such that the product $F_1(x)\ F_2(y)$ exists, then $\mathscr{D}$ is a normal domain in $\mathscr{X} \times \mathscr{Y}$ and it is easy to check that the operator U, given by formula (6), is of type (V).

Remark. If $\mathscr{X} = \mathscr{Y}$ and if U is a symmetric operator of type (V), defined on the set $\mathscr{D}$ of all elements $(x, y) \in \mathscr{X} \times \mathscr{Y}$ for which xy exists, then U can be represented in the form

$$U(x, y) = F(xy),$$

where F is an (*o*)-continuous, multiplicative, linear operator.

Bibliographical comments

The first results on linear operators on ordered vector spaces were obtained by L. V. Kantorovitch [3], [6], [7], [9], [15], [17].

Proposition 2 from Section 5.1.1 belongs to F. Papangelou [1].

The theorem on the extension of the linear positive operators (from Section 5.1.2) has been proved by L. V. Kantorovitch [10].

Theorem 1, from Section 5.1.3 (on the space of regular operators) has been proved for the case of functionals, by F. Riesz [2], and then generalized for operators, by L. V. Kantorovitch. Proposition 5 from the same section has been proved by R. Cristescu [14]. The theorem from Section 5.2.2, in which it is shown that $(\mathscr{X}, \mathscr{Y})_0^r$ is a component of $(\mathscr{X}, \mathscr{Y})^r$, was proved by A. G. Pinsker, under the assumption that $\mathscr{X}$ and $\mathscr{Y}$ are complete vector lattices (see L. V. Kantorovitch, B. Z. Vulich, A. G. Pinsker [1]). A proof which is valid also for the case of an arbitrary vector lattice $\mathscr{X}$, was given in the paper by R. Cristescu [15].

The (ω)-continuous regular operators have been studied by several authors. A. G. Pinsker [7] studied the linear operators which are continuous with respect to (o)-continuous transfinite sequences, and called them completely continuous operators. One can show that if $\mathscr{X}$ and $\mathscr{Y}$ are complete vector lattices and if U is a regular operator, continuous with respect to the (o)-convergent transfinite sequences, then U is (ω)-continuous.

If U is a regular (o)-continuous operator, defined on a σ-complete vector lattice $\mathscr{X}$, with a strong unit element, and if U takes values in a complete vector lattice $\mathscr{Y}$, then U can be represented as an integral.

More precisely, one can prove that

$$U(x) = \int_{-\infty}^{\infty} \lambda \mathrm{d} U_0 (g^x(\lambda)),$$

where U_0 is a function defined on $\mathscr{B}(\mathscr{X})$, taking values in $\mathscr{Y}$ and having the following properties: (i) $U_0(e_1 \vee e_2) = U_0(e_1) + U_0(e_2)$, if $e_1 \perp e_2$; (ii) the set $\{U_0(e);\ e \in \mathscr{B}(\mathscr{X})\}$ is bounded. The integral is defined as in Section 5.11, whereas $g^x(\lambda)$ has the meaning of the same section. This result belongs to B. Z. Vulich [11].

Let $\mathscr{X}$ be a complete vector lattice with unit, such that there exists an (o)-continuous linear functional f_0, for which $x > \mathbf{0}$ implies $f_0(x) > 0$. In the space $\mathscr{X}_o^*$ of the (o)-continuous linear functionals, f_0 can be taken as a unit. For any bounded element f of $\mathscr{X}_o^*$, there exists a unique bounded element $y \in \mathscr{X}$, such that $f(x) = f_0(xy)$, $(\forall x \in \mathscr{X})$. This result belongs to B. Z. Vulich (see L. V. Kantorovitch, B. Z. Vulich, A. G. Pinsker [1]). A similar representation for the (ω)-continuous linear functionals (with the use of the multiplication of the elements in the maximal extension of a complete vector lattice) was given by B. Z. Vulich and G. Ya. Lozanovski [1].

The results from Section 5.2.4, on the extension of the continuous, positive, linear operators, belong to L. V. Kantorovitch [15].

The notion of a normal functional (and more generally, that of a normal operator) was introduced by M. A. Ermolin [1].

The lemmas from Section 5.2.5 were given by A. I. Veksler [2]. Proposition 2 from the same section can be found (for the case of operators), in B. Z. Vulich's book [16]. The almost integral operators were studied by G. Ya. Lozanovski [2] and J. Synacke [1].

The results from Section 5.2.7 concerning functionals and operators on spaces of vector sequences belong to R. Cristescu [33, 36, 37]. For other results, one can see D. Serban [1].

The disjunctive operators, as well as the multiplicative operators, were introduced by B. Z. Vulich [5].

Proposition 3 from Section 5.3.1 belongs to J. Mařik [1]. Propositions 4 and 5 from the same section were given by B. Z. Vulich for the multiplicative operators and were presented by the author for the disjunctive operators.

Translators have been introduced by H. Nakano [9].

The results given in Section 5.4 belong to the author.

The regular bilinear and the (o)-continuous bilinear operators were studied by R. Cristescu [9], who generalized various results on the bilinear functionals, which were obtained by H. Nakano [16].

The biadditive and positive operators, defined on normal domains, were studied (under the name of product operators) by R. Cristescu [7]. The theorem on the representation of operators of type (V) from Section 5.4.2, belongs to the author.

6

Topological ordered vector spaces

In an ordered vector space one can consider various topologies determined by the order relation. But it is also convenient to consider the case in which in an ordered vector space a topology is given from the beginning. In this case one requires not only that the topology be compatible with the structure of vector space, but that it be also compatible with the order relation, in one sense or another.

6.1 Topological ordered vector spaces

6.1.1 The notion of topological ordered vector space

Let $\mathscr{X}$ be an ordered vector space. We shall denote by $\mathscr{X}_+$ the set of the positive elements of $\mathscr{X}$.

Definition 1. A subset E of $\mathscr{X}$ is said to be *full* if $a \leqslant b$ and $a, b \in E$ implies $[a, b] \in E$.

If A is any subset of $\mathscr{X}$, then the set

$$\mathrm{fl}(A) = \bigcup_{\substack{x,y\in A\\ x\leqslant y}} [x, y]$$

is the smallest full set which contains A.

Indeed, it is obvious that $A \subset \mathrm{fl}(A)$. Let $a, b \in \mathrm{fl}(A)$ such that $a \leqslant b$. Then there exist elements a_1, a_2, b_1, b_2, which belong to A, such that $a_1 \leqslant a \leqslant a_2$ and $b_1 \leqslant b \leqslant b_2$. It follows that $a_1 \leqslant a \leqslant b \leqslant b_2$ and, since $[a_1, b_2] \subset \mathrm{fl}(A)$, from $a \leqslant x \leqslant b$ we infer that $x \in \mathrm{fl}(A)$. This shows that the set $\mathrm{fl}(A)$ is full. If B is a full set and if $A \subset B$, then it is obvious that $\mathrm{fl}(A) \subset B$.

The set $\mathrm{fl}(A)$ is called the *full hull* of the set A.

Also the following equality holds

$$\mathrm{fl}(A) = (A + \mathscr{X}_+) \cap (A - \mathscr{X}_+). \tag{1}$$

Indeed, if $z \in \mathrm{fl}(A)$, then there exist $x, y \in A$, such that $x \leqslant z \leqslant y$. It follows that there exist $x_1, y_1 \geqslant 0$, such that $z = x + x_1 = y - y_1$, whence $z \in A + \mathscr{X}_+$ and

$z \in A - \mathscr{X}_+$. Conversely, if $z = x + x_1 = y - y_1$, where $x_1, y_1 \geqslant 0$ and $x, y \in A$, then $x \leqslant z \leqslant y$, hence $z \in \text{fl}(A)$.

From equality (1) it follows that if A is an equilibrated set (or a convex set), hen $\text{fl}(A)$ is equilibrated (or convex) too.

Definition 2. One calls a *topological ordered vector space* any ordered vector space $\mathscr{X}$, which, at the same time, is a topological[1]) vector space (with a topology τ) and which satisfies the following condition: there exists a fundamental system $\mathscr{W}$ of neighbourhoods of the origin, such that each $W \in \mathscr{W}$ have the property

$$0 \leqslant x \leqslant y \text{ and } y \in W \Rightarrow x \in W. \tag{2}$$

Remark. Condition (2) is equivalent to the following condition

$$A \subset \mathscr{X}_+ \text{ and } A \cap W = \varnothing \Rightarrow (A + \mathscr{X}_+) \cap W = \varnothing. \tag{3}$$

Indeed, let us assume that (2) is satisfied and let c be an element of $(A + \mathscr{X}_+) \cap W$. Consequently $c = a + b$, where $a \in A$ and $b \geqslant 0$, $c \in W$. From $0 \leqslant a \leqslant c$ and $c \in W$ it follows that $a \in W$. Consequently, $A \cap W \neq \varnothing$.

Conversely, let (3) be satisfied and let $0 \leqslant x_0 \leqslant y$, where $y \in W$. If $x_0 \notin W$, then from (3) it follows that

$$(x_0 + \mathscr{X}_+) \cap W = \varnothing.$$

But the element y can be written as $y = x_0 + a$, where $a \geqslant 0$ and this implies $y \notin W$, which is a contradiction.

Proposition 1. *Let $\mathscr{X}$ be an ordered vector space which, at the same time, is a topological vector space. In order that $\mathscr{X}$ be a topological ordered vector space it is necessary and sufficient that a fundamental system of full neighbourhoods of the origin should exist.*

Proof. The sufficiency of the condition is obvious. Let us establish its necessity.

Let E be an arbitrary neighbourhood of the origin. Let E_1 be a neighbourhood of the origin, such that $E_1 + E_1 \subset E$. Let W be a neighbourhood of the origin which satisfies condition (2) and such that $W \subset E_1$. Let W_1 be an equilibrated neighbourhood of the origin, such that $W_1 + W_1 \subset W$. Let us define $V = \text{fl}(W_1)$. The set V is a full neighbourhood of the origin. If $x \in V$, then there exist $a, b \in W_1$, such that $a \leqslant x \leqslant b$. It follows that

$$0 \leqslant x - a \leqslant b - a \in W_1 + W_1 \subset W \subset E_1$$

and, therefore, $x - a \in E_1$. We then have

$$x \in a + E_1 \subset W_1 + E_1 \subset W + E_1 \subset E_1 + E_1 \subset E,$$

and this implies $V \subset E$.

---

[1]) In this chapter by a topological vector space we shall mean a Hausdorff topological vector space.

The proposition is proved.

A topology τ in an ordered vector space $\mathscr{X}$, for which a fundamental system of full neighbourhoods of the origin exists, is called a *locally full topology*.

In an ordered vector space $\mathscr{X}$, in which a topology τ is also given, we shall say that a set, which is bounded with respect to the order relation, is (o)-bounded, whereas a set, which is bounded with respect to the topology τ, will be said to be (τ)-bounded. We shall denote by $x_\delta \xrightarrow{\tau} x$ or $x = (\tau)\text{-}\lim x_\delta$ the convergence with respect to the topology τ. The phrases "(τ)-closed set", "(τ)-continuous function", etc., have an obvious meaning.

Proposition 2. *If $\{x_\delta\}_{\delta\in\Delta}$, $\{a_\delta\}_{\delta\in\Delta}$ and $\{b_\delta\}_{\delta\in\Delta}$ are generalized sequences of elements of a topological ordered vector space, such that $a_\delta \leqslant x_\delta \leqslant b_\delta$ for any $\delta\in\Delta$, and if $a_\delta \xrightarrow{\tau} 0$ and $b_\delta \xrightarrow{\tau} 0$, then $x_\delta \xrightarrow{\tau} 0$.*

Indeed, if W is a full neighbourhood of the origin, then there exists a $\delta_w \in \Delta$, such that a_δ, $b_\delta \in W$ for any $\delta \geqslant \delta_w$. From $a_\delta \leqslant x_\delta \leqslant b_\delta$ it follows that $x_\delta \in W$ for any $\delta \geqslant \delta_w$.

Proposition 3. *If $\{a_j\}_{j\in J}$, $\{b_j\}_{j\in J}$ are (τ)-bounded systems of elements of a topological ordered vector space, such that $a_j \leqslant b_j$, for any $j \in J$, then the set*

$$E = \bigcup_{j\in J} [a_j, b_j]$$

is (τ)-bounded.

Indeed, let $\{x_n\}_{n\in\mathbb{N}}$ be a sequence of elements of E and $\{\lambda_n\}_{n\in\mathbb{N}}$ a sequence of positive numbers, which converges to zero. If $a_{j_n} \leqslant x_n \leqslant b_{j_n}$, then

$$\lambda_n a_{j_n} \leqslant \lambda_n x_n \leqslant \lambda_n b_{j_n}$$

and since $\lambda_n a_{j_n} \xrightarrow{\tau} \mathbf{0}$, $\lambda_n b_{j_n} \xrightarrow{\tau} \mathbf{0}$, from proposition 2 it follows that $\lambda_n x_n \xrightarrow{\tau} \mathbf{0}$. This shows that the set E is (τ)-bounded.

In particular, in a topological ordered vector space, any (o)-bounded set is (τ)-bounded.

Definition 3. Let $\mathscr{X}$ be an ordered vector space, which, at the same time, is a topological vector space with the topology τ. Let $\mathscr{V}$ be the set of all the neighbourhoods of the origin for the topology τ. The topology in $\mathscr{X}$, which has the system $\mathscr{W} = \{\text{fl}(V);\ V\in\mathscr{V}\}$ as a fundamental system of neighbourhoods of the origin, is called *locally full topology associated to* τ, and is denoted by $\text{fl}(\tau)$.

Remark. One can show that $\text{fl}(\tau) \leqslant \tau$, $\text{fl}(\text{fl}(\tau)) = \text{fl}(\tau)$ and that from $\tau_1 \leqslant \tau_2$ it follows that $\text{fl}(\tau_1) \leqslant \text{fl}(\tau_2)$.

Proposition 4. *Let $\mathscr{X}$ be an ordered vector space which is also a topological vector space (with the topology τ). The closure of the set $\mathscr{X}_+$ in the topology τ coincides with the closure of $\mathscr{X}_+$ in the topology* $\text{fl}(\tau)$.

Proof. If V is a neighbourhood of the origin for the topology τ, then

$$\mathscr{X}_+ + V \subset \mathscr{X}_+ + \text{fl}(V) \subset \mathscr{X}_+ + (\mathscr{X}_+ + V) = \mathscr{X}_+ + V;$$

hence

$$\mathscr{X}_+ + V = \mathscr{X}_+ + \mathrm{fl}(V). \tag{4}$$

If $\mathscr{V}$ is the set of all neighbourhoods of the origin for the topology τ, then from equality (4) it follows that

$$\bigcap_{V \in \mathscr{V}} (\mathscr{X}_+ + V) = \bigcap_{V \in \mathscr{V}} (\mathscr{X}_+ + \mathrm{fl}(V)). \tag{5}$$

One has only to remark that the left-hand member of equality (5) is the closure of $\mathscr{X}_+$ in the topology τ, whereas the right-hand member is the closure of $\mathscr{X}_+$ in the topology $\mathrm{fl}(\tau)$.

Proposition 5. *Let $\mathscr{X}$ be an ordered vector space which is also a topological vector space (with the topology τ). If the set $\mathscr{X}_+$ is (τ)-closed, then the following hold:*

(i) *if $\{x_\delta\}_{\delta \in \Delta}$ and $\{y_\delta\}_{\delta \in \Delta}$ are generalized sequences of elements of $\mathscr{X}$, such that $x_\delta \leqslant y_\delta$ $(\forall \delta \in \Delta)$ and if $x_\delta \xrightarrow{\tau} x$, $y_\delta \xrightarrow{\tau} y$, then $x \leqslant y$;*

(ii) *if $\{x_\delta\}_{\delta \in \Delta}$ is a generalized sequence of elements of $\mathscr{X}$ and if $x_\delta \uparrow$ (or $x_\delta \downarrow$), and $x_\delta \xrightarrow{\tau} x$, then $x_\delta \uparrow x$ ($x_\delta \downarrow x$, respectively).*

Proof. (i) If $x_\delta \leqslant y_\delta$ $(\forall \delta \in \Delta)$, then

$$y_\delta - x_\delta \in \mathscr{X}_+, \quad (\forall \delta \in \Delta),$$

whence $y - x \in \mathscr{X}_+$, i.e., $x \leqslant y$.

(ii) Let $x_\delta \uparrow$ and $x_\delta \xrightarrow{\tau} x$. Let δ be any index from Δ. If $\delta' \geqslant \delta$, then $x_{\delta'} \geqslant x_\delta$. By tending to the limit with respect to the set of indices $\Delta' = \{\delta'; \delta' \geqslant \delta\}$, it follows that $x \geqslant x_\delta$. On the other hand, if $z \geqslant x_\delta$ $(\forall \delta \in \Delta)$, then, by tending to the (τ)-limit with respect to $\delta \in \Delta$, we get $z \geqslant x$. It follows that $x = \bigvee_{\delta \in \Delta} x_\delta$. The case $x_\delta \downarrow$ can be treated similarly.

Corollary. *Let $\mathscr{X}$ be a directed vector space and τ a topology in $\mathscr{X}$, compatible with the structure of vector space. If the set $\mathscr{X}_+$ is (τ)-closed, then $\mathscr{X}$ is an Archimedean space.*

Indeed, if $x > \mathbf{0}$, then $\frac{1}{n} x \downarrow_{n \in \mathbb{N}}$ and $\frac{1}{n} x \xrightarrow{\tau} \mathbf{0}$, hence $\frac{1}{n} x \downarrow_{n \in \mathbb{N}} \mathbf{0}$.

Definition 4. Two topological ordered vector spaces $\mathscr{X}$, $\mathscr{Y}$ are *isomorphic* if there exists a one-to-one mapping f of $\mathscr{X}$ onto $\mathscr{Y}$, with the following properties:

(i) $f(\alpha x_1 + \beta x_2) = \alpha f(x_1) + \beta f(x_2)$, for any $x_1, x_2 \in \mathscr{X}$ and $\alpha, \beta \in \mathbb{R}$;

(ii) $x_1 \leqslant x_2$ $(x_1, x_2 \in \mathscr{X}) \Leftrightarrow f(x_1) \leqslant f(x_2)$;

(iii) f is bicontinuous with respect to the topologies considered in $\mathscr{X}$ and in $\mathscr{Y}$.

6.1.2 Locally convex ordered spaces

Definition 1. One calls a *locally convex ordered space* any topological ordered vector space in which the topology is locally *convex* [2]).

Proposition 1. *In any locally convex ordered space there exists a fundamental system $\mathscr{W}$ of neighbourhoods of the origin, consisting of equilibrated, convex and full sets.*

Proof. If V is any neighbourhood of the origin, then there exists a full neighbourhood $V_1 \subset V$ of the origin. There exists also an equilibrated and convex neighbourhood V_2 (of the origin), such that $V_2 \subset V_1$. By putting $W = \mathrm{fl}(V_2)$, we get a neighbourhood of the origin, which is an equilibrated, convex and full set, as one can easily check. But $W \subset V_1$, because V_1 is full.

It follows that $W \subset V$ and the proposition is proved.

Definition 2. A semi-norm p, defined on an ordered vector space $\mathscr{X}$, is said to be a *monotonous semi-norm*, if $0 \leqslant x \leqslant y$ $(x, y \in \mathscr{X})$ implies $p(x) \leqslant p(y)$.

Proposition 2. *Let $\mathscr{X}$ be an ordered vector space and τ a locally convex topology in $\mathscr{X}$. In order that $\mathscr{X}$ be a locally convex ordered space it is necessary and sufficient that the topology τ be defined by a family of monotonous semi-norms.*

Proof. If $\mathscr{X}$ is a locally convex ordered space, then there exists a fundamental system $\mathscr{W}$ of neighbourhoods of the origin, consisting of equilibrated, convex and full sets. For any $W \in \mathscr{W}$, let us consider the corresponding Minkowski semi-norm p_W. If $0 \leqslant x \leqslant y$ and if $y \in \mu W$, where $\mu > 0$, then $x \in \mu W$, whence $p_W(x) \leqslant p_W(y)$.

The sufficiency of the condition follows from the fact that if p is a monotonous semi-norm on $\mathscr{X}$, then, for any number $\varepsilon > 0$, the set

$$W(p; \varepsilon) = \{x; p(x) < \varepsilon\}$$

has property (2) from Section 6.1.1.

In the next proposition, by $x_\delta \xrightarrow{w} x$ we shall mean the weak convergence, in the weakened topology associated to a locally convex topology τ.

Proposition 3. *Let $\mathscr{X}$ be a locally convex ordered space (with the topology τ) and $\{x_\delta\}_{\delta \in \Delta}$ a generalized sequence of elements of $\mathscr{X}$. If $0 \leqslant x_\delta \downarrow$ and $x_\delta \xrightarrow{w} 0$, then $x_\delta \xrightarrow{\tau} 0$.*

Proof. As one knows (Section 2.4.4, remark 2) in any Hausdorff locally convex space, the (τ)-closure $\bar{A}$ of a convex set A coincides with its closure in the weakened topology.

Let

$$A = \mathrm{co}\,\{x_\delta; \delta \in \Delta\}.$$

---

[2]) We recall that the topology is assumed to be Hausdorff.

From $x_\delta \xrightarrow{w} 0$ we infer that $0 \in \bar{A}$. Let W be a full neighbourhood of the origin. There exists an element $z \in A \cap W$ and this element can be written as

$$z = \sum_{i=1}^{m} \alpha_i x_{\delta_i}$$

where $\alpha_i > 0$ and $\Sigma\alpha_i = 1$. Let $\delta_0 \in \Delta$ be such that $\delta_0 \geqslant \delta_i$ $(i = 1, 2, \ldots, m)$. If $\delta \geqslant \delta_0$, then

$$0 \leqslant x_\delta = \sum_{i=1}^{m} \alpha_i x_\delta \leqslant \sum_{i=1}^{m} \alpha_i x_{\delta_i} = z;$$

since $z \in W$ and W is full, we infer that $x_\delta \in W$. The proposition is proved.

Proposition 4. *If $\{\mathscr{X}_j\}_{j \in J}$ is a family of locally convex ordered spaces, then the product ordered vector space $\mathscr{X}$, endowed with the product topology, is a locally convex ordered space.*

Proof. As one knows, if $\mathscr{P}_j$ is a family of semi-norms on $\mathscr{X}_j$ defining the topology τ_j of $\mathscr{X}_j$, and if for any $p_j \in \mathscr{P}_j$ and any element $x = \{x_j\}_{j \in J}$ we define

$$q_j(x) = p_j(x_j),$$

then the set of all semi-norms q_j defines the topology τ of $\mathscr{X}$. If the seminorms p_j are monotonous, then the semi-norms q_j are also monotonous, as one can immediately show if one takes into account the definition of the order relation in $\mathscr{X}$. It follows that the topology τ is locally full.

Examples (1) Let T be a locally compact space and $C_0(T)$ the vector space of all continuous real functions, defined on T and having compact supports.

Let S be an arbitrary compact subset of T and let $C_0(T; S)$ be the subspace of $C_0(T)$ consisting of all functions whose supports are contained in S. We organize $C_0(T; S)$ as a Banach space whose topology τ_S is given by the norm

$$\|x\|_S = \max_{t \in S} |x(t)|.$$

Let τ be the inductive limit of the topologies τ_S. With the usual order relation (i.e., $x \leqslant y \Leftrightarrow x(t) \leqslant y(t), \forall t \in T$) and with the topology τ, the vector space $C_0(T)$ becomes a locally convex ordered space.

(2) Let $\mathscr{D}(\mathbb{R})$ be the Schwartz space of the indefinitely differentiable real functions, with compact supports [3]). With the order relation $x \leqslant y \Leftrightarrow x(t) \leqslant y(t)$, $\forall t \in \mathbb{R}$, the vector space $\mathscr{D}(\mathbb{R})$ is an ordered vector space, but the usual topology in Schwartz' space cannot be defined by a family of monotonous semi-norms. It follows that the topology of $\mathscr{D}(\mathbb{R})$ is not locally full.

---

[3]) See R. Cristescu [31].

6.1.3 Conjugated spaces

If $\mathscr{X}$ is an ordered vector space, we shall denote by $\mathscr{X}^{\#}$ the set off all (o)-bounded linear functionals defined on $\mathscr{X}$ and by $\mathscr{X}^r$ the set of all regular [4]) linear functionals defined on $\mathscr{X}$, whereas $\mathscr{X}_o^*$ will stand for the set of all (o)-continuous linear functionals defined on $\mathscr{X}$.

If $\mathscr{X}$ is a topological vector space (with the topology τ), we shall denote by $\mathscr{X}_\tau^*$ the set of all (τ)-continuous linear functionals defined on $\mathscr{X}$.

Proposition 1. *Let $\mathscr{X}$ be an ordered vector space and τ a (Hausdorff) locally convex topology in $\mathscr{X}$. If $\mathscr{X}_\tau^* \subset \mathscr{X}^{\#}$, then any ($o$)-bounded subset of $\mathscr{X}$ is(τ)-bounded. Conversely, if $\mathscr{X}$ is, moreover, a directed space and if any (o)-bounded set of positive elements of $\mathscr{X}$ is (τ)-bounded, then $\mathscr{X}_\tau^* \subset \mathscr{X}^{\#}$.*

Proof. Let us assume that $\mathscr{X}_\tau^* \subset \mathscr{X}^{\#}$. Let A be an (o)-bounded subset of $\mathscr{X}$. If $f \in \mathscr{X}_\tau^*$, then $f(A)$ is bounded, because f is an (o)-bounded functional. It follows that the set A is bounded for the weakened topology of $\mathscr{X}$ and, therefore, it is (τ)-bounded (Section 2.4.4, proposition 2).

Conversely, let us assume that $\mathscr{X}$ is moreover, a directed space and let $\mathscr{X}_\tau^* \not\subset \mathscr{X}^{\#}$. Let $f \in \mathscr{X}_\tau^*$ be such that $f \notin \mathscr{X}^{\#}$. There then exists an ($o$)-bounded set $A \subset \mathscr{X}$, such that $f(A)$ be not bounded. This implies that there exists a sequence $\{x_n\}_{n\in\mathbb{N}}$ of elements of A, such that the sequence $\{f(x_n)\}_{n\in\mathbb{N}}$ be not bounded. Let $a \in \mathscr{X}$ $(a \geqslant \mathbf{0})$ be such that $A \subset [-a, a]$. The sequence $\{f(a - x_n)\}_{n\in\mathbb{N}}$ is not bounded and, consequently, the sequence $\{a - x_n\}_{n\in\mathbb{N}}$ is not (τ)-bounded, because f is linear and (τ)-continuous. But

$$\mathbf{0} \leqslant a - x_n \leqslant 2a,$$

and this implies that $\{a - x_n\}_{n\in\mathbb{N}}$ is an (o)-bounded sequence of positive elements. The proposition is proved.

Lemma. *If τ is a locally convex topology in an ordered vector space, then for any element x_0, which does not belong to the (τ)-closure of the set $\mathscr{X}_+$, there exists a (τ)-continuous, positive, linear functional f, such that $f(x_0) < 0$.*

Proof. There exists a closed hyperplane which strictly separates x_0 from the (τ)-closure $\overline{\mathscr{X}}_+$ of $\mathscr{X}_+$ (Section 2.4.1, proposition 3). Then there exists a (τ)-continuous linear functional f on $\mathscr{X}$, such that

$$f(x_0) < \inf_{x \geqslant 0} f(x). \tag{1}$$

If $x \geqslant \mathbf{0}$, then for any number $\alpha > 0$ we have $\alpha x \geqslant 0$, consequently

$$f(x_0) < f(\alpha x) = \alpha f(x),$$

---

[4]) If $\mathscr{X}$ is a vector lattice, then $\mathscr{X}^{\#} = \mathscr{X}^r$.

whence $f(x) \geqslant 0$, i.e., f is a positive functional. From inequality (1) we then infer that $f(x_0) < 0$.

Proposition 2. *Let τ be a locally convex topology in the ordered vector space $\mathscr{X}$. In order that the set $\mathscr{X}_+$ be (τ)-closed, it is necessary and sufficient that the following condition be satisfied: if $f(x_0) \geqslant 0$, for any (τ)-continuous, positive, linear functional f on $\mathscr{X}$, then $x_0 \geqslant \mathbf{0}$.*

Proof. If the set $\mathscr{X}_+$ is (τ)-closed and if $f(x_0) \geqslant 0$ for any f, such that $0 \leqslant f \in \mathscr{X}_\tau^*$, then $x_0 \geqslant \mathbf{0}$, as one infers from the lemma.

Conversely, if the condition from the statement of the proposition is satisfied, then

$$\mathscr{X}_+ = \bigcap_{0 \leqslant f \in \mathscr{X}_\tau^*} \{x;\ f(x) \geqslant 0\}, \tag{2}$$

and this shows that $\mathscr{X}_+$ is (τ)-closed.

Remarks. (1) Let τ be a locally convex topology in the ordered vector space $\mathscr{X}$, such that the set $\mathscr{X}_+$ be (τ)-closed. If $x_n \xrightarrow{w} x$, $y_n \xrightarrow{w} y$ and $x_n \leqslant y_n$ ($\forall n \in \mathbb{N}$), then $x \leqslant y$.

This is an immediate corollary of proposition 2. It also follows from the fact that the weak closure of $\mathscr{X}_+$ coincides with the (τ)-closure of $\mathscr{X}_+$.

(2) If $\mathscr{X}$ is an ordered vector space and if τ is a locally convex topology in $\mathscr{X}$, such that $\mathscr{X}_+$ be (τ)-closed, then for any element $x_0 > 0$ there exists a positive functional $f \in \mathscr{X}_\tau^*$, such that $f(x_0) > 0$.

Indeed, since $-x_0 \notin \mathscr{X}_+ = \overline{\mathscr{X}}_+$, there exists (by virtue of the lemma) a positive functional $f_0 \in \mathscr{X}_\tau^*$, such that $f_0(-x) \neq 0$; consequently $f_0(x_0) \neq 0$.

Theorem. *If $\mathscr{X}$ is a locally convex ordered space, then any (τ)-continuous linear functional f on $\mathscr{X}$ can be represented as a difference of two (τ)-continuous, positive, linear functionals. In particular, $\mathscr{X}_\tau^* \subset \mathscr{X}^r$.*

Proof. Let $f \in \mathscr{X}_\tau^*$. In order to prove the theorem, we shall need to show that f is majorized by a (τ)-continuous, positive, linear functional. We can assume, from the beginning, that there exists an element $x_0 > \mathbf{0}$ such that $f(x_0) > 0$, since, otherwise, we would infer that $f \leqslant \mathbf{0}$.

Let x be any positive element of $\mathscr{X}$. Since $\mathscr{X}$ is a locally convex ordered space, the interval $[0, x]$ is a (τ)-bounded set and, therefore, the set $\{f(x');\ 0 \leqslant x' \leqslant x\}$ is bounded. Let us define on $\mathscr{X}_+$ the function g given by the formula

$$g(x) = \sup_{0 \leqslant x' \leqslant x} f(x').$$

It is obvious that $g(x) \geqslant 0$, for any $x \in \mathscr{X}_+$. It is easy to show that

$$g(x + y) \geqslant g(x) + g(y),$$

for any x, $y \in \mathscr{X}_+$ and also that

$$g(\alpha x) = \alpha\, g(x),$$

for any $x \in \mathscr{X}_+$ and any number $\alpha \geqslant 0$. The function g is (τ)-continuous at the origin. Indeed, otherwise, we could find a generalized sequence $\{x_\delta\}_{\delta \in \Delta_0}$ of elements of $\mathscr{X}_+$, which is (τ)-convergent to 0, and a number $\varepsilon_0 > 0$, such that

$$g(x_\delta) \geqslant \varepsilon_0, \qquad \delta \in \Delta_0.$$

For any index δ there exists an element $y_\delta \in \mathscr{X}$, which satisfies the following conditions

$$0 \leqslant y_\delta \leqslant x_\delta, \tag{3}$$

$$f(y_\delta) \geqslant \frac{\varepsilon_0}{2}, \qquad \delta \in \Delta_0. \tag{4}$$

But $x_\delta \xrightarrow{\tau} 0$ and from inequalities (3) if follows that $y_\delta \xrightarrow{\tau} 0$, which contradicts (4).

Let us now consider the set

$$G = \{x \geqslant \mathbf{0}; \quad g(x) \geqslant 1\}.$$

This set is convex and $\mathbf{0} \notin \overline{G}$. Consequently, there exists a (τ)-continuous, linear functional h, defined on $\mathscr{X}$, such that $h(x) \geqslant 1$, if $x \in G$ (Section 2.4.1, proposition 3).

For any element $x \geqslant \mathbf{0}$ we have $h(x) \geqslant g(x)$. Indeed, if $x > \mathbf{0}$ and if $g(x) \neq 0$, then $\dfrac{1}{g(x)}\, x \in G$ and, therefore,

$$h\left(\frac{1}{g(x)}\, x\right) \geqslant 1,$$

hence $h(x) \geqslant g(x)$.

If $x \geqslant \mathbf{0}$ and $g(x) = \mathbf{0}$, then take an element $x_0 \in G$ and let us observe that

$$g(nx + x_0) \geqslant g(nx) + g(x_0) = ng(x) + g(x_0) = g(x_0) \geqslant 1,$$

i.e., $nx + x_0 \in G$. It follows that $h(nx + x_0) \geqslant 1$, which implies

$$h(x) \geqslant \frac{1 - h(x_0)}{n}$$

for any $n \in \mathbb{N}$. It follows that $h(x) \geqslant 0$, hence $h(x) \geqslant g(x)$ in this case too. In particular, h is a positive functional. From the definition of g we infer that $h \geqslant f$. The theorem is proved.

Definition. A topology τ in an ordered vector space $\mathscr{X}$ is said to be *(o)-continuous* if $x_n \underset{n \in \mathbb{N}}{\downarrow} 0$ implies $x_n \xrightarrow{\tau} \mathbf{0}$.

Proposition 3. *If $\mathscr{X}$ is a locally convex ordered space (with the topology τ), then the inclusion $\mathscr{X}_\tau^* \subset \mathscr{X}_o^*$ is equivalent to the (o)-continuity of τ.*

Proof. Assume that the topology τ is (o)-continuous. If $f \in \mathscr{X}_\tau^*$, then f is regular (by virtue of the preceding theorem). If $0 \leqslant f \in \mathscr{X}_\tau^*$ and if $x_n \downarrow 0$, then $x_n \xrightarrow{\tau} 0$ and, therefore, $f(x_n) \to 0$. It follows that $f \in \mathscr{X}_o^*$ and, consequently, $\mathscr{X}_\tau^* \subset \mathscr{X}_o^*$.

Conversely, let $\mathscr{X}_\tau^* \subset \mathscr{X}_o^*$. If $x_n \downarrow 0$, then $f(x_n) \to 0$, for any $f \in \mathscr{X}_\tau^*$. With proposition 3 from Section 6.1.2 we infer that $x_n \xrightarrow{\tau} 0$ and, therefore, τ is (o)-continuous.

6.2 Topological vector lattices

6.2.1 The notion of topological vector lattice

Let $\mathscr{X}$ be a vector lattice. For any element $x \in \mathscr{X}$, the set

$$Z(x) = \{y;\ |y| \leqslant |x|\}$$

is equilibrated and convex.

A set E of elements of $\mathscr{X}$ is said to be solid (Section 3.3.1, definition 2), if $x \in E$ implies $Z(x) \subset E$.

Any solid set is equilibrated.

Let A be any subset of $\mathscr{X}$ and let us define

$$\mathrm{so}(A) = \bigcup_{x \in A} Z(x).$$

The set so(A) is the smallest solid set which contains the set A. It is called the *solid hull* of A.

It is obvious that if $A \subset B$, then so(A) $\subset$ so(B).

Definition. One calls a *topological vector lattice* any vector lattice $\mathscr{X}$ which, at the same time, is a topological vector space [5]) (with a given topology τ) in which there exists a fundamental system W of solid neighbourhoods of the origin.

A topology τ in a vector lattice $\mathscr{X}$, for which there exists a fundamental system of solid neighbourhoods of the origin, is called a *locally solid topology*. Any topological vector lattice is a topological ordered vector space.

Two topological vector lattices are said to be *isomorphic*, if they are isomorphic as topological ordered vector spaces (see Section 6.1.1, definition 4).

---

[5]) We recall that the space is assumed to be separated (Hausdorff).

Proposition 1. *If A is a (τ)-bounded subset of a topological vector lattice, then the set so(A) is also (τ)-bounded.*

Proof. Let W be a solid neighbourhood of the origin. Since the set A is (τ)-bounded, there exists a number $\lambda > 0$, such that $A \subset \lambda W$. If $x \in A$ then $x \in \lambda W$ and, since W is solid, it follows that $Z(x) \subset \lambda W$. Hence we further infer that so $(A) \subset \subset \lambda W$ and, therefore, the set so(A) is (τ)-bounded.

Corollary. *The set $\mathcal{G}$ of all solid and (τ)-bounded subsets of a topological vector lattice $\mathcal{X}$ is a fundamental system of (τ)-bounded sets.*

Proposition 2. *Let $\{x_\delta\}_{\delta\in\Delta}$ and $\{y_\delta\}_{\delta\in\Delta}$ be generalized sequences of elements of a topological vector lattice. Then the following properties hold.*

(i) *if $|x_\delta| \leqslant |y_\delta|$, $\forall \delta \in \Delta$, and if $y_\delta \xrightarrow{\tau} 0$, then $x_\delta \xrightarrow{\tau} 0$;*

(ii) *if $x_\delta \xrightarrow{\tau} x$ and $y_\delta \xrightarrow{\tau} y$, then $x_\delta \vee y_\delta \xrightarrow{\tau} x \vee y$ and $x_\delta \wedge y_\delta \xrightarrow{\tau} x \wedge y$.*

Proof. (i) This immediately follows from the existence of a fundamental system of solid neighbourhoods of the origin.

(ii) The following inequalities hold

$$|(x_\delta \vee y_\delta) - (x \vee y)| \leqslant |x_\delta - x| + |y_\delta - y|,$$

$$|(x_\delta \wedge y_\delta) - (x \wedge y)| \leqslant |x_\delta - x| + |y_\delta - y|$$

and we have only to take into account property (i).

Corollary 1. *In a topological vector lattice $\mathcal{X}$, the set $\mathcal{X}_+$ of its positive elements is (τ)-closed.*

This immediately follows from (ii).

Corollary 2. *Any topological vector lattice is an Archimedean space.*

This immediately follows from corollary 1 (see Section 6.1.1, proposition 5, corollary).

Proposition 3. *Let $\mathcal{X}$ be a vector lattice which is, at the same time, a topological vector space with respect to a topology τ. In order that $\mathcal{X}$ be a topological vector lattice, it is necessary and sufficient that there should exist a fundamental system of full neighbourhoods of the origin and that the mapping $x \to x_+$ be (τ)-continuous.*

Proof. The necessity follows from the definition of the topological vector lattice and from proposition 2.

Let us now prove the sufficiency.

Let V be an arbitrary neighbourhood of the origin. By hypothesis, there exists a full neighbourhood $W_0 \subset V$ of the origin. There also exists an equilibrated neighbourhood of the origin $W_1 \subset W_0$. If we denote $V_1 = \mathrm{fl}(W_1)$, then V_1 is equilibrated, full and $V_1 \subset V$.

Let V_2 be a neighbourhood of the origin, such that

$$V_2 + V_2 \subset V_1.$$

Since the mapping $x \mapsto x_+$ is (τ)-continuous, there exists an equilibrated neighbourhood V_3 of the origin, such that

$$x \in V_3 \Rightarrow x_+ \in V_2. \tag{1}$$

Let us now state that

$$x \in V_3 \Rightarrow Z(x) \subset V_1. \tag{2}$$

Indeed, since V_3 is equilibrated, from $x \in V_3$ it follows that $-x \in V_3$. From implication (1) it follows that, on the one hand, we have $x_+ \in V_2$, and on the other

$$x_- = (-x)_+ \in V_2.$$

It follows that

$$|x| = x_+ + x_- \in V_2 + V_2 \subset V_1,$$

i.e.,

$$x \in V_3 \Rightarrow |x| \in V_1.$$

If $y \in Z(x)$, then

$$-|x| \leqslant y \leqslant |x|,$$

and, since V_1 is full and $|x| \in V_1$, $-|x| \in V_1$, it follows that $y \in V_1$. Therefore, (2) holds.

Let us now define $W = \mathrm{so}(V_3)$. Then W is a solid neighbourhood of the origin and $W \subset V$ (by virtue of implication (2)).

The proposition is proved.

Remark. The continuity of the mapping $x \mapsto x_+$ is equivalent to the continuity of the mappings $(x, y) \mapsto x \vee y$, $(x, y) \mapsto x \wedge y$.

6.2.2 Locally convex vector lattices

One calls a *locally convex vector lattice* any topological vector lattice in which the considered topology is locally convex.

Proposition 1. *In a locally convex vector lattice there exists a fundamental system $\mathscr{W}$ of neighbourhoods of the origin, consisting of convex, solid sets.*

Proof. Let V be an arbitrary neighbourhood of the origin. There exists a convex neighbourhood $V_1 \subset V$ of the origin. There also exists a solid neighbourhood $V_2 \subset V_1$. Let us define

$$W = \{x; Z(x) \subset V_1\}.$$

The set W is solid. Indeed, if $x \in W$ and $|y| \leqslant |x|$, then $Z(x) \subset V_1$ and $Z(y) \subset Z(x)$; consequently, $Z(y) \subset V_1$ and, therefore, $y \in W$.

The set W is convex, because $x, y \in W$ and $0 \leqslant \alpha \leqslant 1$ imply that

$$Z(\alpha x + (1-\alpha) y) \subset \alpha Z(x) + (1-\alpha) Z(y) \subset \alpha V_1 + (1-\alpha) V_1 \subset V_1$$

and this implies that $\alpha x + (1-\alpha) y \in W$.

Finally, from $V_2 \subset W$ it follows that W is a neighbourhood of the origin and we obviously have $W \subset V$.

Definition. A semi-norm p, defined on a vector lattice $\mathscr{X}$, is said to be a *solid semi-norm* if $|x| \leqslant |y|$, $(x, y \in \mathscr{X})$ implies $p(x) \leqslant p(y)$.

Proposition 2. *Let $\mathscr{X}$ be a vector lattice and τ a locally convex (Hausdorff) topology in $\mathscr{X}$. Then $\mathscr{X}$ is a locally convex vector lattice iff the topology τ is determined by a set of solid semi-norms,*

The proof is similar to that given for the case of the ordered locally convex spaces (Section 6.1.2, proposition 2).

Proposition 3. *Let $\mathscr{X}$ be a vector lattice and τ a (Hausdorff) locally convex topology in $\mathscr{X}$. Then $\mathscr{X}$ is a locally convex vector lattice iff the following condition is satisfied: if $\{x_\delta\}_{\delta \in \Delta}$, $\{y_\delta\}_{\delta \in \Delta}$ are generalized sequences of elements of $\mathscr{X}$, such that $|x_\delta| \leqslant |y_\delta|$, $(\forall \delta \in \Delta)$ and if $y_\delta \xrightarrow{\tau} 0$, then $x_\delta \xrightarrow{\tau} 0$.*

Proof. The necessity of the condition is contained in proposition 2 from Section 6.2.1.

Let us prove the sufficiency.

Let $\mathscr{P}$ be a set of semi-norms which determines the topology τ of the space $\mathscr{X}$. For any $p \in \mathscr{P}$ (and any $x \in \mathscr{X}$) the set $\{p(z);\ 0 \leqslant z \leqslant |x|\}$ is bounded [6]).

Let us define

$$p'(x) = \sup_{0 \leqslant z \leqslant |x|} p(z). \tag{1}$$

We have

$$p'(x+y) = \sup_{0 \leqslant z \leqslant |x+y|} p(z) \leqslant \sup_{0 \leqslant z \leqslant |x|+|y|} p(z)$$

$$= \sup_{\substack{0 \leqslant x' \leqslant |x| \\ 0 \leqslant y' \leqslant |y|}} p(x'+y') \leqslant \sup_{0 \leqslant x' \leqslant |x|} p(x') + \sup_{0 \leqslant y' \leqslant |y|} p(y') = p'(x) + p'(y).$$

If $\lambda \neq 0$, then

$$p'(\lambda x) = \sup_{0 \leqslant z \leqslant |\lambda x|} p(z) = \sup_{0 \leqslant x' \leqslant |x|} p(|\lambda| x') = |\lambda| \sup_{0 \leqslant x' \leqslant |x|} p(x') = |\lambda| p'(x)$$

It is also obvious that $p'(\mathbf{0}) = 0$. Consequently, p' is a semi-norm on the space $\mathscr{X}$. It is obvious that p' is a solid semi-norm.

---

[6]) For any element $a \geqslant 0$, the interval $[\mathbf{0}, a]$ is (τ)-bounded. Indeed, if $\mathbf{0} \leqslant x_n \leqslant a$ and $0 \leqslant \lambda_n \to 0$ $(\lambda_n \in \mathbb{R})$ then $\mathbf{0} \leqslant \lambda_n x_n \leqslant \lambda_n a$ and $\lambda_n a \xrightarrow{\tau} \mathbf{0}$. With the condition from the statement of the proposition, it follows that $\lambda_n x_n \xrightarrow{\tau} \mathbf{0}$.

Let us denote by $\mathscr{P}'$ the set of all semi-norms of the form (1), where $p \in \mathscr{P}$. Let τ' be the topology determined by $\mathscr{P}'$ and let us show that $\tau' = \tau$.

Let us first observe that

$$p(x) = p(x_+ - x_-) \leqslant p(x_+) + p(x_-) \leqslant p'(x) + p'(x) = 2p'(x),$$

and, consequently, for any $p \in \mathscr{P}$ there exists a $p' \in \mathscr{P}'$, such that

$$p(x) \leqslant 2p'(x), \ \forall x \in \mathscr{X}.$$

It follows that $\tau \leqslant \tau'$. It is, therefore, sufficient to show that any semi-norm $p' \in \mathscr{P}'$ is (τ)-continuous (because from this result we then infer that $\tau' \leqslant \tau$).

Let us assume that p' is not (τ)-continuous. Then there exists a generalized sequence $\{x_\delta\}_{\delta \in \Delta_0}$ of elements of $\mathscr{X}$, and a number $\varepsilon_0 > 0$ such that $x_\delta \xrightarrow{\tau} 0$, but $p'(x_\delta) \geqslant \geqslant \varepsilon_0$, for any $\delta \in \Delta_0$. By taking into account the definition of p', there exist elements y_δ, such that

$$0 \leqslant y_\delta \leqslant |x_\delta|, \ \delta \in \Delta_0, \tag{2}$$

$$p(y_\delta) \geqslant \frac{\varepsilon_0}{2}, \ \delta \in \Delta_0. \tag{3}$$

From $x_\delta \xrightarrow{\tau} 0$ and from (2), by taking into account the hypothesis, it follows that $y_\delta \xrightarrow{\tau} 0$, whence $p(y_\delta) \to 0$, and this is in contradiction with (3).

It follows that p' is (τ)-continuous and the proposition is proved.

Proposition 4. *If $\{\mathscr{X}_j\}_{j \in J}$ is a family of locally convex vector lattices, then the product vector lattice $\mathscr{X} = \prod_{j \in J} \mathscr{X}_j$, endowed with the product topology, is a locally convex vector lattice.*

The proof is similar to that given for the case of the locally convex ordered vector spaces (Section 6.1.2, proposition 4), by taking into account Proposition 2.

Proposition 5. *Let $\mathscr{X}$ be a locally convex vector lattice (with the topology τ) and $\mathscr{X}_0$ a vector sublattice of $\mathscr{X}$. If f_0 is a (τ)-continuous* [7]*, positive, linear functional, defined on $\mathscr{X}_0$, then f_0 can be extended to $\mathscr{X}$ as a (τ)-continuous, positive, linear functional.*

Proof. There exists a (τ)-continuous, solid semi-norm p on $\mathscr{X}$, such that

$$|f_0(x)| \leqslant p(x), \ \ x \in \mathscr{X}_0.$$

If we put

$$q(x) = p(x_+), \ \ x \in \mathscr{X},$$

[7]) We denote by τ the topology induced on $\mathscr{X}_0$, too.

then we get a positively homogeneous, subadditive functional on the space $\mathscr{X}$, and we have

$$f_0(x) \leqslant q(x), \ x \in \mathscr{X}_0.$$

With the well-known Hahn-Banach theorem there exists a linear functional f, defined on $\mathscr{X}$, which extends f_0, and for which the following inequalities hold

$$-q(-x) \leqslant f(x) \leqslant q(x), \ x \in \mathscr{X}.$$

Since for $x \geqslant 0$ we obviously have $q(-x) = 0$, it follows that f is a positive functional. We then have

$$|f(x)| \leqslant f(|x|) \leqslant q(|x|) = p(|x|) = p(x),$$

and this implies that f is (τ)-continuous.

Remark. In Section 3.1.4 we defined a topology in an arbitrary ordered vector space $\mathscr{X}$, topology which we denoted by θ, and called the topology of (o)-boundedness. This topology generally is not separated (Hausdorff).

If $\mathscr{X}$ is a vector lattice, then the topology θ is locally solid.

Indeed, if V is a convex neighbourhood of the origin for the topology θ, and if we denote

$$W = \{x; \ Z(x) \subset V\},$$

then W is convex and solid. If a is an arbitrary element of $\mathscr{X}$, then the set $Z(a)$ is (o)-bounded and, therefore, it is absorbed by V. If $Z(a) \subset \lambda V$, then the inclusion $Z(a) \subset \lambda W$ also holds.

On the other hand, if A is an (o)-bounded subset of $\mathscr{X}$, then there exists an $a \in \mathscr{X}$, such that $A \subset Z(a)$.

It follows that W is absorbing any (o)-bounded set and, therefore W is a neighbourhood of the origin for the topology θ. We have now only to observe that $W \subset V$.

6.2.3 Quasi-normed vector lattices

Definition. A quasi-norm q, defined on a vector lattice $\mathscr{X}$, is said to be a solid *quasi-norm*, if $|x| \leqslant |y|$, $(x, y \in \mathscr{X})$, implies $q(x) \leqslant q(y)$. Any vector lattice, which is a quasi-normed vector space, with respect to a solid quasi-norm, is called a *quasi-normed vector lattice.*

Proposition 1. *Any metrizable locally convex vector lattice is a quasi-normed vector lattice.*

Proof. If $\mathscr{X}$ is a metrizable locally convex vector lattice then the topology τ of $\mathscr{X}$ is determined by a countable system $\{p_n\}_{n \in \mathbb{N}}$ of seminorms. For any $n \in \mathbb{N}$ (and any $x \in \mathscr{X}$), by putting

$$\varphi_n(x) = \sup_{0 \leqslant z \leqslant |x|} p_n(z)$$

we get a solid semi-norm. The system of seminorms $\{\varphi_n\}_{n\in\mathbb{N}}$ determines the same topology τ (see the proof of proposition 3 from Section 6.2.2).
If we put

$$q(x) = \sum_{n=1}^{\infty} \frac{1}{2^n} \frac{\varphi_n(x)}{1+\varphi_n(x)},$$

then q is a solid quasi-norm which determines the same topology τ.

Proposition 2. *Any σ-complete quasi-normed*[8]) *vector lattice, whose topology τ is (o)-continuous*[9]), *is a space of (o)-countable type.*

Proof. Let us assume that there exists an uncountable set A of (non-zero) mutually orthogonal elements, which, moreover, is (o)-bounded. Let q be the quasi-norm of the considered space and, for any $n \in \mathbb{N}$, let us define

$$A_n = \left\{x \in A;\ q(x) > \frac{1}{n}\right\}.$$

Since

$$A = \bigcup_{n\in\mathbb{N}} A_n,$$

it follows that there exists an $n_0 \in \mathbb{N}$, such that the set A_{n_0} is not countable. Let $\{x_n\}_{n\in\mathbb{N}}$ be a bounded sequence of distinct elements of A_{n_0}. If the sequence $\{x_n\}_{n\in\mathbb{N}}$ was (o)-bounded, then one would infer that $x_n \xrightarrow{o} 0$ and, therefore $q(x_n) \to 0$. But

$$q(x_n) > \frac{1}{n_0},\quad n \in \mathbb{N},$$

and this is a contradiction.

6.2.4 Bornological vector lattices

If $\mathscr{X}$ is a locally convex vector lattice (with a topology τ) and if $\mathscr{X}$ is bornological for its topology, then $\mathscr{X}$ is said to be a *bornological vector lattice.*

Proposition 1. *Let $\mathscr{X}$ be a vector lattice which, at the same time, is a bornological space with respect to a given topology τ. If the following property holds: if $\{x_n\}_{n\in\mathbb{N}}$ and $\{y_n\}_{n\in\mathbb{N}}$ are sequences of elements of $\mathscr{X}$, such that $|x_n| \leqslant |y_n|$, $(\forall_n \in \mathbb{N})$, and if $y_n \xrightarrow{\tau} \mathbf{0}$, then $x_n \xrightarrow{\tau} \mathbf{0}$. Then $\mathscr{X}$ is a bornological vector lattice.*

---

[8]) i.e., a quasi-normed vector lattice, which is a σ-complete vector lattice.
[9]) See Section 6.3.1.

Proof. If p is a (τ)-continuous semi-norm on $\mathscr{X}$, and if we define

$$p'(x) = \sup_{0 \leqslant z \leqslant |x|} p(z),$$

then p' is sequentially (τ)-continuous (see the proof of proposition 3 from Section 6.2.2).

Let A be a (τ)-bounded set of elements of $\mathscr{X}$. For any sequence $\{x_n)_{n \in \mathbb{N}}$ of elements of A and any sequence $\{\lambda_n\}_{n \in \mathbb{N}}$ of positive numbers which is converging to zero, we have $\lambda_n x_n \xrightarrow{\tau} 0$. It follows that $\lambda_n p'(x_n) \to 0$, hence the set $p'(A)$ is bounded. Since the space $\mathscr{X}$ is bornological, the semi-norm p' is (τ)-continuous.

As in the proof of the proposition already mentioned, one infers that $\mathscr{X}$ is a locally convex vector lattice, i.e. a bornological vector lattice.

Lemma. *Let $\mathscr{X}$ be a locally convex vector lattice, $\mathscr{X}_0$ a normal subspace of $\mathscr{X}$ and p a semi-norm defined on $\mathscr{X}_0$, which is bounded on any topologically bounded subset of $\mathscr{X}_0$.*

Then, by the formula

$$p'(x) = \sup_{\substack{0 \leqslant z \leqslant |x| \\ z \in \mathscr{X}_0}} p(z), \qquad x \in \mathscr{X}, \tag{1}$$

one defines a solid semi-norm on $\mathscr{X}$, which is bounded on any topologically bounded subset of $\mathscr{X}$.

Proof. Since any interval (with respect to the order relation) of the space $\mathscr{X}$ is topologically bounded, the subset $\{z;\ \mathbf{0} \leqslant z \leqslant |x|,\ z \in \mathscr{X}_0\}$ of $\mathscr{X}_0$ is topologically bounded and, therefore, its image by the semi-norm p is bounded. Formula (1) defines a solid semi-norm on $\mathscr{X}$. The verification of this fact can be done as in the proof of proposition 3 from Section 6.2.2, by taking into account the fact that if $0 \leqslant z \leqslant |x| + |y|$ and $z \in \mathscr{X}_0$, then $z = x' + y'$, where $0 \leqslant x' \leqslant |x|$, $0 \leqslant y' \leqslant |y|$, and $x', y' \in \mathscr{X}_0$.

Let A be a topologically bounded subset of $\mathscr{X}$. The set so(A) is topologically bounded too, and, therefore, the set $B = \text{so}(A) \cap \mathscr{X}_0$ is bounded for the topology of $\mathscr{X}_0$. If $p(z) \leqslant \lambda$ for $z \in B$, then from the conditions $x \in \text{so}\,(A)$, $0 \leqslant z \leqslant |x|$ and $z \in \mathscr{X}_0$ it follows that $p(z) \leqslant \lambda$ and, therefore, $p'(x) \leqslant \lambda$. Consequently, p' is bounded on so (A) and, a fortiori, on A.

Proposition 2. *If $\mathscr{X}$ is a bornological vector lattice, then any normal subspace $\mathscr{X}_0$ of $\mathscr{X}$ also is a bornological vector lattice.*

Proof. Let p be a semi-norm on $\mathscr{X}_0$, which is bounded on any topologically bounded subset of $\mathscr{X}_0$.

Let p' be the semi-norm given by formula (1). Since p' is bounded on any topologically bounded subset of $\mathscr{X}$, and since $\mathscr{X}$ is bornological, p' is continuous for the topology of $\mathscr{X}$. From

$$p(x) \leqslant 2\, p'(x), \qquad x \in \mathscr{X}_0,$$

we infer that p is continuous $\mathscr{X}_0$. Consequently, $\mathscr{X}_0$ is a bornological space.

In what follows we shall denote by $\tilde{\mathscr{X}}$ the Dedekind extension of $\mathscr{X}$.

Definition. If $\mathscr{X}$ is a locally convex vector lattice and if τ is its topology, then one calls the *natural extension* of τ to $\tilde{\mathscr{X}}$ the topology $\tilde{\tau}$ determined in $\tilde{\mathscr{X}}$ by the family of all semi-norms given by

$$\tilde{p}(y) = \inf_{\substack{x \geqslant |y| \\ x \in \mathscr{X}}} p(x), \quad y \in \tilde{\mathscr{X}}, \tag{2}$$

where p is any solid (τ)-continuous semi-norm on $\mathscr{X}$.

It is obvious that the topology $\tilde{\tau}$ induces on $\mathscr{X}$ the topology τ and that any of the semi-norms $\tilde{p}$ is solid. One also immediately observes that the topology $\tilde{\tau}$ is separated.

Proposition 3. *If $\mathscr{X}$ is a bornological vector lattice with the topology τ, then $\tilde{\mathscr{X}}$ is a bornological vector lattice for the natural extension $\tilde{\tau}$ of τ.*

Proof. Let q be a semi-norm defined on $\tilde{\mathscr{X}}$, which is assumed to be bounded on any ($\tilde{\tau}$)-bounded subset of $\tilde{\mathscr{X}}$. If we denote

$$q'(y) = \sup_{0 \leqslant z \leqslant |y|} q(z), \quad y \in \tilde{\mathscr{X}},$$

we get a solid semi-norm on $\tilde{\mathscr{X}}$, which is bounded on any ($\tilde{\tau}$)-bounded subset of $\tilde{\mathscr{X}}$ (see the preceding lemma, in which $\mathscr{X}$ and $\mathscr{X}_0$ are replaced by $\tilde{\mathscr{X}}$). The semi-norm q' is, in particular, bounded on any (τ)-bounded subset of $\mathscr{X}$. Since the space $\mathscr{X}$ is bornological, the semi-norm q' is (τ)-continuous on $\mathscr{X}$. There then exists a (τ)-continuous solid semi-norm p on the space $\mathscr{X}$, such that

$$q'(x) \leqslant p(x), \quad x \in \mathscr{X}.$$

Hence we infer that

$$q'(y) \leqslant \tilde{p}(y), \quad y \in \tilde{\mathscr{X}},$$

as we can easily see. Since

$$q(y) = q(y_+ - y_-) \leqslant q(y_+) + q(y_-) \leqslant 2q'(y),$$

it follows that

$$q(y) \leqslant 2\tilde{p}(y), \quad y \in \tilde{\mathscr{X}},$$

and this shows that q is ($\tilde{\tau}$)-continuous on $\tilde{\mathscr{X}}$. Consequently, $\tilde{\mathscr{X}}$ is a bornological space.

Theorem. *Let $\mathscr{X}$ be a bornological vector lattice with the topology τ. Let τ' be a topology on $\tilde{\mathscr{X}}$, for which $\tilde{\mathscr{X}}$ is a bornological vector lattice and which induces the topology τ on $\mathscr{X}$. Then τ' coincides with $\tilde{\tau}$ iff the following condition is satisfied: if $0 \leqslant f \in \mathscr{X}_\tau^*$ and if g is a positive linear extension of f to $\tilde{\mathscr{X}}$, then g is (τ')-continuous.*

Proof. The necessity. Let $0 \leqslant f \in \mathscr{X}_\tau^*$ and let g be a positive linear extension of f to $\tilde{\mathscr{X}}$. Since the functional f is linear and (τ)-continuous, there exists a solid, (τ)-continuous seminorm p on the space $\mathscr{X}$, such that

$$|f(x)| \leqslant p(x), \; x \in \mathscr{X}. \tag{3}$$

Let y be any positive element of $\tilde{\mathscr{X}}$. If $x \in \mathscr{X}$, and $x \geqslant y$, then

$$g(y) \leqslant g(x) = f(x) \leqslant p(x),$$

whence

$$g(y) \leqslant \tilde{p}(y). \tag{4}$$

Since the semi-norm $\tilde{p}$ is solid and since g is linear and positive, from (4) we infer that for any y we have

$$|g(y)| \leqslant g(|y|) \leqslant \tilde{p}(|y|) = \tilde{p}(y),$$

hence g is a $(\tilde{\tau})$-continuous functional.

The sufficiency. Since τ' induces on $\mathscr{X}$ the topology τ, for any (τ')-continuous, solid semi-norm p' on the space $\tilde{\mathscr{X}}$ there exists a (τ)-continuous, solid semi-norm p on $\mathscr{X}$, such that

$$p'(x) \leqslant p(x), \; x \in \mathscr{X}. \tag{5}$$

If $|y| \leqslant x$, where $y \in \tilde{\mathscr{X}}$ and $x \in \mathscr{X}$, then from (5) and from the fact that p' is solid, it follows that

$$p'(y) \leqslant p'(x) \leqslant p(x),$$

whence, by taking into account the definition of $\tilde{p}$, it immediately follows that

$$p'(y) \leqslant \tilde{p}(y).$$

We have thus shown that $\tau' \leqslant \tilde{\tau}$, whence we infer that

$$\tilde{\mathscr{X}}_{\tau'}^* \subset \tilde{\mathscr{X}}_{\tilde{\tau}}^*. \tag{6}$$

We shall now prove the opposite inclusion. Let $g \in \tilde{\mathscr{X}}^*_{\tilde{\tau}}$. Then $g = g_+ - g_-$, where g_+, g_- are $(\tilde{\tau})$-continuous (by the theorem from Section 6.1.3). Let us denote by f_1, f_2 the restriction of g_+, g_-, respectively to the space $\mathscr{X}$. Obviously, the functionals f_1, f_2 are linear, positive and (τ)-continuous on $\mathscr{X}$. Since g_+ is an extension of f_1 to $\tilde{\mathscr{X}}$, which is also linear and positive, g_+ is, by hypothesis, (τ')-continuous. Analogously, g_- is (τ')-continuous. Consequently, $g \in \tilde{\mathscr{X}}^*_{\tau'}$, and this implies that

$$\tilde{\mathscr{X}}^*_{\tilde{\tau}} \subset \tilde{\mathscr{X}}^*_{\tau'}.$$

From this inclusion and from (6) we infer that

$$\tilde{\mathscr{X}}^*_{\tau'} = \tilde{\mathscr{X}}^*_{\tilde{\tau}}.$$

Since $\tilde{\mathscr{X}}$ is a Mackey space for each of the topologies τ', $\tilde{\tau}$, it follows that $\tau' = \tilde{\tau}$.

Corollary. *Let $\mathscr{X}$ be a bornological vector lattice, with the topology τ, having the following property: if $0 \leqslant f \in \mathscr{X}^*_\tau$, then the extension of f to a positive linear functional on $\tilde{\mathscr{X}}$ is unique. Then there exists a unique topology in $\tilde{\mathscr{X}}$, for which $\tilde{\mathscr{X}}$ is a bornological vector lattice and which induces the topology τ on $\mathscr{X}$.*

6.2.5 Locally convex vector lattices with (ω)-continuous topologies

Definition. A locally convex topology τ in a vector lattice $\mathscr{X}$ is said to be (ω)-*continuous* if $x_\delta \underset{\delta \in \Delta}{\downarrow} \mathbf{0}$ implies $x_\delta \xrightarrow[\delta \in \Delta]{\tau} \mathbf{0}$.

Proposition 1. *Let $\mathscr{X}$ be a locally convex vector lattice whose topology τ is (o)-continuous* [10]). *If $\mathscr{X}$ is of (o)-countable type, then the topology τ is (ω)-continuous.*

Proof. If $x_\delta \underset{\delta \in \Delta}{\downarrow} \mathbf{0}$ in $\mathscr{X}$, then $x_\delta \underset{\delta \in \Delta}{\downarrow} \mathbf{0}$ in $\tilde{\mathscr{X}}$, too (here $\tilde{\mathscr{X}}$ denotes the Dedekind extension of $\mathscr{X}$). But if $\mathscr{X}$ is of (o)-countable type, so is $\tilde{\mathscr{X}}$ (Section 4.2.2, proposition 1). There then exists an increasing sequence of indices $\delta_n \in \Delta$ $(n \in \mathbb{N})$, such that $x_{\delta_n} \downarrow \mathbf{0}$ (Section 4.2.2, Proposition 2, corollary). Since τ is (o)-continuous, it follows that $x_{\delta_n} \xrightarrow{\tau} \mathbf{0}$. If φ is a (τ)-continuous solid semi-norm on the space $\mathscr{X}$, then from $x_\delta \downarrow$ and $x_{\delta_n} \xrightarrow{\tau} \mathbf{0}$ it follows that $\varphi(x_\delta) \to 0$. Consequently, $x_\delta \xrightarrow{\tau} \mathbf{0}$, i.e., the topology τ is (ω)-continuous. The proposition is proved.

In the following proposition we shall denote by $\mathscr{X}^*_\omega$ the set of all (ω)-continuous, linear functionals on $\mathscr{X}$.

---

[10]) See Section 6.1.3.

Proposition 2. *Let $\mathscr{X}$ be a locally convex vector lattice, whose topology τ is (ω)-continuous. If the Dedekind extension $\tilde{\mathscr{X}}$ of $\mathscr{X}$ is a locally convex vector lattice with respect to a topology τ', which induces the topology τ on $\mathscr{X}$, then τ' is also (ω)-continuous.*

Proof. Let $g \in \tilde{\mathscr{X}}^*_{\tau'}$. Let us denote by f_1, f_2 the restrictions of the functionals g_+, g_- to the space $\mathscr{X}$. Since τ is (ω)-continuous we have

$$\mathscr{X}^*_\tau \subset \mathscr{X}^*_\omega,$$

hence

$$0 \leqslant f_1, f_2 \in \mathscr{X}^*_\omega.$$

It follows that $g_+, g_- \in \tilde{\mathscr{X}}^*_\omega$ (Section 5.2.5, lemma 3 and proposition 2), whence we get that $g \in \tilde{\mathscr{X}}^*_\omega$. Consequently, we have the inclusion

$$\tilde{\mathscr{X}}^*_{\tau'} \subset \tilde{\mathscr{X}}^*_\omega,$$

from which we infer that τ' is (ω)-continuous (Section 6.1.2, proposition 3).

Proposition 3. *If $\mathscr{X}$ is a (τ)-complete, locally convex vector lattice whose topology is (ω)-continuous, then $\mathscr{X}$ is a complete vector lattice.*

Proof. By taking into account proposition 2, the natural extension $\tilde{\tau}$ of τ to $\tilde{\mathscr{X}}$ is (ω)-continuous (here τ is the topology of $\mathscr{X}$). Let $\{x_\delta\}_{\delta \in \Delta}$ be a bounded, increasing, generalized sequence of elements of $\mathscr{X}$. There exists a $y \in \tilde{\mathscr{X}}$, such that $x_\delta \uparrow y$ in $\tilde{\mathscr{X}}$. It follows that $x_\delta \xrightarrow{\tilde{\tau}} y$ and, therefore, $\{x_\delta\}_{\delta \in \Delta}$ is a generalized Cauchy sequence (with respect to $\tilde{\tau}$) in $\tilde{\mathscr{X}}$ and, therefore, in $\mathscr{X}$ too (with respect to τ). Since $\mathscr{X}$ is (τ)-complete, it follows that $y \in \mathscr{X}$ and $y = \bigvee_{\delta \in \Delta} x_\delta$ in $\mathscr{X}$.

6.2.6 Spaces of type (KP)

Definition 1. One calls a *space of type* (KP) any locally convex σ-complete vector lattice, whose topology τ is (o)-continuous and which has the property that any increasing sequence of positive elements, which is (τ)-bounded, is also (o)-bounded.

Proposition 1. *A sequence $\{x_n\}_{n \in \mathbb{N}}$ of elements of a space of type* (KP) *is (o)-bounded iff the set E of all elements of the form*

$$|x_n| \vee |x_{n+1}| \vee \ldots \vee |x_{n+m}|, \qquad (n, m \in \mathbb{N}),$$

is (τ)-bounded.

Proof. If the sequence $\{x_n\}_{n \in \mathbb{N}}$ is (o)-bounded, then the set E from the statement of the proposition is (o)-bounded, hence it is (τ)-bounded.

Conversely, if the set E is (τ)-bounded, then by putting

$$y_n = \bigvee_{i=1}^{n} |x_i|, \qquad n \in \mathbb{N},$$

we have $0 \leqslant y_n \uparrow$ and the sequence $\{y_n\}_{n\in\mathbb{N}}$ is (τ),-bounded. From definition 1, the sequence $\{y_n\}_{n\in\mathbb{N}}$ is (o)-bounded and this implies that the sequence $\{x_n\}_{n\in\mathbb{N}}$ is (o)-bounded.

Proposition 2. *A sequence $\{x_n\}_{n\in\mathbb{N}}$ of elements of a space of type (KP) is (o)-convergent to $\mathbf{0}$ iff the following condition is satisfied: for any (τ)-continuous solid semi-norm p on $\mathscr{X}$ and for any number $\varepsilon > 0$, there should exist a natural number $n_0 = n_0(p, \varepsilon)$, such that*

$$p\left(\bigvee_{i=n}^{n+m} |x_i|\right) < \varepsilon, \ (n \geqslant n_0, \quad \forall\, m \in \mathbb{N}). \tag{1}$$

Proof. If $x_n \xrightarrow{o} 0$, then

$$\bigvee_{i=n}^{\infty} |x_i| \underset{n\in\mathbb{N}}{\downarrow} \mathbf{0},$$

whence

$$p\left(\bigvee_{i=n}^{\infty} |x_i|\right) \to 0,$$

for any (τ)-continuous semi-norm p. If, moreover, p is solid then (1) immediately follows.

Conversely, let (1) be satisfied. By denoting $m_0 = n_0\,(p, 1)$ we have

$$\bigvee_{i=1}^{k} |x_i| \leqslant \left(\bigvee_{i=1}^{m_0} |x_i|\right) + \left(\bigvee_{i=m_0}^{k} |x_i|\right), \ k \geqslant m_0,$$

and, since p is solid, we get

$$p\left(\bigvee_{i=1}^{k} |x_i|\right) \leqslant p\left(\bigvee_{i=1}^{m_0} |x_i|\right) + 1, \ (k \geqslant m_0).$$

Since p was any (τ)-continuous, solid semi-norm, it follows that the sequence $\left\{\bigvee_{i=1}^{k} |x_i|\right\}_{k\in\mathbb{N}}$ is (τ)-bounded and, therefore, by taking into account definition (1), it is (o)-bounded. In particular, the sequence $\{x_n\}_{n\in\mathbb{N}}$ is (o)-bounded. Since

$$\bigvee_{i=n}^{n+m} |x_i| \underset{n\in\mathbb{N}}{\uparrow} \bigvee_{i=n}^{\infty} |x_i|,$$

and since p is (o)-continuous, from (1) we infer that

$$p\left(\bigvee_{i=n}^{\infty} |x_i|\right) = \lim_m p\left(\bigvee_{i=n}^{n+m} |x_i|\right) \leqslant \varepsilon$$

if $n \geqslant n_0(p, \varepsilon)$. Since ε was arbitrary, and p was any (τ)-continuous, solid semi-norm, the sequence $\left\{\bigvee_{i=n}^{\infty} |x_i|\right\}_{n\in\mathbb{N}}$ is (τ)-converging to 0; since it is decreasing, this sequence is (o)-converging to 0 (see Section 6.1.1, proposition 5). It follows that $x_n \xrightarrow{o} \mathbf{0}$. The proposition is proved.

Proposition 3. *If $\mathscr{X}$ is a space of type (KP) and of (o)-countable type, then its Dedekind extension $\tilde{\mathscr{X}}$ is a space of type (KP) (and of (o)-countable type) with respect to the natural extension $\tilde{\tau}$ of the topology τ of $\mathscr{X}$.*

Proof. The topology τ of $\mathscr{X}$ is (ω)-continuous (Section 6.2.5, proposition 1) and, therefore, the topology $\tilde{\tau}$ of $\tilde{\mathscr{X}}$ is (ω)-continuous (Section 6.2.5, proposition 2). In particular, the topology $\tilde{\tau}$ is (o)-continuous.

Let now $\{y_n\}_{n\in\mathbb{N}}$ be an increasing, non-bounded sequence of positive elements of $\tilde{\mathscr{X}}$. Since the space $\tilde{\mathscr{X}}$ is of (o)-countable type (Section 6.2.2, proposition 1), there exists a non-bounded, increasing sequence $\{x_n\}_{n\in\mathbb{N}}$ of elements of $\mathscr{X}$, such that [11]) $\mathbf{0} \leqslant x_n \leqslant y_n$. There then exists a (τ)-continuous, solid semi-norm p on $\mathscr{X}$, such that $p(x_n) \to +\infty$, whence we immediately infer that $\tilde{p}(y_n) \to +\infty$ (see the definition of $\tilde{p}$ in section 4).

Definition 2. One calls a *space of type* (LP) any space of type (KP), whose topology τ can be defined by a set of solid seminorms, which are additive on positive arguments.

Proposition 4. *Let $\mathscr{X}$ be a locally convex vector lattice whose topology τ can be defined by a set $\mathscr{P}$ of solid semi-norms, which are additive on positive arguments. If $\mathscr{X}$ is sequentially (τ)-complete, then $\mathscr{X}$ is a space of type (LP).*

Proof. Let $0 \leqslant x_n \uparrow$ and assume that for any $p \in \mathscr{P}$ the sequence $\{p(x_n)\}_{n\in\mathbb{N}}$ is bounded. Since any semi-norm p is solid, the sequence $\{p(x_n)\}_{n\in\mathbb{N}}$ is increasing and, therefore, convergent. Since for $m \geqslant n$ we have

$$p(x_m - x_n) = p(x_m) - p(x_n),$$

it follows that

$$\lim_{m,\,n} p(x_m - x_n) = 0,$$

---

[11]) Since $\tilde{\mathscr{X}}$ is of (o)-countable type, any element y_n can be represented as $x_n = \sup_{m\in\mathbb{N}} x_{nm}$, where $\mathbf{0} \leqslant x_{nm} \in \mathscr{X}$. For any $\nu \in \mathbb{N}$, by putting $z_n = \sup_{k,m=1}^{n} x_{km}$, we have $\mathbf{0} \leqslant z_n \leqslant y_n$ and $z_n \uparrow$. The sequence $\{z_n\}_{n\in\mathbb{N}}$ is not (o)-bounded, because if $z_n \leqslant v$, then $y_n \leqslant \sup_{k,m\in\mathbb{N}} x_{k,m} \leqslant v$.

for any $p \in \mathscr{P}$. Since the space $\mathscr{X}$ is sequentially (τ)-complete, there exists the limit $x = (\tau)\text{-lim } x_n$ and we have $x = \bigvee_{n \in \mathbb{N}} x_n$. Hence we infer that $\mathscr{X}$ is a σ-complete vector lattice. The property of (o)-continuity of the topology τ now immediately follows, by taking into account the fact that $x_n \downarrow \mathbf{0}$ implies

$$\mathbf{0} \leqslant x_1 - x_n \uparrow x_1.$$

From what we have already shown, we deduce that $x_1 = (\tau)\text{-lim } (x_1 - x_n)$ and, therefore $x_n \xrightarrow{\tau} 0$.

Remark. If $\mathscr{X}$ is a complete vector lattice with unit and if there exists a sufficient set of (o)-*continuous linear functionals on $\mathscr{X}$, then $\mathscr{X}$ admits of an enlargement $\mathscr{Y}$, which is a space of type (LP).*

The space $\mathscr{Y}$ can be constructed as follows. The set $\mathscr{F}$ of all (o)-continuous, positive, linear functionals on $\mathscr{X}$ is sufficient. Let $\mathscr{Z}$ be the maximal enlargement of $\mathscr{X}$. For any $f \in \mathscr{F}$ and $x \in \mathscr{Z}$ let us denote

$$A(f; x) = \{f(x');\quad \mathbf{0} \leqslant x' \leqslant |x|,\quad x' \in \mathscr{X}\}.$$

Let $\mathscr{Y}$ be the set of the elements $x \in \mathscr{Z}$, such that $A(f; x)$ is bounded, for any $f \in \mathscr{F}$. Let us define

$$p_f(x) = \sup A(f; x), \qquad f \in \mathscr{F}.$$

One can prove that $\mathscr{Y}$ is an enlargement of $\mathscr{X}$ and, for the topology determined by the family $\{p_f\}_{f \in \mathscr{F}}$ of semi-norms, the space $\mathscr{Y}$ is of type (LP).

6.2.7 Spaces of type (M)

Definition 1. One calls a *locally convex vector lattice of type* (M) any locally convex vector lattice in which there exists a fundamental system of neighbourhoods of the origin, consisting of sub-lattices.

Proposition 1. *In any locally convex vector lattice of type (M) there exists a fundamental system of neighbourhoods of the origin consisting of convex, solid sub-lattices.*

Proof. Let V be any neighbourhood of the origin. There exists a solid neighbourhood of the origin $W_1 \subset V$. There also exists a neighbourhood W_2 of the origin, which is a sub-lattice, such that $W_2 \subset W_1$. Let $W_3 = W_2 \cap (-W_2)$. If $a \in W_3$, then $-a \in W_3$ and, therefore $|a| = a \vee (-a) \in W_3$. Let us put $W = \text{so}\,(W_3)$. If $x, y \in W$, then there exists $a, b \in W_3$, such that $|x| \leqslant a$ and $|y| \leqslant b$. By putting $c = a \vee b$, we have $c \in W_3$ and from

$$|x \vee y| \leqslant |x| \vee |y| \leqslant c$$

it follows that $x \vee y \in W$. Analogously, one can show that $x \wedge y \in W$. Consequently, W is a sub-lattice.

Let $\lambda \in \mathbb{R}$ be such that $0 \leqslant \lambda \leqslant 1$. From

$$|\lambda x + (1 - \lambda)\, y| \leqslant \lambda\, |x| + (1 - \lambda)\, |y| \leqslant c$$

it follows that $\lambda x + (1 - \lambda)\, y \in W$. Consequently, W is a solid, convex sub-lattice and, obviously, $W \subset V$.

Definition 2. A semi-norm p, defined on a vector lattice $\mathscr{X}$, is said to be a *semi-norm of type* (M), if for any positive elements $x, y \in \mathscr{X}$, one has

$$p(x \vee y) = \max\{p(x), p(y)\}. \tag{1}$$

Proposition 2. *Let $\mathscr{X}$ be a vector lattice and τ a (Hausdorff) locally convex topology in $\mathscr{X}$. Then $\mathscr{X}$ is a locally convex vector lattice of type (M) iff the topology τ is determined by a set of solid semi-norms of type (M).*

Proof. If $\mathscr{X}$ is a locally convex vector lattice of type (M), then (with Proposition 1) there exists a fundamental system $\mathscr{W}$ of neighbourhoods of the origin, consisting of solid, convex sub-lattices. If $W \in \mathscr{W}$, then the Minkowski semi-norm p_W, associated to W, is (as one knows) a solid semi-norm. Let now $x, y \geqslant \mathbf{0}$, $x, y \in \mathscr{X}$. Since the semi-norm p is monotonous, we have

$$\max\{p(x), p(y)\} \leqslant p(x \vee y). \tag{2}$$

On the other hand, if

$$\alpha > \max\{p(x), p(y)\},$$

then $\dfrac{1}{\alpha}\, x \in W$ and $\dfrac{1}{\alpha}\, y \in W$, whence

$$\frac{1}{\alpha}(x \vee y) \in W,$$

Consequently, we have

$$\alpha \geqslant p(x \vee y),$$

and from (2) equality (1) immediately follows.

Conversely, let p be a solid semi-norm of type (M) on the space $\mathscr{X}$; then, for any number $\varepsilon > 0$, the set

$$W(p;\, \varepsilon) = \{x;\, p(x) < \varepsilon\}$$

is solid. If $x, y \in W(p;\, \varepsilon)$, then from

$$|x \vee y| \leqslant |x| \vee |y|$$

we get

$$p(x \vee y) \leqslant p(|x| \vee |y|) \doteq p(|x|) \vee p(|y|) = p(x) \vee p(y) < \varepsilon$$

and this implies that $x \vee y \in W(p;\, \varepsilon)$. Hence we immediately infer that $x \wedge y \in W(p;\, \varepsilon)$ and, therefore, $W(p;\, \varepsilon)$ is a sub-lattice.

The proposition is proved.

In the following proposition we shall use the notion of space with M-unit, as defined in Section 4.1.4.

If $\mathscr{X}$ is a space with M-unit, we shall denote by E the total set of unitary elements which has the property: for any $x \in \mathscr{X}$ and $e \in E$ there exists $\xi \in \mathbb{R}$, such that

$$[e]\,|x| \leqslant \xi e.$$

Proposition 3. *Any space $\mathscr{X}$ with M-unit is a locally convex vector lattice of type (M) with respect to the topology τ determined by the family of semi-norms $\{p_e\}_{e \in E}$, where*

$$p_e(x) = \inf\{\xi;\ [e]\,|x| \leqslant \xi e\}. \tag{3}$$

If $\{x_n\}_{n \in \mathbb{N}}$ is an (o)-bounded sequence of elements of $\mathscr{X}$ and if $x_n \xrightarrow{\tau} x$, then $x_n \xrightarrow{o} x$.

Proof. It is easily verified that formula (3) gives a solid semi-norm of type (M).

Let us check the last assertion of the proposition. It is obvious that we can limit ourselves to the case $x = \mathbf{0}$.

From $x_n \xrightarrow{\tau} \mathbf{0}$ and from the inequality

$$[e]\,|x_n| \leqslant p_e(x_n)\, e$$

it follows that

$$(o)\text{-}\lim_n\, [e]\,|x_n| = \mathbf{0}.$$

Consequently, we have

$$[e]\,\overline{\lim}\,|x_n| = \mathbf{0}, \qquad \forall\, e \in E. \tag{4}$$

Since the set E is total, for any element $a \geqslant \mathbf{0}$ the following equality holds

$$\sup_{e \in E}\, [e]a = a.$$

If we put $a = \overline{\lim}\, x_n$, then from (4) we deduce that $\overline{\lim}\,|x_n| = 0$ and, therefore, $x_n \xrightarrow{o} 0$.

Remark. If in the preceding proposition the space $\mathscr{X}$ is complete as a vector lattice, then the property stated for (usual) sequences holds for generalized sequences too.

6.2.8 Conjugated spaces

In this section we shall always denote by τ the topology of a locally convex vector lattice.

Proposition 1. *If $\mathscr{X}$ is a locally convex vector lattice, then $\mathscr{X}^*_\tau$ is a normal sub-space of $\mathscr{X}^r$.*

Proof. As one knows (Section 6.1.3, theorem), we have the inclusion $\mathscr{X}_\tau^* \subset \mathscr{X}^r$. Let $g \in \mathscr{X}^r$ and $f \in \mathscr{X}_\tau^*$ be such that $|g| \leqslant |f|$. Let $x_\delta \xrightarrow[\delta \in \Delta]{\tau} 0$. For any number $\varepsilon > 0$ and any $\delta \in \Delta$ there exists a $y_\delta \in \mathscr{X}$, such that

$$|y_\delta| \leqslant |x_\delta|, \qquad \delta \in \Delta, \tag{1}$$

$$|f|(|x_\delta|) \leqslant f(y_\delta) + \varepsilon, \qquad \delta \in \Delta. \tag{2}$$

From (1) it follows that $y_\delta \xrightarrow{\tau} 0$ and, since f is (τ)-continuous, it follows that $f(y_\delta) \to 0$. From (2), by taking into account the fact that ε is arbitrary, we deduce that $|f|(|x_\delta|) \to 0$. From the inequalities

$$0 \leqslant g_+((x_\delta)_+) \leqslant |f|(|x_\delta|)$$

it then follows that $g_+((x_\delta)_+) \to 0$. Analogously, $g_+((x_\delta)_-) \to 0$, whence we get $g_+(x_\delta) \to 0$. Consequently, we have $g_+ \in \mathscr{X}_\tau^*$. Analogously, one can show that $g_- \in \mathscr{X}_\tau^*$. It follows that $g = g_+ - g_- \in \mathscr{X}_\tau^*$.

Proposition 2. *Let $\mathscr{X}$ be a vector lattice and $\mathscr{F}$ a sufficient set of regular functionals defined on $\mathscr{X}$. If $\mathscr{F}$ is a normal sub-space of $\mathscr{X}^r$ then there exists a topology τ in $\mathscr{X}$, for which $\mathscr{X}$ is a locally convex lattice, such that $\mathscr{X}_\tau^* = \mathscr{F}$.*

Proof. We shall consider on $\mathscr{X}$ the topology τ determined by the system of all (solid) semi-norms p_f of the form

$$p_f(x) = f(|x|), \qquad \mathbf{0} < f \in \mathscr{F}.$$

If $f \in \mathscr{F}$, then f_+, $f_- \in \mathscr{F}$ and from the inequalities

$$|f_+(x)| \leqslant p_{f_+}(x),$$

$$|f_-(x)| \leqslant p_{f_-}(x),$$

we infer that f_+ and f_- are (τ)-continuous. Consequently, $f \in \mathscr{X}_\tau^*$.

Let $\mathbf{0} < f \in \mathscr{X}_\tau^*$. If V is a neighbourhood of the origin, then there exists a positive functional $g \in \mathscr{F}$, such that

$$g(|x|) \leqslant 1 \Rightarrow x \in V.$$

Since f is (τ)-continuous, by putting

$$V = \{x; |f(x)| \leqslant 1\},$$

we get a neighbourhood of the origin for the topology τ. It follows that there exists a positive functional $g \in \mathscr{F}$, such that

$$g(|x|) \leqslant 1 \Rightarrow |f(x)| \leqslant 1.$$

Let $\mathbf{0} < x_0 \in \mathscr{X}$. If $\alpha > g(x_0)$, then $g\left(\frac{1}{\alpha} x_0\right) < 1$, whence we get $f(x_0) \leqslant \alpha$. Consequently, $f(x_0) \leqslant g(x_0)$. Since x_0 was an arbitrary positive element, we infer that $f \leqslant g$. Since $f > \mathbf{0}$ and since $\mathscr{F}$ is a normal subspace, from $g \in \mathscr{F}$ we infer that $f \in \mathscr{F}$.

If $f \in \mathscr{X}_\tau^*$ but if f is not positive, then we have to observe that $|f|$ is (τ)-continuous, because $\mathscr{X}_\tau^*$ is a normal subspace of $\mathscr{X}^r$ (proposition 1). But, in this case, from what we have shown, it follows that $|f| \in \mathscr{F}$ and, since $\mathscr{F}$ is a normal subspace of $\mathscr{X}^r$, we also have $f \in \mathscr{F}$. The proposition is proved.

Proposition 3. *If $\mathscr{X}$ is a locally convex vector lattice and if G is a normal subspace of $\mathscr{X}$, then the polar G^0 of G is a component of $\mathscr{X}_\tau^*$.*

Proof. The following equality holds

$$G^0 = \{f;\ |f|(x) = 0, \quad \text{for} \quad \mathbf{0} \leqslant x \in G\}. \tag{3}$$

Indeed, since G is a vector subspace, we have

$$G^0 = \{f;\ f(x) = 0 \quad \text{for} \quad x \in G\}.$$

If $\mathbf{0} \leqslant x \in G$, then from $|x'| \leqslant x$ it follows that $x' \in G$ and, since

$$|f|(x) = \sup_{|x'| \leqslant x} f(x'),$$

we infer that $|f|(x) = 0$ if $f \in G^0$. Conversely, let $|f|(x) = 0$, for $\mathbf{0} \leqslant x \in G$. Then from $x \in G$ and from the inequality

$$|f(x)| \leqslant |f|(|x|)$$

we infer that $f(x) = 0$ (because $|x| \in G$). Equality (3), therefore, holds.

Now it is obvious that G^0 is a normal subspace of $\mathscr{X}_\tau^*$.

If $\{f_j\}_{j \in J}$ is a bounded system of positive elements of G^0, directed for the order relation in $\mathscr{X}_\tau^*$, and if $\mathbf{0} \leqslant x \in G$, then from

$$(\bigvee_{j \in J} f_j)(x) = \bigvee_{j \in J} f_j(x) = 0$$

we infer that $\bigvee_{j \in J} f_j \in G^0$. Since G^0 is a normal subspace of $\mathscr{X}_\tau^*$, we infer that $\bigvee_{j \in J} f_j \in G^0$, for any bounded from above system $\{f_j\}_{j \in J}$ of elements of G^0.

Since the space $\mathscr{X}_\tau^*$ is a complete vector lattice (see proposition 1), the set G^0 is a component of $\mathscr{X}_\tau^*$ (Section 3.3.3, proposition 5).

Remark. One can show that if E is a normal subspace of $\mathscr{X}$, then

$$(E^0)^\perp \subset (E^\perp)^0,$$

and, by taking into account the preceding proposition, we infer that any $f \in \mathscr{X}_\tau^*$ can be written as $f = f_1 + f_2$, where $f_1 \in E^0$, $f_2 \in (E^\perp)^0$, but this decomposition is not unique.

Proposition 4. *If $\mathscr{X}$ is a sequentially (τ)-complete bornological vector lattice, then $\mathscr{X}_\tau^* = \mathscr{X}^r$.*

Proof. We have only to prove the inclusion $\mathscr{X}^r \subset \mathscr{X}_\tau^*$. In order to do this, it is enough to show that any positive linear functional is (τ)-continuous.

If there existed a positive linear functional f on $\mathscr{X}$, which would not be (τ)-continuous, then there would exist a (τ)-bounded subset A, consisting of positive elements, on which f is not bounded.

Let then $x_n \in A$ be such that $f(x_n) > n^2$. Let us put

$$y = (\tau)\text{-}\sum_{n=1}^{\infty} \frac{1}{n^2} x_n.$$

The series is (τ)-convergent. Indeed, by putting

$$y_n = \sum_{j=1}^{n} \frac{1}{j^2} x_j,$$

we have, for any $m > n$,

$$y_m - y_n = \sum_{j=n+1}^{m} \frac{1}{j^2} x_j.$$

If p is a (τ)-continuous semi-norm on $\mathscr{X}$, then there exists a $\lambda > 0$, such that $p(x_j) \leqslant \lambda$, for any $j \in \mathbb{N}$; consequently, we have the following inequalities

$$p(y_m - y_n) \leqslant \sum_{j=n+1}^{m} \frac{1}{j^2} p(x_j) \leqslant \lambda \sum_{j=n+1}^{m} \frac{1}{j^2}.$$

It follows that $\{y_n\}_{n \in \mathbb{N}}$ is a Cauchy sequence and, since the space $\mathscr{X}$ is sequentially (τ)-complete, there exists the limit $y = (\tau)\text{-}\lim y_n$ and $y_n \uparrow y$.

It follows that

$$f(y) \geqslant f(y_n) = \sum_{j=1}^{n} \frac{1}{j^2} f(x_j) > n,$$

for any $n \in \mathbb{N}$, and this is a contradiction. The proposition is proved.

Remark. If $\mathscr{X}$ is a locally convex vector lattice, then the inclusion $\mathscr{X}_\tau^* \subset \mathscr{X}_o^*$ is equivalent to the (o)-continuity of the topology τ, whereas the inclusion $\mathscr{X}_\tau^* \subset \mathscr{X}_\omega^*$ is equivalent to the (ω)-continuity of τ.

If $\mathscr{X}$ is a sequentially (τ)-complete bornological vector lattice and if the topology τ is (o)-continuous, then

$$\mathscr{X}_\tau^* = \mathscr{X}_o^* = \mathscr{X}_r^*,$$

whereas if τ is (ω)-continuous, then

$$\mathscr{X}_\tau^* = \mathscr{X}_\omega^* = \mathscr{X}_o^* = \mathscr{X}_r^*.$$

6.2.9 Normed vector lattices

Definition 1. One calls a *normed vector lattice* any vector lattice on which a solid norm is given (i.e. such that $|x_1| \leqslant |x_2| \Rightarrow \|x_1\| \leqslant \|x_2\|$). If, moreover, $\mathscr{X}$ is a Banach space, then $\mathscr{X}$ is said to be a *Banach vector lattice* (or a Banach lattice).

If $\mathscr{X}$ is a normed vector lattice, we shall denote by τ the norm topology.

Definition 2. A sequence $\{x_n\}_{n\in\mathbb{N}}$ of elements of an Archimedean vector lattice is said to (ρ^*)-converge to an element x if from any subsequence $\{x_{j_n}\}_{n\in\mathbb{N}}$ of the given sequence one can select another subsequence $\{x_{j_{i_n}}\}_{n\in\mathbb{N}}$ which (ρ)-converges to x. One writes, in this case, $x_n \xrightarrow{\rho^*} x$.

Proposition 1. *In a Banach vector lattice one has $x_n \xrightarrow{\tau} x$ iff $x_n \xrightarrow{\rho^*} x$.*

Proof. It is enough to consider the case $x = \mathbf{0}$. If $\|x_n\| \to 0$, then there exists an increasing sequence of natural numbers $\{j_n\}_{n\in\mathbb{N}}$ such that $n^3\|x_{j_n}\| \to 0$. By defining

$$r = (\tau)\text{-}\sum_{n=1}^{\infty} n|x_{j_n}|,$$

it follows that

$$|x_{j_n}| \leqslant \frac{1}{n} r, \; n \in \mathbb{N},$$

consequently, $x_{j_n} \xrightarrow{\rho} \mathbf{0}$. It is obvious that the given sequence may be replaced by any subsequence; therefore, $x_n \xrightarrow{\rho^*} \mathbf{0}$.

Conversely, let us assume that $x_n \xrightarrow{\rho^*} \mathbf{0}$. If $\{x_{j_n}\}_{n\in\mathbb{N}}$ is any subsequence of the given sequence, and if $x_{j_{i_n}} \xrightarrow{\rho} \mathbf{0}$, then we obviously have $\|x_{j_{i_n}}\| \to 0$. It follows that $\|x_n\| \to \mathbf{0}$, i.e., $x_n \xrightarrow{\tau} \mathbf{0}$.

Proposition 2. *In any Banach vector lattice, which is separable for the norm topology, there exist total elements.*

Proof. Let us consider a countable set E, which is dense in $\mathscr{X}$ (for the topology τ). If $\mathbf{0} \leqslant a \in \mathscr{X}$ and if $a \perp E$, then $a = \mathbf{0}$, as one can immediately see.

Let us represent the set E as a sequence $\{x_n\}_{n\in\mathbb{N}}$ and define

$$u = (\tau)\text{-}\sum_{n=1}^{\infty} \frac{1}{2^n\|x_n\|}|x_n|.$$

If $\mathbf{0} \leqslant a \in \mathscr{X}$ and $a \perp u$, then $a \perp x_n$, for any $n \in \mathbb{N}$, whence $a = 0$. Consequently, u is total.

Proposition 3. *Any locally convex vector lattice is isomorphic* [12]) *to a topological vector sublattice of a product of normed vector lattices.*

Proof. Let $\mathscr{X}$ be a locally convex vector lattice and $\{p_j\}_{j\in J}$ a family of seminorms, which determines the topology of $\mathscr{X}$. Let $G_j = \overset{-1}{p_j}(0)$ $(j \in J)$ and $\mathscr{Z}_j = \mathscr{X}/G_j$. As we have shown in Section 2.2.2, by defining

$$q_j(\dot{x}) = p_j(x), \quad \dot{x} \in \mathscr{Z}_j$$

we get a norm on $\mathscr{Z}_j$. By denoting

$$\mathscr{Z} = \prod_{j\in J} \mathscr{Z}_j$$

and considering in $\mathscr{Z}$ the product topology, we infer that the space $\mathscr{X}$ is isomorphic [13]), as a topological vector space, to a subspace $\mathscr{Z}_0$ of $\mathscr{Z}$. The isomorphism is given by the mapping l defined by $l(x) = \{l_j(x)\}_{j\in J}$, where l_j is the natural mapping of $\mathscr{X}$ onto $\mathscr{Z}_j$.

We shall define in each space $\mathscr{Z}_j$ the (vector space) order relation by defining $\dot{a} \leqslant \dot{b}$ to mean that there exist $a' \in \dot{a}$ and $b' \in \dot{b}$ such that $a' \leqslant b'$. It is easily verified that by defining $c = a \vee b$, we have $\dot{c} = \dot{a} \vee \dot{b}$, whence one can infer that $\mathscr{X}$ is a vector lattice.

If $|\dot{a}| \leqslant |\dot{b}|$, then there exists an $e \in G$, such that $|a| \leqslant |b| + e$, whence we immediately infer that $q_j(\dot{a}) \leqslant q_j(\dot{b})$.

Consequently, each $\mathscr{Z}_j$ is a normed vector lattice. Endowed with the product order relation, $\mathscr{Z}$ is a locally convex vector lattice.

Since

$$l_j(a \vee b) = l_j(a) \vee l_j(b), \quad j \in J,$$

the relation $a \leqslant b$ is equivalent to $l(a) \leqslant l(b)$.

Indeed, if $a \leqslant b$, then $l_j(a) \leqslant l_j(b)$, for any $j \in J$, hence $l(a) \leqslant l(b)$. In order to prove the converse, it is enough to show that if $l(a) \geqslant \mathbf{0}$, then $a \geqslant \mathbf{0}$. Let then $l_j(a) \geqslant \mathbf{0}$, for any $j \in J$. If $a_- \neq \mathbf{0}$, then there exists a $j_0 \in J$, such that $l_{j_0}(a_-) \neq \mathbf{0}$. But, in this case

$$(l_{j_0}(a))_- = l_{j_0}(a_-) \neq \mathbf{0}$$

and this is a contradiction. The proposition is proved.

Proposition 4. *Let $\mathscr{X}$ be a normed vector lattice, $\hat{\mathscr{X}}$ its completion as a normed vector space and E the closure of $\mathscr{X}_+$ in $\hat{\mathscr{X}}$. Then $\hat{\mathscr{X}}$ is a Banach vector lattice with respect to the order relation $y_1 \leqslant y_2 \Leftrightarrow y_2 - y_1 \in E$, whereas $\mathscr{X}$ is a vector sublattice of $\hat{\mathscr{X}}$.*

[12]) As a topological ordered vector space.

[13]) In Section 2.2.2 we completed the space $\mathscr{Z}_j$ to a Banach space.

Proof. It is obvious that if $y_1, y_2 \in E$, then $y_1 + y_2 \in E$, whereas if $y \in E$ and $\mathbf{0} \leqslant \alpha \in \mathbb{R}$, then $\alpha y \in E$.

Let now $y \in E$ and $-y \in E$. Consequently, $y = (\tau)\text{-lim } a_n$ and $-y = (\tau)\text{-lim } b_n$, where $a_n, b_n \geqslant \mathbf{0}$. From $\mathbf{0} \leqslant a_n \leqslant a_n + b_n$ we get $\|a_n\| \leqslant \|a_n + b_n\|$ and, therefore $y = \mathbf{0}$. One thus obtains an order relation in $\hat{\mathscr{X}}$, by defining $y_1 \leqslant y_2$ to be equivalent to $y_2 - y_1 \in E$.

If we put $h_0(x) = x_+$, then h_0 is a (τ)-continuous mapping of $\mathscr{X}$ into $\mathscr{X}$ (Section 6.2.2, proposition 2). It has a unique continuous extension to $\hat{\mathscr{X}}$, with values in $\hat{\mathscr{X}}$. From $h(\mathscr{X}) \subset \mathscr{X}_+$ one obtains $h(\hat{\mathscr{X}}) \subset E$. Let $y \in \mathscr{Y}$ and let us show that $h(y) = y \vee \mathbf{0}$ in $\hat{\mathscr{X}}$.

Let $y = (\tau)\text{-lim } x_n$, where $x_n \in \mathscr{X}_+$. We have $h(x_n) = (x_n)_+$ in $\mathscr{X}$, hence $h(x_n) - x_n \geqslant \mathbf{0}$ and this implies that $h(y) - y \in E$. But $h(y) \in E$ and, therefore, $h(y) \geqslant y$ and $h(y) \geqslant \mathbf{0}$. Let $z \geqslant y$ and $z \geqslant \mathbf{0}$ in $\hat{\mathscr{X}}$. We have $z \in E$ and $z - y \in E$, hence $z = (\tau)\text{-lim } z_n$, $z - y = (\tau)\text{-lim } v_n$, where $z_n, v_n \in \mathscr{X}_+$. It follows that $y = (\tau)\text{-lim } y_n$, where $y_n = z_n - v_n$. But $z_n - h(y_n) \in \mathscr{X}_+$, because $z_n \geqslant \mathbf{0}$ and $z_n \geqslant y_n$.

It follows that $z - h(y) \in E$, i.e., $z \geqslant h(y)$ in $\hat{\mathscr{X}}$. Consequently, $h(y) = y \vee \mathbf{0}$ in $\hat{\mathscr{X}}$, and thus we have shown that $\hat{\mathscr{X}}$ is a vector lattice. It is now obvious that $\mathscr{X}$ is a vector sublattice of $\hat{\mathscr{X}}$.

If $\mathbf{0} \leqslant y_1 \leqslant y_2$ in $\hat{\mathscr{X}}$, then $y_1 = (\tau)\text{-lim } a_n$, where $a_n \in \mathscr{X}_+$ and $y_2 - y_1 = (\tau)\text{-lim } b_n$, where $b_n \in \mathscr{X}_+$. It follows that $y_2 = (\tau)\text{-lim } x_n$, where $x_n = a_n + b_n$. From $\mathbf{0} \leqslant a_n \leqslant x_n$ we get that $\|a_n\| \leqslant \|x_n\|$ and, therefore $\|y_1\| \leqslant \|y_2\|$.

If $y \in \hat{\mathscr{X}}$ and $y = (\tau)\text{-lim } x_n$, where $x_n \in \mathscr{X}$, then $|y| = (\tau)\text{-lim } |x_n|$ (because the (τ)-continuity of the mapping $y \mapsto y_+$ implies the (τ)-continuity of the mapping $y \mapsto |y|$). It follows that

$$\||y|\| = \lim \||x_n|\| = \lim \|x_n\| = \|y\|.$$

In the following proposition we shall assume that the completion $\hat{\mathscr{X}}$ of a normed vector lattice $\mathscr{X}$ is endowed with the order relation given in the statement of the preceding proposition.

Proposition 5. *If $\mathscr{X}$ is a normed vector lattice, in which any (τ)-bounded, increasing sequence is a (τ)-Cauchy sequence, then in $\hat{\mathscr{X}}$ any (τ)-bounded, increasing sequence is (τ)-convergent.*

Proof. Let $\{y_n\}_{n\in\mathbb{N}}$ be a (τ)-bounded, increasing sequence in $\hat{\mathscr{X}}$. Let us consider an arbitrary number $\varepsilon > 0$ and, for any $n \in \mathbb{N}$, let $x_n \in \mathscr{X}$, be such that

$$\|y_n - x_n\| \leqslant \frac{\varepsilon}{2^n}.$$

If we put

$$x'_n = \bigvee_{i=1}^{n} x_i, \; n \in \mathbb{N},$$

then $x_n' \in \mathscr{X}$ and $x_n' \uparrow$. We have

$$x_n' - y_n = \bigvee_{i=1}^{n} (x_i - y_n) \leqslant \bigvee_{i=1}^{n} (x_i - y_i) \leqslant \sum_{i=1}^{n} |y_i - x_i|$$

$$y_n - x_n' \leqslant y_n - x_n \leqslant \sum_{i=1}^{n} |y_i - x_i|,$$

and, therefore,

$$|y_n - x_n'| \leqslant \sum_{i=1}^{n} |y_i - x_i|.$$

It follows that

$$\|y_n - x_n'\| \leqslant \sum_{i=1}^{n} \|y_i - x_i\| \leqslant \sum_{i=1}^{n} \frac{\varepsilon}{2^i} \leqslant \varepsilon,$$

for any $n \in \mathbb{N}$.

In particular, the sequence $\{y_n - x_n'\}_{n \in \mathbb{N}}$ is (τ)-bounded and, since $\{y_n\}_{n \in \mathbb{N}}$ is (τ)-bounded, it follows that the sequence $\{x_n'\}_{n \in \mathbb{N}}$ is (τ)-bounded. Since the latter is also increasing, it is a (τ)-Cauchy sequence, by hypothesis. Let then n_0 be such that

$$\|x_m' - x_n'\| \leqslant \varepsilon, \quad m, n \geqslant n_0.$$

For $m, n \geqslant n_0$ one then infers that

$$\|y_m - y_n\| \leqslant \|y_m - x_m'\| + \|x_m' - x_n'\| + \|x_n' - y_n\| \leqslant 3\varepsilon,$$

and this shows that $\{y_n\}_{n \in \mathbb{N}}$ is a (τ)-Cauchy sequence in $\hat{\mathscr{X}}$. The proposition is proved.

Definition 3. One calls a *space of type* (KB) any complete vector lattice, which, at the same time, is a normed vector lattice and a space of type (KP), with respect to the norm topology.

Proposition 6. *Any space $\mathscr{X}$ of type (KB) is a Banach space.*

Proof. Let $\{x_n\}_{n \in \mathbb{N}}$ be a (norm) Cauchy sequence of elements of $\mathscr{X}$. Let us consider a subsequence $\{x_{j_n}\}_{n \in \mathbb{N}}$, such that the series with the terms $\|x_{j_n} - x_{j_{n+1}}\|$ be convergent. There then exists a y such that $x_{j_1} - x_{j_n} \xrightarrow{o} y$, because we can define

$$y = \sum_{n=1}^{\infty} (x_{j_n} - x_{j_{n+1}}).$$

It follows that $x_j - x_{j_n} \xrightarrow{\tau} y$ and, by denoting $x = x_{j_1} - y$, we have $x_{j_n} \xrightarrow{\tau} x$. Since $\{x_n\}_{n\in\mathbb{N}}$ is a (norm) Cauchy sequence, it follows that $x_n \xrightarrow{\tau} x$.

Proposition 7. *In a space of type (KB) a set E is (o)-bounded iff the set of all elements of the form* $\bigvee_{i=1}^{n} |x_i|$, *where* $x_i \in E$ *is* (τ)-*bounded.*

Proof. Let E be an (o)-bounded set. There then exists a y, such that $|x| \leqslant y$, for any $x \in E$. Therefore, for any finite system of elements $x_i \in E$ we have $\bigvee_{i=1}^{n} |x_i| \leqslant y$, whence

$$\left\|\bigvee_{i=1}^{n} |x_i|\right\| \leqslant \|y\|.$$

Conversely, let E be such that the set of all elements of the form $\bigvee_{i=1}^{n} |x_i|$, where $x_i \in E$, be (τ)-bounded. We can assume that the elements of E are positive (because, otherwise, we could consider the set $\{|x|;\ x \in E\}$). We can also suppose that E contains the least upper bounds of all its finite subsets (otherwise, we can add them to E).

Let now $x_1 \in E$. By induction, we can construct a sequence $\{x_n\}_{n\in\mathbb{N}}$ of elements of E, which is increasing and such that

$$\|x_{n+1} - x_n\| \geqslant \sup_{x_n \leqslant x \in E} \|x - x_n\| - \frac{1}{n}.$$

Since $\mathbf{0} \leqslant x_n \uparrow$ and $\{x_n\}_{n\in\mathbb{N}}$ is (τ)-bounded, there exists an element y, such that $x_n \uparrow y$. If $x \in E$, then

$$\|x \vee x_n - x_n\| \leqslant \sup_{x_n \leqslant x' \in E} \|x' - x_n\| \leqslant \|x_{n+1} - x_n\| + \frac{1}{n} \leqslant \|y - x_n\| + \frac{1}{n}.$$

By tending to the limit with respect to n, we get that

$$x \vee y - y = \mathbf{0},$$

and, therefore, $x \leqslant y$. Since $y = \bigvee_{n\in N} x_n$, it follows that $y = \sup E$.

Proposition 8. *In a space of type (KB), the convergence* $x_n \xrightarrow{\tau} x$ *is equivalent to* $x_n \xrightarrow{o^*} x$.

Proof. If $x_n \xrightarrow{\tau} \mathbf{0}$, then, by taking into account proposition 1, it follows that $x_n \xrightarrow{o^*} \mathbf{0}$. Conversely, if $x_n \xrightarrow{o^*} \mathbf{0}$, then from any subsequence $\{x_{j_n}\}_{n\in\mathbb{N}}$ one can choose another subsequence $\{x_{j_{i_n}}\}_{n\in\mathbb{N}}$, such that $x_{j_{i_n}} \xrightarrow{o} \mathbf{0}$, and therefore, $x_{j_{i_n}} \xrightarrow{\tau} \mathbf{0}$. It follows that $x_n \xrightarrow{\tau} \mathbf{0}$.

Proposition 9. *Any space $\mathscr{X}$ of type (KB) is a σ-regular space.*

Proof. If $x_n \xrightarrow{o} \mathbf{0}$, then

$$\lim \| \bigvee_{i=1}^{n+m} |x_i| \| = 0,$$

uniformly with respect to m (Section 6.2.6, proposition 2). There then exists an increasing sequence $\{j_n\}_{n\in\mathbb{N}}$ of natural numbers, such that

$$\| |x_{j_n}| \vee |x_{j_n+1}| \vee \ldots \vee |x_{j_n+m}| \| < \frac{1}{n^3}$$

for any, $n, m \in \mathbb{N}$. Let us define

$$\lambda_i = \begin{cases} 1 \text{ if } & 1 \leqslant i < j_1 \\ n \text{ if } & j_n \leqslant i < j_{n+1}, \ n \in \mathbb{N}. \end{cases}$$

For $i \geqslant j_n$, we have [14])

$$\| \bigvee_{h=0}^{k} |\lambda_{i+h}\, x_{i+h}| \| \leqslant \sum_{m=n}^{\infty} m \, \| \, |x_{j_m}| \vee |x_{j_m+1}| \vee \ldots \vee |x_{j_{m+1}-1}| \, \| < \sum_{m=n}^{\infty} \frac{1}{m^2}$$

whence $\lambda_n x_n \xrightarrow{o} \mathbf{0}$. Consequently, $\mathscr{X}$ is a regular space (Section 4.2.1, proposition 1).

On the other hand, for any sequence $\{x_n\}_{n\in\mathbb{N}}$ of elements there exists a sequence $\{\mu_n\}_{n\in\mathbb{N}}$ of strictly positive numbers, such that the series with the terms $\mu_n \|x_n\|$ be convergent. There then exists the sum

$$x = (o)\text{-}\sum_{n=1}^{\infty} \mu_n x_n,$$

whence we get $\mu_n x_n \xrightarrow{o} \mathbf{0}$. Consequently, the space $\mathscr{X}$ is σ-regular (Section 4.2.1, proposition 3).

Definition 4. One calls a *space of type* (*L*) any space of type (KB) in which the norm is solid and additive on positive arguments.

Proposition 10. *Any Banach vector lattice, in which the norm is* (*solid and*) *additive on positive arguments, is a space of type* (*L*).

This is a consequence of proposition 4 from Section 6.2.6.

With definition 2 from Section 6.2.7, a norm $\|\cdot\|$ on a vector lattice $\mathscr{X}$ is of type (*M*) if for any $x, y \geqslant \mathbf{0}$

$$\|x \vee y\| = \max\{\|x\|, \|y\|\}.$$

---

[14] If $j_n \leqslant j_m \leqslant i < j_{m+1}$, then $\lambda_i = m$ and, therefore, $|\lambda_i x_i| = m|x_i| \leqslant m(|x_{j_m}| \vee \ldots \vee |x_{j_{m+1}-1}|)$

Definition 5. One calls a *normed vector lattice of type* (M) any vector lattice on which a solid norm of type (M) has been defined.

Proposition 11. *Any Archimedean vector lattice, which has a strong unit u, is a normed vector lattice of type* (M) *with respect to the norm*

$$\|x\| = \inf\{\xi;\ |x| \leqslant \xi u\}. \tag{1}$$

In this space the convergence with respect to the norm implies the (ρ)*-convergence.*

This assertion is easy to check.

We shall say that the norm given by formula (1) is *the norm associated to the unit.*

Proposition 12. *With respect to the norm associated to the unit any* σ*-complete vector lattice with a strong unit is a Banach space.*

Proof. If

$$\|x_n - x_m\| < \varepsilon,\ (n, m \geqslant n_\varepsilon)$$

then

$$|x_n - x_m| < \varepsilon u\,(n, m \geqslant n_\varepsilon).$$

There then exists the limit $x = (o)\text{-}\lim\limits_n x_n$ (Section 3.2.6, proposition 1), and

$$|x_n - x| \leqslant \varepsilon u, \quad (n \geqslant n_\varepsilon).$$

It follows that

$$\|x_n - x\| \leqslant \varepsilon, \quad (n \geqslant n_\varepsilon),$$

and, therefore, $x_n \xrightarrow{\tau} x$.

Remark. In a normed vector space an element z is said to be an *extremal point* of the bal $\bar{S}(\mathbf{0}; 1) = \{x;\ \|x\| \leqslant 1\}$ if z cannot be written as $z = \alpha x + (1 - \alpha)y$, where $x, y \in \bar{S}(\mathbf{0}; 1)$, $x \neq y$, and $0 < \alpha < 1$.

In a σ-complete vector lattice $\mathscr{X}$ with a strong unit u, the ball $\bar{S}(0; 1)$ (with respect to the norm (1)) has the element u as an extremal point. If, moreover, $\mathscr{X}$ is a complete vector lattice, then the family of all closed balls in $\mathscr{X}$ has the binary intersection property.

Examples. (1) Let (c) be the set of all convergent sequences of real numbers. This set becomes a vector lattice if the operations and the order relation are defined as in (s) (Section 3.4.2). Defining

$$\|x\| = \sup_{n \in \mathbb{N}} |\xi_n|,$$

(where $x = \{\xi_n\}_{n\in\mathbb{N}}$), the space (c)-becomes a Banach vector lattice.

(2) The space (l^p) (where $p \geqslant 1$) (see Section 3.4.2) is a space of type (KB) with respect to the norm

$$\|x\| = \left(\sum_{n=1}^{\infty} |\xi_n|^p\right)^{1/p}.$$

If $p = 1$, it is a space of type (L).

(3) The space $C(T)$ of all continuous real functions, defined on a compact space T, the operations and the order relation being defined as in $F(T)$ (see Section 3.4.3), and in which the norm is given by

$$\|x\| = \max_{t\in T} |x(t)|$$

is a Banach vector lattice. The norm convergence in this space coincides with the (ρ)-convergence (and means the uniform convergence on T).

(4) The space $V(T)$ of all real functions of bounded variation, defined on an interval $T = [\alpha, \beta]$ (Section 3.4.3) is a space of type L with respect to the norm

$$\|x\| = |x(\alpha)| + \operatorname*{var}_{\alpha}^{\beta} x(t).$$

6.3 Linear operators defined on topological vector lattices

6.3.1 (τo)-Bounded linear operators

Let $\mathscr{Y}$ be a complete vector lattice.

Definition. An operator U, which is defined on a topological vector space $\mathscr{X}$, and takes values in $\mathscr{Y}$, is said to be (τo)-*bounded* if for any (τ)-bounded set $A \subset \mathscr{X}$ the set $U(A)$ is (o)-bounded.

If U is a (τo)-bounded linear operators, then for any (τ)-bounded subset $A \subset \mathscr{X}$ we shall write

$$|U|_A = \sup_{x\in A} |U(x)|.$$

If U_1, U_2 and U are (τo)-bounded linear operators, and $\alpha \in \mathbb{R}$, then $U_1 + U_2$ and αU are (τo)-bounded operators and

$$|U_1 + U_2|_A \leqslant |U_1|_A + |U_2|_A,$$

$$|\alpha U|_A = |\alpha|\,|U|_A.$$

These relations are easy to prove.

In the following part of this section we shall suppose that $\mathscr{X}$ is a topological vector lattice.

We shall denote by $\mathscr{G}$ the set of all solid and (τ)-bounded subsets of $\mathscr{X}$. The set $\mathscr{G}$ is a fundamental system of (τ)-bounded sets, because if A is a (τ)-bounded set, then the set so(A) is (τ)-bounded, too. Consequently, in order that an operator U, which maps $\mathscr{X}$ into $\mathscr{Y}$, be (τo)-bounded it is sufficient that for any $A \in \mathscr{G}$ the set $U(A)$ be (o)-bounded.

We shall denote by $(\mathscr{X}, \mathscr{Y})^{\tau o}$ the set of all (τo)-bounded linear operators which map $\mathscr{X}$ into $\mathscr{Y}$. As in Chapter 5, we shall denote by $(\mathscr{X}, \mathscr{Y})^r$ the space of all regular operators.

Proposition 1. *The set $(\mathscr{X}, \mathscr{Y})^{\tau o}$ is a normal subspace of the space $(\mathscr{X}, \mathscr{Y})^r$.*

Proof. If $U \in (\mathscr{X}, \mathscr{Y})^{\tau o}$, then U is regular. Indeed, if A is an (o)-bounded subset of $\mathscr{X}$, then A is (τ)-bounded and, therefore, $U(A)$ is (o)-bounded. It follows (see Section 5.1.3, theorem 2) that U is regular.

The set $(\mathscr{X}, \mathscr{Y})^{\tau o}$, therefore, is a subset and, obviously, a vector subspace of the space $(\mathscr{X}, \mathscr{Y})^r$.

If $U \in (\mathscr{X}, \mathscr{Y})^{\tau o}$, then $|U| \in (\mathscr{X}, \mathscr{Y})^{\tau o}$.

Indeed, if $A \in \mathscr{G}$ and $x \in A$, then

$$| \, |U|(x)| \leqslant |U| \, (|x|) = \sup_{|x'| \leqslant |x|} |U(x')| \leqslant |U|_A.$$

Let now $U_2 \in (\mathscr{X}, \mathscr{Y})^{\tau o}$ and U_1 be a positive, additive operator such that $U_1 \leqslant U_2$. If $A \in \mathscr{G}$ and $x \in A$, then

$$|U_1(x)| \leqslant U_1(|x|) \leqslant |U_2|_A$$

and therefore, $U_1 \in (\mathscr{X}, \mathscr{Y})^{\tau o}$.

Let now $U_0 \in (\mathscr{X}, \mathscr{Y})^{\tau o}$ and $U \in (\mathscr{X}, \mathscr{Y})^r$ be such that $|U| \leqslant |U_0|$. From what we already showed, it follows that $|U_0|$ is (τo)-bounded and, therefore, $|U|$ is (o)-bounded too. From

$$0 \leqslant U_+, \; U_- \leqslant |U|$$

it follows that the operators U_+, U_- are (τo)-bounded. It follows that $U \in (\mathscr{X}, \mathscr{Y})^{\tau o}$.

Corollary. *The set of all linear, (τo)-bounded operators (which map $\mathscr{X}$ into $\mathscr{Y}$) is a complete vector lattice.*

Proposition 2. *If U_0 is a (τo)-bounded, positive, linear operator, defined on a normal subspace $\mathscr{X}_0$ of the space $\mathscr{X}$ (taking values in $\mathscr{Y}$), then U_0 can be extended to $\mathscr{X}$ as a (τo)-bounded, positive and linear operator.*

Proof. For any positive element $x \in \mathscr{X}$ let us define

$$U_1(x) = \sup_{\substack{0 \leqslant x' \leqslant x \\ x' \in \mathscr{X}_0}} U_0(x').$$

Let us show that

$$U_1(x_1 + x_2) = U_1(x_1) + U_1(x_2), \tag{1}$$

for any positive elements x_1, x_2.

Let $a \in \mathscr{X}_0$ be such that

$$\mathbf{0} \leqslant a \leqslant x_1 + x_2.$$

The element a can then be represented as $a = b + c$, where b, c are such that $\mathbf{0} \leqslant b \leqslant x_1$ and $\mathbf{0} \leqslant c \leqslant x_2$. The elements b, c belong to $\mathscr{X}_0$, because $\mathscr{X}_0$ is a normal subspace. From the relations

$$U_0(a) = U_0(b) + U_0(c) \leqslant U_1(x_1) + U_1(x_2)$$

it follows that

$$U_1(x_1 + x_2) \leqslant U_1(x_1) + U_1(x_2).$$

Let now $b, c \in \mathscr{X}_0$ be such that $\mathbf{0} \leqslant b \leqslant x_1$ and $\mathbf{0} \leqslant c \leqslant x_2$. It follows that $b + c \in \mathscr{X}_0$ and $\mathbf{0} \leqslant b + c \leqslant x_1 + x_2$. From the relations

$$U_0(b) + U_0(c) = U_0(b + c) \leqslant U_1(x_1 + x_2)$$

it follows that

$$U_1(x_1) + U_1(x_2) \leqslant U_1(x_1 + x_2).$$

We have thus proved equality (1).

If we now put

$$U(x) = U_1(x_+) - U_1(x_-),$$

for any element x, we obtain an additive operator on $\mathscr{X}$, which obviously is an extension of U_0. It is also obvious that U is positive.

If $A \in \mathscr{G}$, then there exists a $y_0 \in \mathscr{Y}$, such that

$$x \in A \cap \mathscr{X}_0 \Rightarrow |U_0(x)| \leqslant y_0.$$

If $\mathbf{0} \leqslant x \in A$ and $\mathbf{0} \leqslant x' \leqslant x$, whereas $x' \in \mathscr{X}_0$, then $U_0(x') \leqslant y_0$ and, therefore, $U(x) \leqslant y_0$. For any element $x \in A$ we consequently have

$$|U(x)| \leqslant U(|x|) \leqslant y_0$$

and this shows that the set $U(A)$ is (o)-bounded. Consequently, the operator U is (τo)-bounded.

6.3.2 (po)-Bounded linear operator

Let $\mathscr{X}$ be a locally convex vector lattice, τ its topology and $\mathscr{Y}$ a complete vector lattice.

Definition.[15] An operator U, which maps $\mathscr{X}$ into $\mathscr{Y}$, is said to be *(po)-bounded* if there exists a (τ)-continuous seminorm p on the space $\mathscr{X}$ and an element $y_0 \in \mathscr{Y}$, such that

$$|U(x)| \leqslant p(x)y_0, \tag{1}$$

for any $x \in \mathscr{X}$.

It is obvious that any (po)-bounded operator is (τo)-bounded.

If U is a (po)-bounded operator, then in (1) the seminorm p can be assumed to be not only (τ)-continuous, but also a solid semi-norm.

Lemma [16]) *Let $\mathscr{V}$ be a (real) vector space and P a mapping of $\mathscr{V}$ into a complete vector lattice $\mathscr{Y}$, which satisfies the following conditions*

$$P(v_1 + v_2) \leqslant P(v_1) + P(v_2),$$

$$P(\alpha v) = \alpha P(v), \ (0 \leqslant \alpha \in \mathbb{R}).$$

Let $\mathscr{V}_0$ be a vector subspace of $\mathscr{V}$ and U_0 a linear operator, defined on $\mathscr{V}_0$, taking values in $\mathscr{Y}$, such that

$$U_0(v) \leqslant P(v), \quad v \in \mathscr{V}_0.$$

There then exists a linear operator U, defined on $\mathscr{V}$, taking values in $\mathscr{Y}$, such that $U(v) = U_0(v)$, for $v \in \mathscr{V}_0$ and $U(v) \leqslant P(v)$, for any $v \in \mathscr{V}$.

Proof. Let $w \in \mathscr{V}$ be such that $w \notin \mathscr{V}_0$ and denote by $\mathscr{V}_1$ the linear hull of the set $\mathscr{V}_0 \cup \{w\}$. By denoting

$$a = \bigvee_{v \in \mathscr{V}_0} \{-P(-v-w) - U_0(v)\},$$

$$b = \bigwedge_{v \in \mathscr{V}_0} \{P(v + w) - U_0(v)\},$$

we have $a \leqslant b$. Let $y_0 \in \mathscr{Y}$ be such that $a \leqslant y_0 \leqslant b$. If we define

$$U_1(v + \alpha w) = U_0(v) + \alpha y_0,$$

for any $v \in \mathscr{V}_0$ and any $\alpha \in \mathbb{R}$, we get a linear operator U_1, defined on $\mathscr{V}_1$ and taking values in $\mathscr{Y}$, such that $U_1(v) = U_0(v)$, for $v \in \mathscr{V}_0$, and moreover,

$$U_1(v + \alpha w) \leqslant P(v + \alpha w).$$

---

[15]) The definition makes sense even if $\mathscr{X}$ is an arbitrary locally convex space and $\mathscr{Y}$ a vector lattice.

[16]) This lemma is a generalization (by L. V. Kantorovitch) of the well-known Hahn-Banach theorem (see R. Cristescu [31]). Since the proof is quite similar, we do not insist upon the details.

The lemma will now follow by applying Zorn's lemma.

Theorem. *For any element $x_0 > 0$ of the space $\mathscr{X}$, for any (τ)-continuous solid semi-norm p defined on $\mathscr{X}$, and for any element $y_0 > 0$ of the space $\mathscr{Y}$, there exists a (po)-bounded, positive, linear operator U, defined on $\mathscr{X}$, and taking values in $\mathscr{Y}$, such that*

$$U(x_0) = p(x_0)\, y_0, \tag{2}$$

$$|U(x)| \leqslant p(x)\, y_0, \quad (\forall\, x \in \mathscr{X}). \tag{3}$$

Proof. Let us define

$$\mathscr{X}_0 = \{\lambda x_0; \quad \lambda \in \mathbb{R}\}$$

and consider the linear operator U_0 (which maps $\mathscr{X}_0$ into $\mathscr{Y}$) given by the formula

$$U_0(\lambda x_0) = \lambda p(x_0) y_0.$$

With the preceding lemma, U_0 can be extended to $\mathscr{X}$ as a linear operator U_1 which satisfies the condition

$$|U_1(x)| \leqslant p(x) y_0, \quad \forall x \in \mathscr{X}.$$

Since p is solid, one immediately infers that

$$|U_1|(|x|) \leqslant p(x) y_0,$$

for any x. From the relations

$$p(x_0)y_0 = U_1(x_0) \leqslant |U_1|(x_0) \leqslant p(y_0)\, y_0$$

it follows that

$$|U_1|(x_0) = p(x_0)\, y_0.$$

By defining $U = |U_1|$, one obtains a (po)-bounded, positive, linear operator, which satisfies conditions (2) and (3).

6.3.3 (τo)-bounded operators and (τo)-continuous operators defined on normed spaces

Let $\mathscr{Y}$ be a complete vector lattice.

If $\mathscr{X}$ is a normed vector space, then a linear operator U, which maps $\mathscr{X}$ into $\mathscr{Y}$ is (τo)-bounded if it maps the ball $S = \{x;\ \|x\| \leqslant 1\}$ into an (o)-bounded set.

If U is a (τo)-bounded linear operator (defined on the normed vector space $\mathscr{X}$, taking values in $\mathscr{Y}$), then the element given by the formula

$$\mathbf{|} U \mathbf{|} = \sup_{\|x\| \leqslant 1} |U(x)|$$

is called the *vector norm* of U. The following inequality

$$|U(x)| \leqslant \|x\| \, |U| \tag{1}$$

holds for any $x \in \mathscr{X}$. In particular, U is (τo)-continuous (i.e. $x_n \xrightarrow{\tau} x$ implies $U(x_n) \xrightarrow{o} U(x)$).

Proposition 1. *If U_0 is a (τo)-bounded, linear operator, defined on a vector subspace $\mathscr{X}_0$ of the normed vector space $\mathscr{X}$ (and taking values in $\mathscr{Y}$), then U_0 can be extended, to $\mathscr{X}$ as a linear operator with the same vector norm.*

Proof. By defining

$$P(x) = \|x\| \, |U_0|$$

the hypothesis of the lemma from Section 6.3.2 is satisfied. The operator U_0 can be, therefore, extended to a linear operator U on $\mathscr{X}$, which satisfies the inequality

$$|U(x)| \leqslant \|x\| \, |U_0|.$$

From this inequality we infer that $|U| \leqslant |U_0|$ and since U is an extension of U_0, we immediately infer that $|U| = |U_0|$.

Proposition 2. *If $\mathscr{X}$ is any normed vector space and if $\mathscr{Y}$ is of the bounded type, then in order that a linear operator U (which maps $\mathscr{X}$ into $\mathscr{Y}$) be (τo)-bounded it is necessary and sufficient that it be (τo)-continuous.*

Proof. We have only to prove the sufficiency. Let U be a (τo)-continuous operator and $y_n \in U(S)$, $n \in \mathbb{N}$. Let $x_n \in S$ be such that $y_n = U(x_n)$. If $\alpha_n \to 0$ $(\alpha_n \in \mathbb{R})$, then $\alpha_n x_n \xrightarrow{\tau} \mathbf{0}$, whence

$$\alpha_n U(x_n) = U(\alpha_n x_n) \xrightarrow{o} \mathbf{0}.$$

Since the space $\mathscr{Y}$ is of bounded type, the set $U(S)$ is bounded, hence U is (τo)-bounded.

Proposition 3. *Any (τo)-bounded linear operator, defined on a normed vector lattice $\mathscr{X}$ (and taking values in the complete vector lattice $\mathscr{Y}$) is a regular operator.*

Indeed, if A is an (o)-bounded subset of $\mathscr{X}$, then there exists an $x_0 \in \mathscr{X}$, such that $|x| \leqslant x_0$ for any $x \in A$. From inequality (1) we infer that

$$|U(x)| \leqslant \|x_0\| \, |U|$$

for any $x \in A$ and, therefore, $U(A)$ is (o)-bounded.

Proposition 4. *If $\mathscr{X}$ is a space with a strong unit (and if $\mathscr{Y}$ is an arbitrary complete, vector lattice), then in order that a linear operator U, which maps $\mathscr{X}$ into $\mathscr{Y}$, be (τo)-bounded, it is necessary and sufficient that it be regular.*

Indeed, if U is regular and if u is the unit element of $\mathscr{X}$, then $\|x\| \leqslant 1$ implies $|x| \leqslant u$ and, therefore,

$$|U(x)| \leqslant |U|(|x|) \leqslant |U|(u).$$

Remark. If $\mathscr{X}$ is a space with a strong unit and if U is a (τo)-bounded, linear operator, then

$$\|U\| = |U|(u).$$

This assertion can be easily proved.

6.3.4 (*o*)-Continuous linear operators defined on spaces of type (KB)

Let $\mathscr{X}$ and $\mathscr{Y}$ be spaces of type (KB).

Proposition 1. *In order that a linear operator U, which maps $\mathscr{X}$ into $\mathscr{Y}$, be (o)-continuous, it is necessary and sufficient that a real number μ should exist, such that*

$$\left\| \bigvee_{i=1}^{n} |U(x_i)| \right\| \leqslant \mu \left\| \bigvee_{i=1}^{n} |x_i| \right\|, \tag{1}$$

for any elements $x_1, x_2, \ldots, x_n$ of $\mathscr{X}$.

Proof. If U is (o)-continuous, then it is (τ)-continuous because the (τ)-convergence of sequences in a space of type (KB) coincides with the (o^*)-convergence (Section 6.2.9, proposition 8).

On the other hand, since the space $\mathscr{Y}$ is of bounded type, any (o)-continuous linear operator, which maps $\mathscr{X}$ into $\mathscr{Y}$, is regular (Section 5.2.2, remarks).

If U is an (o)-continuous linear operator, which maps $\mathscr{X}$ into $\mathscr{Y}$, then $|U|$ is (o)-continuous, hence (τ)-continuous. If $x_1, x_2, \ldots, x_n$ are elements of $\mathscr{X}$, then from

$$\bigvee_{i=1}^{n} |U(x_i)| \leqslant |U|\left(\bigvee_{i=1}^{n} |x_i|\right)$$

it follows that

$$\left\| \bigvee_{i=1}^{n} |U(x_i)| \right\| \leqslant \left\| |U|\left(\bigvee_{i=1}^{n} |x_1|\right)\right\| \leqslant \| \, |U| \, \| \left\| \bigvee_{i=1}^{n} |x_i| \right\|.$$

We have thus shown the necessity of the condition.

Conversely, let U be a linear operator which satisfies condition (1). If A is an (o)-bounded subset of $\mathscr{X}$, then there exists a number $\alpha > 0$, such that

$$\left\| \bigvee_{i=1}^{n} |x_i| \right\| \leqslant \alpha,$$

for any elements $x_1, x_2, \ldots, x_n \in A$. It follows that

$$\left\| \bigvee_{i=1}^{n} |U(x_i)| \right\| \leqslant \mu\alpha$$

and, therefore, $U(A)$ is (o)-bounded (see Section 6.2.9, proposition 7). Consequently, the operator U is regular and, since $\mathscr{X}$ is a regular space, U is (o)-continuous (Section 5.2.2, remarks).

Remark. The smallest number μ which satisfies relation (1) is equal to $\| \, |U| \, \|$.

Indeed, since the space $\mathscr{Y}$ is of (o)-countable type, for any $x > 0$ there exists a sequence $\{x_n\}_{n\in\mathbb{N}}$ of elements of $\mathscr{X}$, such that $|x_n| \leqslant x$ and

$$|U|(x) = (o)\text{-}\lim_n \bigvee_{i=1}^{n} |U(x_i)|.$$

It follows that

$$\| \, |U(x)| \, \| = \lim_n \left\| \bigvee_{i=1}^{n} |U(x_i)| \right\| \leqslant \mu \|x\|$$

and this implies that $\| |U| \| \leqslant \mu$.

One can also observe that

$$\| |U| \| = \sup \left\{ \left\| \bigvee_{i=1}^{n} |U(x_i)| \right\|; \ \left\| \bigvee_{i=1}^{n} |x_i| \right\| \leqslant 1 \right\}.$$

Proposition 2. *If $\mathscr{X}$ is a space of type (L), whereas $\mathscr{Y}$ is an arbitrary space of type (KB), then in order that a linear operator U, which maps $\mathscr{X}$ into $\mathscr{Y}$, be (o)-continuous it is necessary and sufficient that it be (τ)-continuous.*

Proof. We have to prove only the sufficiency. Let then U be a (τ)-continuous linear operator, which maps $\mathscr{X}$ into $\mathscr{Y}$. Let a be any positive element of $\mathscr{X}$ and let us define

$$G(a) = \left\{ \sum_{i=1}^{n} |U(x_i)|; \quad 0 \leqslant x_i \in \mathscr{X}, \quad \sum_{i=1}^{n} x_i = a \right\}. \tag{2}$$

It is easy to show that the set $G(a)$ is directed to the right. If $y \in G(a)$, then

$$\|y\| \leqslant \sum_{i=1}^{n} \|U(x_i)\| \leqslant \|U\| \sum_{i=1}^{n} \|x_i\| = \|U\| \, \|a\|.$$

If $y_1, y_2, \ldots, y_n$ are arbitrary elements of $G(a)$, then there exists a $y_0 \in G(a)$, such that

$$\bigvee_{i=1}^{n} y_i \leqslant y_0,$$

whence

$$\left\| \bigvee_{i=1}^{n} y_i \right\| \leqslant \|y_0\| \leqslant \|U\| \cdot \|a\|.$$

The set $G(a)$ is, therefore, (o)-bounded. Let us define

$$P(a) = \sup G(a). \tag{3}$$

If x is an element of $\mathscr{X}$ such that $|x| \leqslant a$, then $x_+ + x_- \leqslant a$ and, therefore

$$|U(x)| \leqslant |U(x_+)| + |U(x_-)| \leqslant P(a).$$

If A is an (o)-bounded subset of $\mathscr{X}$, then there exists an a such that $|x| \leqslant a$, for any $x \in A$. If follows that $U(A)$ is (o)-bounded, hence U is a regular operator. Since the space $\mathscr{X}$ is regular, U is an (o)-continuous operator.

Remark. If $\mathscr{X}$ is a space of type (L), whereas $\mathscr{Y}$ is an arbitrary space of type (KB), then for any (o)-continuous linear operator U, which maps $\mathscr{X}$ into $\mathscr{Y}$, the equality $\| |U| \| = \| U \|$ holds.

Indeed, since the space $\mathscr{Y}$ is of countable type, there exists a sequence $\{y_n\}_{n \in \mathbb{N}}$ of elements of $G(x)$ (see formulae (2) and (3)) such that

$$\bigvee_{n=1}^{\infty} y_n = P(x).$$

Since the set $G(a)$ is directed to the right, one can assume that the sequence $\{y_n\}_{n \in \mathbb{N}}$ is increasing. As we have shown in the proof of proposition 2, we have

$$\|y_n\| \leqslant \| U \| \cdot \| x\|,$$

whence

$$\|P(x)\| \leqslant \|U\| \cdot \|x\|.$$

Since for any $x \geqslant \mathbf{0}$ we have

$$|U|(x) \leqslant P(x)$$

it follows that

$$\| |U|(x) \| \leqslant \|U\| \cdot \|x\|$$

and, therefore, we get the inequality $\| |U| \| \leqslant \| U \|$. The opposite inequality can be immediately proved.

★

Let $\mathscr{X}$ be a space of type (KB) and U a positive translator on $\mathscr{X}$. For any element $a \in \mathscr{X}$ and any natural number n, let $a^n = \frac{1}{n} \sum_{i=0}^{n-1} U^i(a)$ (see Section 5.3.3). One can prove the following ergodic theorem [17]).

[17]) See H. Nakano [9].

If there exists a number $\alpha > 0$, such that

$$\|U^n(x)\| \leqslant \alpha \|x\|, \tag{4}$$

for any $n \in \mathbb{N}$ and $x \in \mathscr{X}$, then $\mathscr{X}$ has property $\mathscr{E}$ with respect to U and for any element v, which is invariant with respect to U, the sequence $\{[v]a^n\}_{n \in \mathbb{N}}$ is individually convergent.

The theorem also holds[18]) if $\mathscr{X}$ is a space of type (KP) and if condition (4) is extended in the following manner: there exists a directed family $\mathscr{P} = \{p_j\}_{j \in J}$ of solid semi-norms which determines the topology of $\mathscr{X}$ and a family $\{\alpha_j\}_{j \in J}$ of positive numbers, such that

$$p_j(U^n(x)) \leqslant \alpha_j \, p_j(x),$$

for any $j \in J$ (and $n \in \mathbb{N}$, $x \in \mathscr{X}$).

6.4 Special spaces

6.4.1 Spaces of Lebesgue measurable functions

(1) Let $T = [\alpha, \beta]$ be an interval of the real line and $S(T)$ the space of classes of real Lebesgue measurable functions, which was defined in Section 4.1.1 (example 3). We shall denote by x (instead of $\dot{x}$) the element of $S(T)$ which corresponds to a function x.

As one knows, $S(T)$ is a vector lattice. We shall show that it is a complete vector lattice.

If $\{x_n\}_{n \in \mathbb{N}}$ is a majorized sequence of elements of $S(T)$, then by putting

$$x(t) = \sup_{n \in \mathbb{N}} x_n(t)$$

we have $x \in S(T)$ and one can easily show that $x = \bigvee_{n \in \mathbb{N}} x_n$; consequently, $S(T)$ is a σ-complete vector lattice.

In order to show that $S(T)$ is a complete vector lattice, we shall consider the following function

$$q(x) = \int_T \frac{x(t)}{1 + x(t)} \, \mathrm{d}t, \qquad (\mathbf{0} \leqslant x \in S(T)).$$

This function has the following properties

(i) $$\mathbf{0} \leqslant x < y \Rightarrow q(x) < q(y),$$

(ii) $$\mathbf{0} \leqslant x_n \xrightarrow{o} x \Rightarrow q(x_n) \to q(x).$$

---

18) See R. Cristescu [11].

Let now A be a majorized set of positive elements of $S(T)$. Let us denote by E the set of all least upper bounds of all finite subsets of A. Let

$$\lambda = \sup_{x \in E} q(x)$$

and let us consider a sequence $\{x_n\}_{n \in \mathbb{N}}$ of elements of E, such that $q(x_n) \to \lambda$. Let us denote

$$y_n = \bigvee_{i=1}^{n} x_i, \qquad n \in \mathbb{N}.$$

The sequence $\{y_n\}_{n \in \mathbb{N}}$ consists of elements of E, is increasing and $q(y_n) \to \lambda$. If z is an upper bound for A and, therefore, for E, then $y_n \leqslant z$. Then there exists the element

$$y = \bigvee_{n=1}^{\infty} y_n$$

and we have $y \leqslant z$ and $q(y) = \lambda$. If $x \in E$, then

$$q(x \vee y) = \lambda,$$

as one can easily see. But then, with (i), one infers that $x \vee y = y$ and this implies that y is an upper bound for E. Consequently, we have

$$y = \sup E = \sup A.$$

This proves that $S(T)$ is a complete vector lattice.

In the space $S(T)$ the (o)-convergence $x_n \xrightarrow{o} x$ is equivalent to $x_n(t) \to x(t)$, almost everywhere, whereas $x_n \xrightarrow{o^*} x$ is equivalent to $x_n(t) \to x(t)$ in measure.

The space $S(T)$ is a σ-regular space, because the proposition on the diagonal sequence (Section 4.2.1, proposition 2) coincides in this case with the well-known Frechet theorem on the almost everywhere convergent sequences.

One can also easily check that $S(T)$ is a space of (o)-countable type, a space of bounded type and a maximal space.

(2) Let us now consider a real number $p \geqslant 1$ and the subset $L^p(T)$ of $S(T)$ consisting of all elements corresponding to the Lebesgue p-integrable functions. Endowed with the algebraic operations and the order relation induced by those of $S(T)$, the set $L^p(T)$ is a normal subspace. In particular, $L^p(T)$ is a complete vector lattice [19]).

In the space $L^p(T)$ the convergence $x_n \xrightarrow{o} x$ means: (i) $x_n(t) \to x(t)$ almost everywhere, (ii) there exists a p-integrable function y such that $| x_n(t) | \leqslant y(t)$, almost everywhere.

---

[19]) If $p = 1$, one writes $L(T)$ instead of $L^1(T)$.

The space $L^p(T)$ is a space of type (KB) with respect to the norm

$$\|x\| = \left(\int_T |x(t)^p| \,\mathrm{d}t\right)^{1/p}.$$

For $p = 1$, one gets a space of type (L).

One can show that in the space $L^p(T)$ the convergence $x_n \xrightarrow{o} x$ is equivalent to

$$\lim_n \int_T \max_{n \leqslant i \leqslant n+m} | x(t) - x_i(t) |^p \,\mathrm{d}t = 0,$$

for any natural number m.

(3) Let us now consider the space $AC(T)$ of all absolutely continuous real functions, defined on T, which vanish at α (Section 3.4.3).

It is a space of type (KB) with respect to the norm

$$\| x \| = \operatorname*{var}_{t \in T} x(t).$$

As one knows, in order that $x \in AC(T)$ it is necessary and sufficient that x have the following form

$$x(t) = \int_\alpha^t y(t) \,\mathrm{d}t, \tag{1}$$

where y is a Lebesgue integrable function, defined on T. In this case we have

$$\operatorname*{var}_{t \in T} x(t) = \int_\alpha^\beta | y(t) | \,\mathrm{d}t.$$

The mapping $x \to y$ of $AC(T)$ onto $L(T)$, given by equality (1), is an isomorphism of ordered vector spaces, as one can easily prove.

This shows that $L(T)$ is isomorphic to $AC(T)$ as a normed vector lattice.

If $p > 1$, we shall consider the space $V_0^p(T)$ of all real functions, defined on T, which vanish at α and are of bounded p-variation on T. (Section 3.4.3). It is a space of type (KB) with respect to the norm

$$\| x \| = (p)\text{-}\operatorname*{var}_{t \in T} x(t).$$

In order that $x \in V_0^p(T)$ it is necessary and sufficient that x be representable in the form (1), where y is a Lebesgue p-integrable function on T. In this case we have

$$(p)\text{-}\operatorname*{var}_{t \in T} x(t) = \left(\int_\alpha^\beta | y(t) |^p \,\mathrm{d}t\right)^{1/p}.$$

The mapping $x \to y$ is an isomorphism between the normed vector lattice $V_0^p(T)$ and $L^p(T)$.

(4) Let $M(T)$ be the subset of the space $S(T)$ consisting of all elements which correspond to essentially bounded functions. If one chooses as a unit in $S(T)$ the element u represented by the function $u(t) \equiv 1$, then $M(T)$ is the space of the bounded elements of $S(T)$. In particular, $M(T)$ is a complete vector lattice with strong unit. The norm associated to the unit is given by the formula

$$\| x \| = \underset{t \in T}{\text{vrai max}} \ | x(t) |.$$

In the space $M(T)$ the convergence $x_n \xrightarrow{o} x$ means: (i) $x_n(t) \to x(t)$, almost everywhere; (ii) there exists a number λ, such that $| x_n(t) | \leqslant \lambda$, almost everwhere (for any $n \in \mathbb{N}$).

One can also show that in the space $M(T)$ the convergence $x_n \xrightarrow{o} x$ is equivalent to

$$\lim_{n \to \infty} \int_T \max_{n \leqslant i \leqslant n+m} | x_i(t) - x(t) | \, y(t) \, dt = 0,$$

for any Lebesgue integrable function y which is almost everywhere positive on T (and for any natural number m).

6.4.2 The concrete representation of type (L) spaces

Let T be a set, $\mathscr{T}$ a σ-algebra of subsets of T, and μ a measure on $\mathscr{T}$. One can consider the space $L(T, \mathscr{T}, \mu)$ of all classes of $(\mathscr{T}, \mu)$-integrable functions, which is defined similarly to the space $L(T)$ from the preceding section (6.4.1) [20]). By using the notation conventions from the same section, the relation $x \leqslant y$ in $L(T, \mathscr{T}, \mu)$, therefore, means that $x(t) \leqslant y(t)$, μ-almost everywhere. With the norm

$$\| x \| = \int_T | x(t) | \, d\mu(t)$$

the space $L(T, \mathscr{T}, \mu)$ is of type (L). The element u, which is represented by the constant function equal to one, is a total element and it can be chosen as a unit element.

Theorem. *Any space of type* (*L*), *with a unit element, is isomorphic* (*as a normed vector lattice* [21]) *to a space* $L(T, \mathscr{T}, \mu)$.

Proof. Let $\mathscr{X}$ be a space of type (L), with the unit u, and let $\mathscr{B}(\mathscr{X})$ be the basis of the space $\mathscr{X}$. Without loss of generality, one can assume that $\| u \| = 1$. Since the

---

[20]) For the theory of integration, see P. Halmos [1] or N. Dinculeanu [1]. The notions of which we make use in this section can also be found in Al. Ghika [4], Chapter 1.

[21]) i.e. isomorphic as an ordered vector space by a mapping which is also an isometry.

set $\mathscr{B}(\mathscr{X})$ is a (complete) Boolean lattice, there exists a totally disconnected compact topological space T, such that $\mathscr{B}(\mathscr{X})$ be isomorphic to the set $\mathscr{A}$ of all clopen subsets of T, the latter being ordered by inclusion (Section 1.2.4). Let $e \mapsto E_e$ be this isomorphism. If, for any subset A of T, we define

$$\mu^*(A) = \inf_{E_e \supset A} \| e \|,$$

the function μ^* has all properties of an exterior measure.

Indeed, it is obvious that $\mu^*(\varnothing) = 0$, and that $A_1 \subset A_2$ implies $\mu^*(A_1) \leqslant \mu^*(A_2)$.

Let now $\{A_n\}_{n \in \mathbb{N}}$ be a sequence of subsets of T. For any number $\varepsilon > 0$ and any $n \in \mathbb{N}$, there exists a set E_{e_n} such that $A_n \subset E_{e_n}$ and

$$\| e_n \| < \mu^*(A_n) + \frac{\varepsilon}{2^n}.$$

If we put

$$e = \bigvee_{n \in \mathbb{N}} e_n,$$

then

$$\bigcup_{n \in \mathbb{N}} A_n \subset E_e,$$

and

$$\mu^*(\bigcup_{n \in \mathbb{N}} A_n) \leqslant \| e \| \leqslant \sum_{n=1}^{\infty} \| e_n \| \leqslant \sum_{n=1}^{\infty} \mu^*(A_n) + \varepsilon.$$

Since ε was arbitrary, it follows that

$$\mu^*\left(\bigcup_{n=1}^{\infty} A_n\right) \leqslant \sum_{n=1}^{\infty} \mu^*(A_n).$$

Let $\mathscr{T}$ be the set of all subsets of T, which are μ^*-measurable, i.e. $A \in \mathscr{T}$ iff

$$\mu^*(B) = \mu^*(B \cap A) + \mu^*(B \cap \complement A),$$

for any B. One easily shows that $\mathscr{A} \subset \mathscr{T}$. The set $\mathscr{T}$ is a σ-algebra whereas the restriction μ of μ^* to $\mathscr{T}$ is a measure on $\mathscr{T}$.

For any set $A \in \mathscr{T}$ there exists a set $E_e \in \mathscr{A}$ which is equivalent to A.

Indeed, there exists a sequence $\{E_{e_n}\}_{n \in \mathbb{N}}$ of sets from $\mathscr{A}$, such that $E_{e_n} \supset A$, for any n, and such that

$$\mu(A) = \lim_n \mu(E_{e_n}).$$

If we define

$$e = \bigwedge_{n \in \mathbb{N}} e_n,$$

then one easily shows that A and E_e are equivalent.

In the space $L(T, \mathscr{T}, \mu)$ by taking as a unit the element represented by the constant function equal to one, the basis of the space will consist of the elements represented by the characteristic functions of the sets belonging to $\mathscr{T}$.

But then it is obvious that $\mathscr{B}(\mathscr{X})$ is isomorphic (as an ordered set) to the basis of the space $L(T, \mathscr{T}, \mu)$ by the mapping $e \mapsto \chi_{E_e}$, where χ_A denotes the characteristic function of the set A. Moreover $\| e \| = \| \chi_{E_e} \|$. The fact that the spaces $\mathscr{X}$ and $L(T, \mathscr{T}, \mu)$ are isomorphic can now be easily proved.

Corollary. *Any separable space of type* (L)*, is isomorphic* (*as a normed vector lattice*) *to a closed vector sublattice of the space* L ([0,1]).

Indeed, if $\mathscr{X}$ is separable as a normed space, then the basis of the space $L(T, \mathscr{T}, \mu)$ is separable with respect to the distance given by the formula

$$d(\chi_{A_1}, \chi_{A_2}) = \mu(A_1 \cup A_2 \cap \complement(A_1 \cap A_2)). \tag{1}$$

On the other hand, we have $\mu(T) = 1$.

As one knows [22], in this case, $\mathscr{T}$ can be indentified with a σ-algebra of Lebesgue measurable subsets of [0,1], whereas μ corresponds to the Lebesgue measure by this isomorphism.

6.4.3 An application to ergodic theory

Let $\mathscr{X}$ be a space of type (L).

Let U be a positive, linear operator, defined on $\mathscr{X}$ and taking values in $\mathscr{X}$, such that $\| U(x) \| = \| x \|$, for any $x \geqslant 0$.

If x is an arbitrary element of $\mathscr{X}$, then

$$\| U(x) \| = \| U(x_+) - U(x_-) \|$$

$$\leqslant \| U(x_+) \| + \| U(x_-) \| = \|x_+\| + \|x_-\| = \| x \|,$$

hence U is (τ)-continuous and $\| U \| = 1$.

An element $x \in \mathscr{X}$ is said to be *invariant* with respect to U if $U(x) = x$.

If $a \in \mathscr{X}$ and n is an arbitrary natural number, we shall define

$$a^n = \frac{1}{n} \sum_{i=0}^{n-1} U^i(a),$$

---

[22]) See, e.g., P. Halmos [1], Section 8.4.1.

where $U^0(a) = a$, whereas $U^i(a) = U(U^{i-1}(a))$, for any $i \in \mathbb{N}$.

Lemma 1. *For any natural number m we have*

$$\lim_{n\to\infty} \| (a - U^m(a))^n \| = 0.$$

Indeed, we have only to observe that if $n, m \in \mathbb{N}$ and $n > m$, then

$$\left\| \frac{1}{n} \sum_{i=0}^{n-1} U^i(a - U^m(a)) \right\| = \frac{1}{n} \left\| \sum_{i=0}^{n-1} U^i(a) - \sum_{i=m}^{n+m-1} U^i(a) \right\| \leqslant \frac{2m}{n} \| a \| .$$

Lemma 2. *If the sequence $\{a^n\}_{n\in\mathbb{N}}$ contains a weakly convergent* [23]) *subsequence, whose limit is b, then b is invariant with respect to U.*

Indeed, let

$$\lim_{i\to\infty} f(a^{n_i} - b) = 0,$$

for any $f \in \mathscr{X}_\tau^*$. Since

$$\lim_{n\to\infty} \| a^n - U(a^n) \| = 0,$$

it follows that

$$\lim_{i\to\infty} f(a^{n_i} - U(a^{n_i})) = 0,$$

for any $f \in \mathscr{X}_\tau^*$. One can now easily infer that $U(b) = b$.

Definition. One calls an *ergodic element* (with respect to U) any element $a \in \mathscr{X}$ such that there exists an element b, invariant with respect to U, such that $a^n \xrightarrow{\tau} b$.

Lemma 3. *The set E of all ergodic elements, with respect to U, is a* (τ)*-closed vector subspace.*

Indeed, the set E is obviously a vector subspace of $\mathscr{X}$. Let $\{a_m\}_{m\in\mathbb{N}}$ be a sequence of elements of E such that $a_m \xrightarrow{\tau} a$.
Let us define

$$b_m = (\tau)\text{-}\lim_n (a_m)^n.$$

Since $\{b_m\}_{m\in\mathbb{N}}$ is a Cauchy sequence, whereas each b_m is invariant, there exists an invariant element b, such that $b_m \xrightarrow{\tau} b$. From the relations

$$\| a^n - b \| \leqslant \| a^n - (a_m)^n \| + \| (a_m)^n - b_m \| + \| b_m - b \|$$
$$\leqslant \| a - a_m \| + \| (a_m)^n - b_m \| + \| a_m - a \| = 2 \| a - a_m \| + \| (a_m)^n - b_m \|$$

---

[23]) i.e., convergent for the weakened topology with respect to the topology τ of $\mathscr{X}$.

it immediately follows that $a^n \xrightarrow{\tau} b$, and, therefore, $a \in E$.

Theorem. *If the sequence* $\{U^n(a)\}_{n \in \mathbb{N}}$ *is* (o)*-bounded, then a is an ergodic element.*

Proof. Let us first suppose that the sequence $\{a^n\}_{n \in \mathbb{N}}$ contains a subsequence $\{a^{n_i}\}_{i \in \mathbb{N}}$ which converges weakly to an element b. By virtue of lemma 1, the element $a - U^m(a)$ is ergodic (for any $m \in \mathbb{N}$). On the other hand, any element of the form $a - a^i$ can be obviously written with the help of an element of the form $a - U^m(a)$. It follows that the element $a - a^{n_i}$ is ergodic (for any $i \in \mathbb{N}$) and, with lemma 3, it follows that $a - b$ is ergodic. From lemma 2 we infer that the element b is ergodic and this shows that the element a is ergodic.

If $\mathscr{X} = L([0,1])$, any (o)-bounded sequence of elements of the space contains a weakly convergent subsequence. One can easily observe that from

$$x_1 \leqslant U^i(a) \leqslant x_2 \quad (\forall\, i \in \mathbb{N})$$

it follows that

$$x_1 \leqslant a^n \leqslant x_2, \quad (\forall n \in \mathbb{N}),$$

and this shows that the sequence $\{a^n\}_{n \in \mathbb{N}}$ is (o)-bounded. Consequently, if $\mathscr{X} = L([0,1])$, then the element a is ergodic.

If $\mathscr{X}$ is an arbitrary space of type (L), let $\mathscr{X}_0$ be the (τ)-closed vector sublattice generated by the elements $U^n(a)$, $n \in \mathbb{N}$. Since $\mathscr{X}_0$ is a (τ)-closed, separable subspace of the space $\mathscr{X}$, it follows that $\mathscr{X}_0$ can be considered as being a subspace of the space $L([0,1])$ (see Section 6.4.2). Since the sequence $\{U^n(a)\}_{n \in \mathbb{N}}$ is (o)-bounded in $\mathscr{X}_0$, from what we have previously shown, it follows that the element a is ergodic.

6.4.4 Spaces of continuous real functions [24]

(1) Let $\mathscr{X}$ be a locally convex lattice of type (M) and τ the topology of the space. We shall denote by $\mathscr{W}$ the set of all equilibrated convex neighbourhoods of the origin.

Let $\mathscr{F}$ be the set of all functionals f defined on $\mathscr{X}$, which, moreover, are linear, (τ)-continuous and satisfy the following condition

$$f(x \vee y) = \max\{f(x), f(y)\},$$

(for any $x, y \in \mathscr{X}$). We shall endow $\mathscr{F}$ with the weak topology, i.e., the topology induced by the topology $\sigma(\mathscr{X}^*_\tau, \mathscr{X})$.

Let $C(\mathscr{F})$ be the space of all continuous real functions, defined on $\mathscr{F}$, endowed with the topology defined by the system of all semi-norms of the form [25]

$$p_W(\varphi) = \sup_{f \in W^0 \cap F} |\varphi(f)|, \ (W \in \mathscr{W}),$$

where W^0 is the polar of W in $\mathscr{X}^*_\tau$, whereas $\varphi \in C(\mathscr{F})$.

---

[24]) In this section some of the results will be given without proofs.

[25]) One can prove that if W is an equilibrated and convex neighbourhood of the origin, in a locally convex space $\mathscr{X}$, then its polar W^0 in $\mathscr{X}^*_\tau$ is a compact set for the topology $\sigma(\mathscr{X}^*_\tau, \mathscr{X})$.

One can prove [26]) that the space $\mathscr{X}$ is isomorphic (as a topological ordered vector space) to a subset of the space $C(\mathscr{F})$.

The isomorphism is given by the correspondence $x \mapsto \varphi_x$, where $\varphi_x(f) = f(x)$.

(2) Let T be a locally compact σ-compact space, i.e., a locally compact space, such that there exists a sequence $\{T_n\}_{n\in\mathbb{N}}$ of relatively compact open sets, with the following properties: $\overline{T}_n \subset T_{n+1}$ and $\bigcup_{n\in\mathbb{N}} T_n = T$.

Let us denote by $C_0(T)$ the set of all continuous real functions, defined on T, and having compact supports. With the usual operations and the natural order relation (i.e., $x \leqslant y$ iff $x(t) \leqslant y(t)$, $t \in T$), the set $C_0(T)$ becomes a vector lattice.

For any continuous real function y on T, such that $y(t) > 0$ for any $t \in T$, let us define

$$W(y) = \{x \in C_0(T);\ |x(t)| \leqslant y(t),\ \forall t \in T\}.$$

We shall consider in $C_0(T)$ the topology τ which has as a fundamental system of neighbourhoods of the origin, the system of all sets of the form $W(y)$. Endowed with the topology τ, $C_0(T)$ is a locally convex vector lattice.

One can observe that $C_0(T)$ can be produced as a strict inductive limit of a sequence of complete metrizable locally convex spaces.

Conversely, let $\mathscr{X}$ be a locally convex vector lattice, which can be represented as a strict inductive limit of a sequence of complete metrizable locally convex spaces. One can prove [27]) that $\mathscr{X}$ is isomorphic to a space $C_0(T)$, iff there exists a sequence $\{e_n\}_{n\in\mathbb{N}}$ of positive elements of $\mathscr{X}$, with the following property: for any element $x \in \mathscr{X}$ there exists an e_n and a number $\lambda > 0$, such that $|x| \leqslant \lambda e_n$.

6.4.5 Examples of regular operators defined on special spaces

If $\mathscr{X}$ and $\mathscr{Y}$ are spaces of type (KB), then an operator U, which maps $\mathscr{X}$ into $\mathscr{Y}$, is regular iff it is linear and (o)-continuous (see the remarks from Section 5.2.2). We shall consider the spaces $L^p([0,1])$ $(p \geqslant 1)$, with the structure of spaces of (KB)-type, as shown in section 1, as well as the space $V([0,1])$ of all real functions of bounded variation on $[0,1]$ (see Section 3.4.3), which is a space of type KB if we endow it with the norm

$$\| x \| = | x(0) | + \operatorname*{var}_{t\in[0,1]} x(t).$$

Let φ be a measurable real function, defined on $[0,1] \times [0,1]$. Let us consider a real number $p \geqslant 1$ and assume that

$$\int_0^1 | \varphi(s, t) |^p \, ds \leqslant \mu, \tag{1}$$

where μ is a given real number, whereas the inequality holds for almost every t.

The operator U, which maps the space $L([0,1])$ into $L^p([0,1])$, given by the formula

$$(U(x))(s) = \int_0^1 \varphi(s, t)\, x(t)\, dt, \tag{2}$$

is regular (and (τ)-continuous), as one can easily prove.

[26]) See G. J. O. Jameson [1].
[27]) See I. Kawai [1].

We shall show that, if $p = 1$, the operator U, given by formula (2), is almost integral [28]). In order to do this, it is enough to suppose that $\varphi(s, t) \geqslant 0$, almost everywhere.

If we define

$$\varphi_n(s, t) = \min\{\varphi(s, t), n\}, \quad n \in \mathbb{N},$$

then any operator U_n given by the formula

$$(U_n(x))(s) = \int_0^1 \varphi_n(s, t)\, x(t)\, dt$$

is almost integral. Indeed, if we denote

$$f(x) = \int_0^1 x(t)\, dt,$$

we have

$$0 \leqslant (U_n(x))(s) \leqslant n f(x),$$

almost everywhere (with respect to s). Consequently,

$$\mathbf{0} \leqslant U_n \leqslant n f u, \tag{3}$$

where u is the element of L ([0,1]) corresponding to the function constantly equal to 1.

From inequalities (3) it obviously follows that U_n is almost integral.

From $\varphi_n(s, t) \uparrow \varphi(s, t)$, it follows that $(U_n(x))(s) \uparrow (U(x))(s)$, almost everywhere, for any $x \geqslant \mathbf{0}$. Consequently, $U_n \uparrow U$, whence we infer that U is almost integral.

If the measurable function φ satisfies condition (1) for $p = 1$, as well as the condition

$$\int_0^1 |\varphi(s, t)|\, dt \leqslant \lambda,$$

where λ is a real number, whereas the inequality holds almost everywhere, then formula (2) defines a regular operator, which maps the space L^2 ([0,1]) into itself.

If the measurable function φ satisfies the conditions

(i) for any s, the function ψ_s, given by the formula $\psi_s(t) = \varphi(s, t)$, is bounded;

(ii) for any t, the function χ_t, given by the formula $\chi_t(s) = \varphi(s, t)$ is of bounded variation and there exists a number α, such that

$$\operatorname*{var}_{s \in [0,1]} \varphi(s, t) \leqslant \alpha, \quad \forall t \in [0,1],$$

then formula (2) defines a regular operator, which maps L ([0,1]) into V([0,1].

[28]) The converse can also be proved (see Lozanovski [2]).

Bibliographical comments

The topological ordered vector spaces have been studied since 1937. We mention the papers by L. V. Kantorovitch [12], [13], as well as the papers by M. G. Krein and his co-workers. In the first papers the considered topologies were derived from a norm, the corresponding spaces being mainly normed vector lattices.

The case of the ordered vector spaces, endowed with locally convex topologies, or even with arbitrary topologies (suitably related to the order relation), has been considered later, and it was also imposed by the subsequent development of the theory of topological vector spaces, mainly of the locally convex ones.

In 1952, G. T. Roberts [1] considered topologies in vector lattices, defined with the help of the fundamental systems of solid neighbourhoods. H. Nakano [14] studies various types of topologies in a complete vector lattice, whereas R. Cristescu [2], [8] considers locally convex vector lattices, in which the topology is given by a system of solid seminorms.

The theory of topological ordered vector spaces has been afterwards developed by H. Schaefer [1] – [5], I. Namioka [1], I. Kawai [1], A. L. Peressini [1], H. Gordon [1], et al.

The definition of the topological ordered vector space, given in this chapter, is equivalent to the definition of the locally full space, given by I. Namioka [1].

Proposition 3 from Section 6.1.2 was proved, for the case of ordinary sequences, by J. D. Weston [1].

The theorem on the representation of the (τ)-continuous linear functionals, as a difference of (τ)-continuous positive linear functionals, has been proved in different ways by F. F. Bonsall [4], J. D. Weston [3], J. Kist [1], H. Schaeffer [2], I. Namioka [1], R. Cristescu [30].

The topological vector lattices have been introduced by G. T. Roberts [1]. He proved the necessity of the condition stated in proposition 3 from Section 6.2.1. The sufficiency of the same condition has been established by I. Namioka [1]. As far as the propositions from Section 6.2.2 are concerned, we mention the fact that proposition 3 is proved by a method which belongs to Kawai [1], the condition given in this proposition being stated by the latter author in the definition of the locally convex vector lattice. Proposition 5 in the same section was given by R. Cristescu [30]. The Dedekind extension of a bornological vector lattice was studied by R. Cristescu [29], who generalized some results of V. A. Solovev [1] concerning the extension of a normed vector lattice. The results from Sections 6.2.5 and 6.2.6 belong to the present author (see R. Cristescu [30]). The notion of a space of type (M) can be found in the paper by G.T.O. Jameson [1], whereas that of a space with M-unit, in the paper by R. Cristescu [30].

The topological completion of a normed vector lattice has been studied by W. A. J. Luxemburg [1], and his results have been generalized for the case of the topological vector lattices by M. Duhoux [2].

Proposition 2 from Section 6.2.8 is due to H. Gordon [1], whereas the proposition and the remark following it belong to I. Kawai [1].

Various types of normed vector lattices were studied by L. V. Kantorovitch, A. G. Pinsker, H. Nakano, M. Nakamura, G. Birkhoff, S. Kakutani, T. Ogasawara, etc.

The spaces of type (KB) were introduced by L. V. Kantorovitch [13].

A special type of ordered vector space, which allows a suitable development of a spectral theory, is that introduced by H. Nakano [10] under the name of modulared space.

A complete vector lattice $\mathscr{X}$ is called a modulared space if a function m is defined on $\mathscr{X}$, taking values in $[0, +\infty]$, which is called a modular function, with the following properties:

(1) for any element $x \in \mathscr{X}$ there exists a number $\xi > 0$, such that $m(\xi x) < +\infty$;

(2) if $m(\xi x) = 0$, for any $\xi > 0$, then $x = \mathbf{0}$;

(3) if $|x| \leqslant |y|$, then $m(x) \leqslant m(y)$;

(4) for any element $x \in \mathscr{X}$ and any numbers $\xi, \eta > 0$ the following inequality holds

$$m\left(\frac{\xi + \eta}{2}\, x\right) \leqslant \frac{1}{2}\,(m(\xi\, x) + m(\eta\, x));$$

(5) if $x \perp y$, then $m(x + y) = m(x) + m(y)$;

(6) if $0 \leqslant x_\delta \underset{\delta \in \Delta}{\uparrow} x$, then $m(x) = \sup_{\delta \in \Delta} m(x_\delta)$.

Nakano's spaces were studied by various authors: J. Musielak and W. Orlicz [1], T. Ando [3], T. Shimogaki [1], [2], etc.

Various types of ordered vector spaces were introduced also by L. V. Kantorovitch [6], [13] (spaces with a metric function) and A. G. Pinsker (spaces with a generalized additive norm) (see L. V. Kantorovitch, B. Z. Vulich, A. G. Pinsker [1]).

G. Biebinger [1], N. Popa [2], [4], and others, studied the ordered topological tensor products.

If τ is a locally convex and locally full topology on a vector lattice $\mathscr{X}$, whereas $\mathscr{W}$ is a fundamental system of neighbourhoods of the origin, consisting of equilibrated, convex and full sets, and if we define $\check{W} = \{x; Z(x) \subset W\}$ (see Section 6.2.1), the system $\{\check{W}; W \in \mathscr{W}\}$ is a fundamental system of neighbourhoods of the origin for a locally convex and locally solid topology $\check{\tau}$. The topology $\check{\tau}$ is the weakest locally convex and locally solid topology, which is finer than τ. It is called the locally solid topology associated to τ. This topology was studied by Yauchuen Wong [1]. The same author gave various conditions equivalent to the semi-reflexivity or the reflexivity of the locally convex vector lattices (Wong [2]).

If $\mathscr{X}$ is an ordered vector space, a subset E of $\mathscr{X}$ is said to be *solid* if it has the following properties: (i) if $x \in E$, then there exists a $y \in E$, such that $-y \leqslant x \leqslant y$; (ii) if $\mathbf{0} \leqslant x \in E$, then $[-x, x] \subset E$. A topology τ in an ordered vector space $\mathscr{X}$ is said to be *locally solid* if there exists a fundamental system of neighbourhoods of the origin, consisting of solid sets. Locally solid topologies in directed vector spaces, or in arbitrary ordered vector spaces, were considered by R. Cristescu [32], Gh. Grigore [5], D. Duhoux [1], Kung-Fu Ng [1]. One can show that if $\mathscr{X}$ is a directed vector space which has the Riesz decomposition property (property stated in proposition 2, Section 3.2.1), then the topology of the (o)-boundedness in $\mathscr{X}$ is the finest locally convex and locally solid topology in $\mathscr{X}$.

The results from Sections 6.3.1 and 6.3.2 are due to the author. The notion of a (τo)-bounded operator (section 1) was introduced by R. Cristescu [28], whereas the notion of a (po)-bounded operator was first defined by L. Tsitsas [1] (who used the term of (τo)-bounded operator). The lemma from Section 6.3.2 is due to L. V. Kantorovitch [1].

The linear and (τ)-continuous operators, which map a locally convex vector lattice into a Banach lattice, and which are "absolutely continuous" with respect to a continuous, positive, linear functional were studied by C. Niculescu [1]. Positively definite distributions, taking values in a locally convex vector lattice, were considered by I. Cioränescu [1].

The individual ergodic theory in spaces of type (KB) was developed by H. Nakano [9]. The results still hold in spaces of type (KP) (R. Cristescu [11]).

Other individual ergodic theorems in complete vector spaces were obtained by K. Winkelbauer [1].

The concrete representation of the spaces of type (L) belongs to S. Kakutani [2]. Another version of the representation theorem for the spaces of type (L) was given by S. Teleman [1].

For some spaces of type (LP), and for other types of locally convex vector lattices, representation theorems were given by Gh. Grigore [1], [2] and by R. Kuller [1].

In connection with the ergodic theorem for spaces of type (L), see G. Birkhoff [7], S. Kakutani [1] and F. Riesz [3].

7

Vector integrals and the representation of some linear operators

In this chapter various types of integrals for functions taking values in a vector lattice are introduced and also for functions taking values in a vector space endowed with a vector norm.

7.1 Spaces with vector norm, bounded linear operators and vector integrals

7.1.1 Vector spaces with vector norm

Let $\mathscr{F}$ be a (real) vector space and $\mathscr{X}$ a vector lattice.

A function p, defined on $\mathscr{F}$ and taking values in $\mathscr{X}$ is said to be a *vector norm* if it has the following properties

(i) $p(f_1 + f_2) \leqslant p(f_1) + p(f_2)$;

(ii) $p(\alpha f) = |\alpha| p(f)$, $\alpha \in R$;

(iii) if $f \neq \mathbf{0}$, then $p(f) \neq \mathbf{0}$.

From (i) and (ii) it follows that the values of a vector norm are positive elements.

If a vector norm (taking values in $\mathscr{X}$) is given on $\mathscr{F}$, then one says that $\mathscr{F}$ is a *space with vector norm.*

In this case the space $\mathscr{F}$ is denoted by $\mathscr{F}_{\mathscr{X}}$ and one writes $\mathbf{|} f \mathbf{|}$ instead of $p(f)$.

If $\mathscr{X} = \mathbb{R}$, one obtains the usual definition of a normed vector space. A vector lattice is also a special case of a space with vector norm if one takes $\mathbf{|} x \mathbf{|} = |x|$, for any element x of the space.

Let $\mathscr{F}_{\mathscr{X}}$ be a space with vector norm. *A* sequence $\{f_n\}_{n \in \mathbb{N}}$ of elements of the space $\mathscr{F}_{\mathscr{X}}$ is said to (v)-converge to an element $f \in \mathscr{F}_{\mathscr{X}}$ if

$$\mathbf{|} f_n - f \mathbf{|} \xrightarrow{o} \mathbf{0}.$$

We shall denote $f_n \xrightarrow{v} f$ or (v)-lim $f_n = f$ in order to describe this fact.

If $f_n \xrightarrow{v} f$, then $\mathbf{|} f_n \mathbf{|} \xrightarrow{o} \mathbf{|} f \mathbf{|}$.

If $\mathscr{X}$ is an Archimedean space and if $\{\alpha_n\}_{n\in\mathbb{N}}$ is a sequence of real numbers which converges to zero, then $\alpha_n f \overset{\vee}{\rightarrow} 0$, for any $f \in \mathscr{F}_{\mathscr{X}}$.

A sequence $\{f_n\}_{n\in\mathbb{N}}$ of elements of the space $\mathscr{F}_{\mathscr{X}}$ is said to be a Cauchy sequence with respect to the vector norm if there exists a sequence $\{x_n\}_{n\in\mathbb{N}}$ of elements of $\mathscr{X}$, such that $x_n \downarrow 0$ and

$$|f_n - f_m| \leqslant x_n,$$

for $m > n$.

The space $\mathscr{F}_{\mathscr{X}}$ is said to be (ν)-*complete* if any Cauchy sequence (for the vector norm) is (ν)-converging to an element of the space.

We shall say that the vector norm of the space $\mathscr{F}_{\mathscr{X}}$ has the *decomposition property* if from the equality $|f| = x_1 + x_2$, where $x_1, x_2 \geqslant \mathbf{0}$, it follows that there exist two elements $f_1, f_2 \in \mathscr{F}_{\mathscr{X}}$, such that $f_1 + f_2 = f$ and $|f_1| = x_1$, $|f_2| = x_2$.

If $\mathscr{X} = \mathbb{R}$, i.e., if $\mathscr{F}$ is a usual normed vector space, then the norm has the decomposition property. Indeed, if $f \in \mathscr{F}$ and $f \neq \mathbf{0}$, whereas $\|f\| = \xi_1 + \xi_2$, where ξ_1, $\xi_2 \geqslant 0$, then by taking $f_1 = \dfrac{\xi_1}{\|f\|} f$ and $f_2 = \dfrac{\xi_2}{\|f\|} f$ we shall have $f_1 + f_2 = f$, and $\|f_1\| = \xi_1$, $\|f_2\| = \xi_2$.

If $\mathscr{F}$ is a vector lattice, then $\mathscr{F}$ is a space with vector norm if we take $\mathscr{X} = \mathscr{F}$ and $|f| = |f|$. If $|f| = x_1 + x_2$, where $x_1, x_2 \geqslant \mathbf{0}$, then by taking

$$f_1 = x_1 \wedge f_+ - x_1 \wedge f_-,$$

$$f_2 = f - f_1 = (f_+ - x_1 \wedge f_+) - (f_- - x_1 \wedge f_-),$$

we shall have

$$f_1 + f_2 = f, \quad |f_1| = x_1, \quad |f_2| = x_2.$$

One calls a *space of type* (B_K) any space $\mathscr{F}_{\mathscr{X}}$ whith vector norm, which is (ν)-complete and whose vector norm has the decomposition property.

A Banach space obviously is a special case of a space of type (B_K). Any σ-complete vector space also is a space of type (B_K) if the absolute value of its elements is taken to be the vector norm.

One can also give other examples of spaces of type (B_K).

In this section we shall consider spaces of type (B_K) whose elements are vector functions.

7.1.2 (ν)-regular operators and (ν)-bounded operators

Let $\mathscr{F}_{\mathscr{X}}$ and $\mathscr{G}_{\mathscr{Y}}$ be vector spaces with vector norm.

Definition 1. An additive operator U, defined on $\mathscr{F}_{\mathscr{X}}$ and taking values in $\mathscr{G}_{\mathscr{Y}}$ is said to be (ν)-*regular* if there exists an additive and positive operator V, which maps $\mathscr{X}$ into $\mathscr{Y}$ such that

$$|U(f)| \leqslant V(|f|), \tag{1}$$

for any $f \in \mathscr{F}_{\mathscr{X}}$. If, moreover, V is (o)-continuous, then U is said to be (ν)-*bounded.*

Proposition 1. *If the space $\mathscr{Y}$ is Archimedean, then any* (ν)*-regular operator U, which maps $\mathscr{F}_{\mathfrak{X}}$ into $\mathscr{G}_{\mathscr{Y}}$, is homogeneous.*

Proof. If $\mathscr{Y}$ is Archimedean, then the operator V, which satisfies condition (1), is homogeneous. Let α be an arbitrary real number and let $\{\lambda_n\}_{n\in\mathbb{N}}$ be a sequence of rational numbers, converging to α. We have

$$|U(\alpha f) - U(\lambda_n f)| = |U(\alpha - \lambda_n) f|$$

$$\leqslant V(|\alpha - \lambda_n| |f|) = |\alpha - \lambda_n| V(|f|).$$

On the other hand

$$|\lambda_n U(f) - \alpha U(f)| = |\lambda_n - \alpha| |U(f)| \leqslant |\lambda_n - \alpha| V(|f|).$$

It follows that

$$|U(\alpha f) - \alpha U(f)| \leqslant 2 |\lambda_n - \alpha| V(|f|)$$

and, therefore,

$$U(\alpha f) = \alpha U(f).$$

Proposition 2. *Let U be a* (ν)*-regular operator which maps $\mathscr{F}_{\mathfrak{X}}$ into $\mathscr{G}_{\mathscr{Y}}$. If $\mathscr{Y}$ is a complete vector lattice, and if the vector norm on $\mathscr{F}_{\mathfrak{X}}$ has the decomposition property, then there exists the smallest positive linear operator V, which satisfies condition* (*1*).

Proof. Let $\{V_j\}_{j\in J}$ be the system of all positive linear operators, which satisfy condition (1). Let $f\in\mathscr{F}_{\mathfrak{X}}$ and $|f| = x$. If $x = \sum_{i-1}^{n} x_i$ where $x_i \geqslant 0$, then there exist $_i \in \mathscr{F}_{\mathfrak{X}}$ $(i = 1, 2, \ldots, n)$, such that $f = \sum_{i=1}^{n} f_i$ and $|f_i| = x_i$.

For any indices $j_1, j_2, \ldots, j_n$ from J the following inequalities hold

$$|U(f)| \leqslant \sum_{i=1}^{n} |U(f_i)| \leqslant \sum_{i=1}^{n} V_{j_i}(|f_i|) = \sum_{i=1}^{n} V_{j_i}(x_i).$$

By denoting

$$V^* = \bigwedge_{j\in J} V_j,$$

it follows that

$$|U(f)| \leqslant V^*(x) = V^*(|f|),$$

and the proposition is proved.

The smallest positive linear operator V, which satisfies condition (1), and whose existence is established by proposition 2, is called the *operator norm* of U and it is denoted by $\| U \|$.

If U is (ν)-bounded, then $\| U \|$ is (o)-continuous.

7.1.3 Continuous vector functions and vector functions of bounded variation

Let T be an interval of the real line and $\mathscr{X}$ a complete vector lattice.

A function f, defined on T and taking values in $\mathscr{X}$, is said to be *(o)-bounded* if there exists an $a \in \mathscr{X}$, such that $| f(t) | \leqslant a$, for any $t \in T$.

We shall denote by $B(T, \mathscr{X})$ the set of all (o)-bounded functions which are defined on T and take values in $\mathscr{X}$. On this set we shall consider the usual operations of addition of two functions and multiplication of a function by a real number. Also, we shall consider the following vector norm

$$\| f \| = \sup_{t \in T} | f(t) |,$$

with values in $\mathscr{X}$.

It is not difficult to check that $B(T, \mathscr{X})$ becomes a space of type (B_K).

If $f \in B(T, \mathscr{X})$, whereas $0 < \eta \in \mathbb{R}$, we shall define

$$v(f; \eta) = \sup_{|t_1 - t_2| \leqslant \eta} | f(t_1) - f(t_2) |.$$

We shall denote by $C(T, \mathscr{X})$ the subset of $B(T, \mathscr{X})$ which contains any function f satisfying the condition of continuity:

$$(o)\text{-}\lim_{\eta_n \to 0} v(f; \eta_n) = 0.$$

Endowed with the same operations and vector norms as in $B(T, \mathscr{X})$, the set $C(T, \mathscr{X})$ is also a space of type (B_K).

Let now $\mathscr{Z}$ be a complete vector lattice.

A function g, defined on T and taking values in $\mathscr{Z}$, is said to be of *(o)-bounded variation,* if the set of all elements of the form

$$\sum_i | g(t_{i+1}) - g(t_i) |,$$

corresponding to all finite partitions of the interval T, is (o)-bounded. We shall denote by $(o)\text{-}\operatorname{var}_{t \in T} g(t)$ the least upper bound of this set.

In particular, we shall consider the case in which $\mathscr{Z} = (\mathscr{X}, \mathscr{Y})_o^r$, where $\mathscr{Y}$ is a complete vector lattice.

7.1.4 The integral representation of the bounded linear operators defined on the space C(T, $\mathscr{X}$)

Let T be an interval of the real line and $\mathscr{X}$, $\mathscr{Y}$ two complete vector lattices.

Let G be a function defined on T and taking values in the space $(\mathscr{X}, \mathscr{Y})_o^r$, of (o)-bounded variation.

We shall denote

$$V_G = (o)\text{-}\underset{t\in T}{\text{var}}\, G(t).$$

If $f \in C(T, \mathscr{X})$ and D is a finite partition of the interval T (determined by the points t_i), let us form the sum

$$s = \sum_i (G(t_{i+1} - G(t_i)) (f(\theta_i)), \tag{1}$$

where the "intermediate points" $\theta_i \in [t_i, t_{i+1}]$ are chosen arbitrarily.

Let us denote by $\eta(D)$ the norm of the partition

$$\eta(D) = \max (t_{i+1} - t_i).$$

Let us consider a sequence $\{D_n\}_{n\in\mathbb{N}}$ of partitions of T, such that $\eta_n = \eta(D)_n \to 0$. For any $n \in \mathbb{N}$ let us denote by s_n the sum of the form (1), corresponding to the partition D_n.

If, for any $n \in \mathbb{N}$, the partition D_{n+1} is finer than the partition D_n, then one can show that

$$|s_n - s_{n+m}| \leqslant V_G(v(f, \eta_n)),\ n, m \in \mathbb{N},$$

where the notations from Section 7.1.3 have been used. Since $V_G \in (\mathscr{X}, \mathscr{Y})_o^r$, whereas $v(f, \eta_n) \xrightarrow{o} 0$, there exists the limit $y = (o)\text{-}\lim_n s_n$. As for the case of the real functions, one can prove that if $\{D_n'\}_{n\in\mathbb{N}}$ is a sequence of arbitrary partitions, such that $\eta(D_n') \to 0$, then for any choice of the intermediate points, which are used when forming the sums s_n', we have $y = (o)\text{-}\lim_n s_n'$.

We shall write

$$y = \int_T \mathrm{d}G(t)\,(f(t)),$$

and we shall say that y is the Stieltjes integral of f with respect to G.

One can easily show that if $f_1, f_2 \in C(T, \mathscr{X})$, then

$$\int_T \mathrm{d}G(t)\,(f_1(t) + f_2(t)) = \int_T \mathrm{d}G(t)\,(f_1(t)) + \int_T \mathrm{d}G(t)\,(f_2(t)), \tag{2}$$

whereas for $f \in C(T, \mathscr{X})$ and $\lambda \in \mathbb{R}$ we have

$$\int_T dG(t)(\lambda f(t)) = \lambda \int_T dG(t)(f(t)). \tag{3}$$

For any $f \in C(T, \mathscr{X})$ the inequality

$$\left| \int_T dG(t)(f(t)) \right| \leqslant V_G (\sup_{t \in T} |f(t)|) \tag{4}$$

also holds, and its verification is easy to carry out.

If, in particular, $\mathscr{X} = \mathbb{R}$, then we shall have

$$G(t)(f(t)) = f(t)\, g(t),$$

where g is a function defined on T and taking values in $\mathscr{Y}$, of bounded variation. In this case the integral of f with respect to g will be denoted by

$$\int_T f(t)\, dg(t).$$

Theorem. *The general form of the* (v)*-bounded linear operators, defined on the space* $C(T, \mathscr{X})$ *and taking values* [1] *in* $\mathscr{Y}$, *is given by the formula*

$$U(f) = \int_T dG(t)(f(t)), \tag{5}$$

where G *is a function of bounded variation, taking values in* $(\mathscr{X}, \mathscr{Y})_0^r$. *The function* G *can be chosen so that the equality*

$$\mathbf{|}U\mathbf{|} = V_G \tag{6}$$

hold.

Proof. If U is of the form (5), then from relations (2), (3), (4) we infer that U is a (v)-bounded linear operator and

$$\mathbf{|}U\mathbf{|} \leqslant V_G. \tag{7}$$

Conversely, let U be an arbitrary (v)-bounded linear operator, defined on $C(T, \mathscr{X})$ and taking values in $\mathscr{Y}$. Then we have

$$|U(f)| \leqslant \mathbf{|}U\mathbf{|}(\mathbf{|}f\mathbf{|}),$$

[1]) The space $\mathscr{Y}$ is considered to be a space of type (B_K), with the vector norm $\mathbf{|}y\mathbf{|} = |y|$.

where $|U| \in (\mathscr{X}, \mathscr{Y})_o^r$. The operator U can be extended to the space $B(T, \mathscr{X})$, such that the linearity and the operator norm be conserved (see Section 6.3.2, lemma). We shall denote by U the considered extension.

If $T = [\alpha, \beta]$, and $t \in (\alpha, \beta]$ we shall denote by χ_t the characteristic function of the interval $[\alpha, t]$ (defined on T); by χ_α we shall denote the identically vanishing function, defined on T.

For any $t \in T$, we shall denote by $G(t)$ the operator defined on $\mathscr{X}$ and taking values in $\mathscr{Y}$, given by the formula

$$G(t)(x) = U(\chi_t x).$$

It is easily verified that for any $t \in T$ the operator $G(t)$ is regular and (o)-continuous.

Since for any positive element $x \in \mathscr{X}$ and any partition D of the interval T, given by a finite system of points t_i, the following inequality

$$(\sum_i |G(t_{i+1}) - G(t_i)|)(x) \leqslant |U|(x)$$

holds, it follows that G (as a function defined on T and taking values in $(\mathscr{X}, \mathscr{Y})_o^r$), is of bounded variation and

$$V_G \leqslant |U|. \tag{8}$$

Let now $f \in C(T, \mathscr{X})$ and D be a partition of the interval T. With the preceding notations we have

$$\sup_{t', t'' \in [t_i, t_{i+1}]} |f(t') - f(t'')| \leqslant v(f; \eta). \tag{9}$$

Let us define

$$h(t) = \begin{cases} f(t_0), & \text{if} \quad t_0 \leqslant t \leqslant t_1 \\ f(t_i), & \text{if} \quad t_i < t \leqslant t_{i+1}, \end{cases} \qquad (i \geqslant 1).$$

It is obvious that $h \in B(T, \mathscr{X})$ and from (9) we get

$$|h - f| \leqslant v(f; \eta),$$

whence

$$|U(h) - U(f)| \leqslant |U|(v(f; \eta)). \tag{10}$$

On the other hand we have

$$h = \sum_i (\chi_{t_{i+1}} - \chi_{t_i}) f(t_i),$$

and, therefore,

$$U(h) = \sum_i (G(t_{i+1}) - G(t_i))(f(t_i)). \tag{11}$$

From (10) and (11) we get formula (5), whereas from inequalities (7) and (8), equality (6) follows. The theorem is proved.

Corollary [2]). *The general form of the (τo)-bounded linear operators, defined on the Banach space $C(T)$ and taking values in $\mathcal{Y}$, is given by the formula*

$$U(f) = \int_T f(t)\,\mathrm{d}g(t),$$

where g is a function defined on T, taking values in $\mathcal{Y}$, and of bounded variation.

7.1.5 Hellinger integrals

Let $\mathcal{V}$ and $\mathcal{Z}$ be complete vector lattices and $\mathcal{F}_{\mathcal{V}}$, $\mathcal{G}_{\mathcal{Z}}$ spaces of type (B_K).

We shall denote by $\mathcal{M}(\mathcal{F}_{\mathcal{V}}, \mathcal{G}_{\mathcal{Z}})$ the set of all (ν)-bounded linear operators, which map $\mathcal{F}_{\mathcal{V}}$ into $\mathcal{G}_{\mathcal{Z}}$.

Let T be an interval of the real line and G a function defined on T and taking values in $\mathcal{M}(\mathcal{F}_{\mathcal{V}}, \mathcal{G}_{\mathcal{Z}})$.

Let us consider a function f, defined on T, taking values in $\mathcal{F}_{\mathcal{V}}$ and let D be a partition of T determined by the points [3]) t_i $(i = 0, 1, 2, \ldots, m)$. Let us denote

$$G_i = G(t_{i+1}) - G(t_i),$$

$$f_i = f(t_{i+1}) - f(t_i),$$

$$\theta_i = t_{i+1} - t_i$$

and let us consider the sum

$$s_D = \sum_{i=0}^{m-1} \frac{1}{\theta_i} G_i(f_i).$$

Let $\eta(D)$ be the norm of the partition D. If there exists an element $h \in \mathcal{G}_{\mathcal{Z}}$, such that

$$h = (\nu)\text{-lim}\, s_{D_n},$$

---

[2]) We consider the Banach space $C(T)$ of all real continuous functions, defined on T, with the usual norm.

[3]) By a partition determined by the points t_i $(i = 0, 1, 2, \ldots, m)$ we shall mean a finite indexed family $\{t_i\}$ of points of $T = [\alpha, \beta]$, such that $\alpha = t_0 < t_1 < t_2 < \ldots < t_{m-1} < t_m = \beta$.

for any sequence $\{D_n\}_{n\in\mathbb{N}}$ of partitions, such that $\eta(D_n) \to 0$, then we shall write

$$h = \int_T \frac{1}{\mathrm{d}t} \{\mathrm{d}G(t)\,(\mathrm{d}f(t))\}$$

and we shall say that h is the Hellinger integral of f with respect to G.

One easily verifies that if the Hellinger integrals of the functions f_1, f_2, with respect to G, exist, then, for any real numbers, α, β the Hellinger integral of the function $\alpha f_1 + \beta f_2$, with respect to G, exists too and the equality

$$\int_T \frac{1}{\mathrm{d}t} \{\mathrm{d}G(t)\,(\alpha f_1(t) + \beta f_2(t))\}$$
$$= \alpha \int_T \frac{1}{\mathrm{d}t} \{\mathrm{d}G(t)\,(f_1(t))\} + \beta \int_T \frac{1}{\mathrm{d}t} \{\mathrm{d}G(t)(f_2(t))\},$$

holds.

By definition, the function G satisfies the Lipschitz condition if there exists a positive operator $H \in (\mathscr{V}, \mathscr{X})_o^r$, such that

$$|G(t') - G(t'')| \leqslant |t' - t''|\, H, \tag{1}$$

for any $t', t'' \in T$.

A function f, defined on T and taking values in $\mathscr{F}_{\mathscr{V}}$, is said to be *of bounded variation* (with respect to the vector norm) if the set of all elements of the form

$$\sum_i |f(t_{i+1}) - f(t_i)|,$$

corresponding to all finite partitions of the interval T, is (o)-bounded in $\mathscr{V}$. We shall denote by $(\nu)\text{-}\operatorname{var}_{t\in T} f(t)$ the least upper bound of this set.

If G satisfies the Lipschitz condition (1), if f is a function of bounded variation and if the Hellinger integral of f with respect to G exists, then

$$\left|\int_T \frac{\mathrm{d}}{\mathrm{d}t} \{\mathrm{d}G(t)\,(\mathrm{d}f(t))\}\right| \leqslant H\ ((\nu)\text{-}\operatorname*{var}_{t\in T} f(t)). \tag{2}$$

Indeed, with the preceding notations, we have

$$|s_D| \leqslant \sum_{i=1}^{m-1} \frac{1}{\theta_i} |G_i(f_i) \leqslant \sum_{i=1}^{m-1} \frac{1}{\theta_i} |G_i| (|f_i|)$$
$$\leqslant \sum_{i=1}^{m-1} H(|f_i|) = H\left(\sum_{i=1}^{m-1} |f(t_{i+1}) - f(t_i)|\right)$$
$$\leqslant H((\nu)\text{-}\operatorname*{var}_{t\in T} f(t)),$$

whence inequality (2) immediately follows.

7.1.6 The integral representation of the bounded linear operators defined on the space L(T, $\mathscr{X}$)

Let $\mathscr{X}$ be a (real) Banach space and $\mathscr{Y}_{\mathscr{Z}}$ a space of type (B_K), where $\mathscr{Z}$ is a complete vector lattice.

A linear operator U, which maps $\mathscr{X}$ into $\mathscr{Y}_{\mathscr{Z}}$, is a (v)-bounded operator if there exists an element $z \in \mathscr{Z}$, such that

$$| U(x) | \leqslant \|x\| \, z, \tag{1}$$

for any $x \in \mathscr{X}$.

The smallest element z which satisfies condition (1), is called the (vector) norm of U and it is denoted by $| U |$.

Let $T = [\alpha, \beta]$ be an interval of the real line.

A function G, defined on T and taking values in $\mathscr{M}(\mathscr{X}, \mathscr{Y}_{\mathscr{Z}})$ satisfies the Lipschitz condition, if there exists an $l \in \mathscr{Z}$, such that

$$| G(t') - G(t'') | \leqslant | t' - t'' | \, l. \tag{2}$$

A function f, defined on T and taking values in $\mathscr{X}$, is said to be a *polygonal function* if there exists a partition D of T, given by the points t_j $(j = 0, 1, \ldots, m)$ and the elements $c_j \in \mathscr{X}$, $(j = 0, 1, \ldots, m)$, such that

$$f(t) = \frac{t - t_i}{t_{i+1} - t_i} (c_{i+1} - c_i) + c_i$$

for $t \in [t_i, t_{i+1}]$, $i = 0, 1, \ldots, m - 1$. If g is a function defined on T and taking values in $\mathscr{X}$, such that $g(t_i) = c_i$, then the polygonal function f will be denoted by g_D.

Lemma 1. *If G satisfies the Lipschitz condition* (2) *and if f is a polygonal function, then the integral of f with respect to G exists and the equality* [4])

$$\int \frac{1}{\mathrm{d}t} \{\mathrm{d}G(t) \, (\mathrm{d}f(t))\} = \Sigma \frac{1}{\theta_i} G_i (c_{i+1} - c_i)$$

holds.

The proof can easily be made.

We shall now consider functions defined on T, taking values in $\mathscr{X}$, and Bochner integrable. We shall denote by $L(T, \mathscr{X})$ the Banach space of the Bochner integrable functions [5]), endowed with the usual norm, i.e.

$$\|f\| = \int_T \|f(t)\| \, \mathrm{d}t.$$

[4]) We shall use the notations from Section 7.1.5, i.e., $G_i = G(t_{i+1}) - G(t_i)$, $\theta_i = t_{i+1} - t_i$.

[5]) i.e., of the classes of Bochner integrable functions, each class containing the functions which differ from one another on a null-set.

In connection with the Bochner integral and the space $L(T, \mathscr{X})$, see M. Nicolescu [1].

If f is a Bochner integrable function, defined on $T = [\alpha, \beta]$, we shall denote

$$f^*(t) = \int_\alpha^t f(\xi)\, d\xi. \tag{3}$$

In the following lemma we shall denote by $\eta(D)$ the norm of a partition D of T. We shall consider the variation with respect to the norm of the space $\mathscr{X}$.

Lemma 2. *If f is a Bochner integrable function and $\{D_n\}_{n\in\mathbb{N}}$ a sequence of partitions of the interval T, such that $\eta(D_n) \to 0$, then*

$$\lim_{n\to\infty} \{\operatorname*{var}_{t\in T} (f^*(t) - f^*_{D_n}(t))\} = 0. \tag{4}$$

Proof. Let D be a partition of T determined by the points t_i $(i = 0, 1, \ldots, m)$. Let us denote

$$e_i = \frac{1}{t_{i+1} - t_i} \int_{t_i}^{t_{i+1}} f(t)\, dt, \; i = 0, 1, \ldots, m-1,$$

and let k_D be the simple function, defined on T, which is equal to e_i on the interval $[t_i, t_{i+1}]$. We have the following equality

$$f^*_D(t) = \int_\alpha^t k_D(\xi)\, d\xi.$$

If $\|f(t)\| \leqslant \lambda$, then $\|k_D(t)\| \leqslant \lambda$ and, therefore,

$$\|f(t) - k_D(t)\| \leqslant 2\lambda.$$

On the other hand, by considering a sequence $\{D_n\}_{n\in\mathbb{N}}$ of partitions, as in the statement of the lemma, we have

$$\lim_n \|f(t) - k_{D_n}(t)\| = 0$$

almost everywhere. It follows that

$$\lim_n \|f - k_{D_n}\| = 0$$

and, therefore, equality (4) holds if f is bounded (in norm).

If f is not bounded, then let us consider an arbitrary number $\varepsilon > 0$ and let λ be such that by putting $A = \{t;\, \|f(t)\| > \lambda\}$, we have

$$\int_A \|f(t)\|\, dt < \varepsilon.$$

If we denote by χ_E the characteristic function of a set E and if we define $g(t)=f(t)\chi_{\complement A}(t)$, $h(t)=f(t)\chi_A(t)$, then we have $f=g+h$ and g is bounded on T.

The proof of the lemma now immediately follows.

Lemma 3. *If G satisfies the Lipschitz condition* (2), *whereas f is Bochner integrable, then there exists the Hellinger integral of f^* with respect to G and*

$$\left|\int_T \frac{1}{\mathrm{d}t}\{\mathrm{d}G(t)(\mathrm{d}f^*(t))\}\right| \leqslant (\operatorname*{var}_{t\in T} f^*(t))\, l. \tag{5}$$

Proof. Let D be a partition of T, determined by the points t_i $(i=0,1,\ldots,m)$ and denote

$$s_D=\sum_{i=0}^{m-1}\frac{1}{\theta_i}G_i(x),$$

where

$$x_i=f^*(t_{i+1})-f^*(t_i).$$

We have

$$|s_D|\leqslant\sum_{i=0}^{m-1}\left|\frac{1}{\theta_i}G_i(x_i)\right|\leqslant\sum_{i=0}^{m}\|x_i\|l\leqslant(\operatorname*{var}_{t\in T}f^*(t))l.$$

On the other hand

$$s_D=\int_T\frac{1}{\mathrm{d}t}\{\mathrm{d}G(t)(\mathrm{d}f_D^*)(t))\}.$$

Consequently, if D' and D'' are two partitions, then

$$|s_{D'}-s_{D''}|\leqslant\operatorname*{var}_{t\in T}(f_{D'}^*(t)-f_{D''}^*(t))l. \tag{6}$$

If $\{D_n\}_{n\in\mathbb{N}}$ is a sequence of partitions such that $\eta(D_n)\to 0$, then, by using inequality (6) and taking into account lemma 2, we infer that

$$(\rho)\text{-}\lim_{n,m\to\infty}|s_{D_n}-s_{D_m}|=0.$$

Therefore, the limit $y=(\nu)\text{-}\lim s_{D_n}$ exists and it is easy to check that y does not depend on the chosen sequence of partitions. Inequality (5) will follow from the inequality

$$|s_D|\leqslant(\operatorname*{var}_{t\in T}f^*(t))l,$$

which was already established.

Theorem. *The general form of the* (ν)*-bounded linear operators which map the space* $L(T, \mathscr{X})$ *into the space* $\mathscr{Y}_{\mathscr{Z}}$ *is given by the formula*

$$U(f) = \int_T \frac{1}{\mathrm{d}t} \{\mathrm{d}G(t)\,(\mathrm{d}f^*(t))\}, \tag{7}$$

where G is a function, defined on T and taking values in $\mathscr{M}(\mathscr{X}, \mathscr{Y}_{\mathscr{Z}})$, *which satisfies the Lipschitz condition, whereas* f^* *is given by formula* (3). *If the operator U is given, then there exists a G, such that* $|U|$ *be the smallest element* $l \in \mathscr{Z}$ *which satisfies the Lipschitz condition* (2).

Proof. If $f \in L(T, \mathscr{X})$, whereas G satisfies condition (2), then by virtue of lemma 2 the Hellinger integral of f^* with respect to G exists. Formula (7) then determines a linear operator which maps $L(T, \mathscr{X})$ into $\mathscr{Y}_{\mathscr{Z}}$. From inequality (5) we infer the inequality

$$|U(f)| \leqslant \|f\| l,$$

whence we infer that U is (ν)-bounded and

$$|U| \leqslant l. \tag{8}$$

Let U be a (ν)-bounded linear operator, which maps $L(T, \mathscr{X})$ into $\mathscr{Y}_{\mathscr{Z}}$.

For any $t \in T$, let $G(t)$ be the operator defined on $\mathscr{X}$ (and taking values in $\mathscr{Y}_{\mathscr{Z}}$), given by the following formula

$$G(t)(x) = U(\chi_t . x),$$

where χ_t is the characteristic function of the interval $[\alpha, t]$, whereas

$$(\chi_t . x)(s) = \chi_t(s) . x, \qquad s \in T.$$

It is easy to check that $G(t)$ is linear and (ν)-bounded.

The function G (i.e., the mapping $t \mapsto G(t)$ of T into $\mathscr{M}(\mathscr{X}, \mathscr{Y}_{\mathscr{Z}})$) satisfies the Lipschitz condition.

Indeed, if $t' < t''$, then

$$\begin{aligned} |(G(t') - G(t''))\,(x)| &= |U\{(\chi_{t''} - \chi_{t'})x\}| \\ &\leqslant \|(\chi_{t''} - \chi_{t'})\,x\|\,|U| = (t'' - t')\,\|x\|\,|U|, \end{aligned}$$

and this implies that

$$|G(t') - G(t'')| \leqslant |t' - t''|\,|U|. \tag{9}$$

Let now f be a function defined on T, taking values in $\mathscr{X}$, of the form

$$f(t) = \sum_{i=0}^{m-1} \chi_{A_i}(t) . a_i, \tag{10}$$

where $a_i \in \mathscr{X}$, $A_i = [t_i, t_{i+1}]$, χ_{A_i} is the characteristic function of A_i, whereas t_i ($i = 0, 1, \ldots, m$) are the points of a partition of T.

One observes that

$$a_i = \frac{1}{t_{i+1} - t_i}(f^*(t_{i+1}) - f^*(t_i))$$

whence it immediately follows

$$U(f) = \sum_{i=0}^{m-1} \frac{1}{t_{i+1} - t_i}(G(t_{i+1}) - G(t_i))(f^*(t_{i+1}) - f^*(t_i)).$$

Since the function f^* is polygonal, the last sum equals the integral of this function with respect to G (lemma 1).

If f is an arbitrary Bochner integrable function, then there exists a sequence $\{f_n\}_{n\in\mathbb{N}}$ of functions of the form (10), such that $f = \lim f_n$ in the space $L(T, \mathscr{X})$. In this case $U(f) = (\nu)\text{-}\lim U(f_n)$, because

$$|U(f_n) - U(f)| \leqslant \|f_n - f\| \, |U|.$$

On the other hand, by defining

$$U_0(f) = \int_T \frac{1}{\mathrm{d}t}\{\mathrm{d}G(t)\,(\mathrm{d}f^*(t))\},$$

the operator U_0 is linear and (ν)-bounded (as we saw in the first part of the proof of the theorem), and we have $U(f_n) = U_0(f_n)$. From $U_0(f_n) \overset{\nu}{\to} U_0(f)$ it follows that $U(f) = U_0(f)$, hence (7) holds.

From (8) and (9) it follows that $|U|$ is the smallest element which satisfies condition (2). The theorem is proved.

Remark. Let $\mathscr{X}$ be a Banach vector lattice and $\mathscr{Y}$ a complete vector lattice. The space $L(T, \mathscr{X})$ becomes a Banach vector lattice if we define $f \geqslant \mathbf{0}$ to mean that $f(t) \geqslant 0$ almost everywhere.

One can easily prove the following proposition:

In order that a (τo)-bounded linear operator, which maps $L(T, \mathscr{X})$ into $\mathscr{Y}$ be positive, it is necessary and sufficient that in the representation (7) of U the function G be increasing.

7.2 Integrals with respect to vector measures

7.2.1 The vector measure and almost uniform convergence

Let T be an arbitrary set and $\mathscr{T}$ a σ-algebra of subsets of T.

Let $\mathscr{X}, \mathscr{Y}, \mathscr{Z}$ be σ-complete vector lattices, such that $\mathscr{X}$ and $\mathscr{Y}$ be σ-regular spaces, whereas $\mathscr{Z}$ be a regular space.

We shall assume that a positive bilinear operator P, defined on $\mathscr{X} \times \mathscr{Y}$ and taking values in $\mathscr{Z}$, is given.

A function $\mathbf{m}$, which is defined on $\mathscr{T}$ and takes values in $\mathscr{Y}$ is said to be a *measure* if it has positive values and if for any sequence $\{A_n\}_{n\in\mathbb{N}}$ of mutually disjoint subsets of $\mathscr{T}$ the equality

$$\mathbf{m}(\bigcup_{n\in\mathbb{N}} A_n) = o\text{-}\sum_{n=1}^{\infty} \mathbf{m}(A_n)$$

holds.

We shall assume that a measure $\mathbf{m}$, defined on $\mathscr{T}$ and taking values in $\mathscr{Y}$, is given.

We shall now consider functions which map T into $\mathscr{X}$.

A sequence $\{f_n\}_{n\in\mathbb{N}}$ of functions, defined on T and taking values in $\mathscr{X}$, is said to (o)*-converge point-wise* on a subset A of T, to a function f, if $f_n(t) \xrightarrow{o} f(t)$, for any $t \in A$.

The sequence $\{f_n\}_{n\in\mathbb{N}}$ is said to (o)*-converge uniformly* on a subset A of T, to a function f, if there exists a sequence $\{r_n\}_{n\in\mathbb{N}}$ of elements of $\mathscr{X}$, such that $r_n \downarrow \mathbf{0}$ and

$$|f_n(t) - f(t)| \leqslant r_n,$$

for any $t \in A$.

We shall write $f_n \xrightarrow{u} f$ on A.

Since the space $\mathscr{X}$ is regular, we have $f_n \xrightarrow{u} f$ on A iff there exists an element $v \in \mathscr{X}$, having the property that for any number $\varepsilon > 0$ a number $n_\varepsilon \in \mathbb{N}$ exists, such that

$$|f_n(t) - f(t)| \leqslant \varepsilon v, \ (n \geqslant n_\varepsilon),$$

for any $t \in A$.

We shall say that the sequence of functions $\{f_n\}_{n\in\mathbb{N}}$ is (o)-converging $(\mathscr{T}, \mathbf{m})$*-almost uniformly* on T, to the function f, if there exists a sequence $\{A_n\}_{n\in\mathbb{N}}$ of sets from $\mathscr{T}$, such that $m(\complement A_n) \xrightarrow{o} \mathbf{0}$, anf $f_n \xrightarrow{u} f$ on A_k (for any $k \in \mathbb{N}$).

If the sequence $\{f_n\}_{n\in\mathbb{N}}$ is (o)-converging $(\mathscr{T}, \mathbf{m})$-almost uniformly on T, to the function f, then the sequence (o)*-converges* $(\mathscr{T}, m)$*-almost everywhere* on T to f, i.e., there exists a set $T_0 \in \mathscr{T}$ on which the sequence (o)-converges pointwise to f, and $\mathbf{m}(\complement T_0) = \mathbf{0}$.

The proof of this assertion offers no difficulties.

A $\mathscr{T}$*-simple function* is, by definition, a function of the form

$$f(t) = \sum_{i=1}^{k} \chi_{A_i}(t)\, x_i,$$

where $\{ A_i; i = 1, 2, \ldots k\}$ is a partition of T consisting of subsets belonging to $\mathscr{T}$, and $x_i \in \mathscr{X}$.

A function $f: T \to \mathscr{X}$ is said to be $(\mathscr{T}, \mathbf{m})$-*measurable* if there exists a sequence $\{f_n\}_{n\in\mathbb{N}}$ of $\mathscr{T}$-simple functions which (o)-converges $(\mathscr{T}, \mathbf{m})$-almost uniformly on T, to f. The sequence $\{f_n\}_{n\in\mathbb{N}}$ will be called an *approximating sequence* for f.

If $\{f_n\}_{n\in\mathbb{N}}$ is a sequence of $(\mathscr{T}, \mathbf{m})$-measurable functions which (o)-converges $(\mathscr{T}, \mathbf{m})$-almost uniformly to a function f, then f is $(\mathscr{T}, \mathbf{m})$-measurable.

In order to prove this proposition it is sufficient to prove the following proposition:

If for any $n \in \mathbb{N}$, the sequence $\{f_n^k\}_{k\in\mathbb{N}}$ is $(\mathscr{T}, \mathbf{m})$-almost uniformly (o)-convergent to f_n, and if the sequence $\{f_n\}_{n\in\mathbb{N}}$ is $(\mathscr{T}, \mathbf{m})$-almost uniformly (o)-convergent to f, then there exists an increasing sequence $\{k_n\}_{n\in\mathbb{N}}$ of natural numbers, such that the sequence $\{f_n^{k_n}\}_{n\in\mathbb{N}}$ be $(\mathscr{T}, \mathbf{m})$-almost uniformly (o)-convergent to f.

Indeed, let $\{\varepsilon_n\}_{n\in\mathbb{N}}$ be an arbitrary sequence of positive numbers let $\{A_n\}_{n\in\mathbb{N}}$ be a sequence of sets from $\mathscr{T}$ and $\{p_n\}_{n\in\mathbb{N}}$ a sequence of elements of $\mathscr{Y}$, such that

$$\mathbf{m}(\complement A_n) \leqslant \varepsilon_n p_n, \quad n \in \mathbb{N},$$

and

$$|f_n^k(t) - f_n(t)| \leqslant v_n^k, \quad t \in A_n,$$

and such that $(o)\text{-}\lim_k v_n^k = \mathbf{0}$, for any $n \in \mathbb{N}$. Let also $\varepsilon_0 > 0$ be an arbitrary number, $A_0 \in \mathscr{T}$ and $p_0 \in \mathscr{Y}$ be such that

$$\mathbf{m}(\complement A_0) \leqslant \varepsilon_0 p_0$$

and

$$|f_n(t) - f(t)| \leqslant v_n, \qquad t \in A_0,$$

where the sequence $\{v_n\}_{n\in\mathbb{N}}$ is such that $v_n \xrightarrow{o} \mathbf{0}$.

Since the space $\mathscr{Y}$ is σ-regular, there exists a sequence $\{\lambda_n; n = 0, 1, 2, \ldots\}$ of strictly positive numbers, such that the sequence $\{\lambda_n p_n; n = 0, 1, \ldots\}$ be bounded. Let then $p \in \mathscr{Y}$ be such that $\lambda_n p_n \leqslant p$.

Let $\varepsilon > 0$ be arbitrary and let us define $\varepsilon_n = \dfrac{\lambda_n}{2^{n+1}}\varepsilon$ and

$$T_0 = \bigcap_{n=0}^{\infty} A_n.$$

We have $m(\complement T_0) \leqslant \varepsilon p$.

On the other hand, since $\mathscr{X}$ is a σ-regular space, there exists an increasing sequence $\{k_n\}_{n\in\mathbb{N}}$ of natural numbers, such that $v_n^{k_n} \xrightarrow{o} \mathbf{0}$.

By denoting

$$w_n = v_n^{k_n} + v_n$$

we have

$$|f_n^{k_n}(t) - f(t)| \leqslant w_n$$

for any $t \in T_0$.

7.2.2 The integral of a vector function with respect to a vector measure

We shall use the same notations as in section 1.

If f is a $\mathscr{T}$-simple function (defined on T and taking values in $\mathscr{X}$), given by

$$f(t) = \sum_{i=1}^{k} \chi_{A_i}(t) \cdot x_i,$$

we shall define the integral of f with respect to the measure $\mathbf{m}$ by the formula

$$\int_T P(f(t), \mathrm{d}\mathbf{m}(t)) = \sum_{i=1}^{k} P(x_i, \mathbf{m}(A_i))\cdot$$

The integral of the function f on an arbitrary set $A \in \mathscr{T}$ will be given by the formula

$$\int_A P(f(t), \mathrm{d}\mathbf{m}(t)) = \sum_{i=1}^{k} P(x_i, \mathbf{m}(A \cap A_i)).$$

A function f: $T \to \mathscr{X}$ is said to be $(\mathscr{T}, \mathbf{m})$-*integrable* on T if it is $(\mathscr{T}, \mathbf{m})$-measurable and if there exists an approximating sequence $\{f_n\}_{n\in\mathbb{N}}$ such that

$$(o)\text{-}\lim_n \left(\sup_{i,j \geqslant n} \int_T P(|f_i(t) - f_j(t)|, \mathrm{d}\mathbf{m}(t)) \right) = \mathbf{0}.$$

The sequence $\{f_n\}_{n\in\mathbb{N}}$ will be called an *integrating sequence* for f.

Lemma 1. *Let $\{f_n\}_{n\in\mathbb{N}}$ be an arbitrary integrating sequence for the function f and $\{S_j\}_{j\in\mathbb{N}}$ a sequence of sets from $\mathscr{T}$, such that* $\mathbf{m}(S_j) \overset{o}{\to} \mathbf{0}$.

Then there exists an element $z_0 \in \mathscr{Z}$ and, for any real number $\varepsilon > 0$, there exists a number $j_\varepsilon \in \mathbb{N}$, such that for $j \geqslant j_\varepsilon$ we have

$$\int_{S_j} P(|f_n(t)|, \mathrm{d}\mathbf{m}(t)) \leqslant \varepsilon\, z_0$$

for any $n \in \mathbb{N}$

Proof. Let

$$f_n(t) = \sum_{i=1}^{k_n} \chi_{A_{ni}}(t)\, x_{ni},$$

and let us define

$$x_n = \bigvee_{i=1}^{k_n} |x_{ni}|.$$

Let $\{\lambda_n\}_{n\in\mathbb{N}}$ be a sequence of strictly positive real numbers, such that $\lambda_n x_n \overset{o}{\to} \mathbf{0}$.

Then there exists an $r \in \mathscr{X}$, such that for any number $\varepsilon > 0$ there exists an $n'_\varepsilon \in \mathbb{N}$, having the property

$$\lambda_n x_n \leqslant \varepsilon r, \quad n \geqslant n'_\varepsilon. \tag{1}$$

Let us define [6])

$$l_n = \bigvee_{i,j \geqslant n} \int_T |f_i - f_j| \, \mathbf{dm},$$

and let z_1 be a convergence regulator of the sequence $\{l_n\}_{n \in \mathbb{N}}$. If we now denote

$$l_n^*(S) = \bigvee_{i,j \geqslant n} \left| \int_S |f_i| \, \mathbf{dm} - \int_S |f_j| \, \mathbf{dm} \right|,$$

then z_1 is a convergence regulator for the sequence $\{l_n^*(S)\}_{n \in \mathbb{N}}$ too. For any number $\varepsilon > 0$, there exists a natural number $n_\varepsilon \geqslant n'_\varepsilon$, such that

$$l_n^*(S) \leqslant \varepsilon z_1, (n \geqslant n_\varepsilon). \tag{2}$$

Let us now write the following inequality

$$\int_S |f_n| \, \mathbf{dm} \leqslant \left| \int_S |f_n| \, \mathbf{dm} - \int_S |f_{n_\varepsilon}| \, \mathbf{dm} \right| + \int_S |f_{n_\varepsilon}| \, \mathbf{dm},$$

which holds for any $n \in \mathbb{N}$ and $S \in \mathscr{T}$. From (2) we infer that

$$\int_S |f_n| \, \mathbf{dm} \leqslant \varepsilon z_1 + P(x_{n_\varepsilon}, \mathbf{m}(S)), \ (n \geqslant n_\varepsilon). \tag{3}$$

On the other hand, if

$$\mathbf{m}(S_j) \leqslant \lambda_{n_\varepsilon} y_0,$$

then from (1) we have

$$P(x_{n_\varepsilon}, \mathbf{m}(S_j)) \leqslant P(x_{n_\varepsilon}, \lambda_{n_\varepsilon} y_0) = P(\lambda_{n_\varepsilon} x_{n_\varepsilon}, y_0) \leqslant \varepsilon P(r, y_0).$$

By denoting

$$z_2 = z_1 + P(r, y_0),$$

from (3) it follows that

$$\int_{S_j} |f_n| \, \mathbf{dm} \leqslant \varepsilon z_2, \ (n \geqslant n_\varepsilon). \tag{4}$$

---

[6]) In this proof we shall use a simplified notation of the integral.

By defining

$$x_0 = \bigvee_{n \in \mathbb{N}} \lambda_n x_n,$$

denoting

$$\mu_{n_\varepsilon} = \min \{\lambda_i; i = 1, 2, \ldots, n_\varepsilon - 1\}$$

and assuming that

$$\mathbf{m}(S_j) \leqslant \varepsilon \mu_{n_\varepsilon} y_0,$$

we infer that

$$\int_{S_j} |f_i| \, \mathrm{d}\mathbf{m} \leqslant P(x_0, \frac{1}{\lambda_i} \mathbf{m}(S_j)) \leqslant \varepsilon \, P(x_0, y_0)$$

for any $i = 1, 2, \ldots, n_\varepsilon - 1$.

By denoting

$$z_0 = z_2 \vee P(x_0, y_0),$$

and

$$\eta(\varepsilon) = \min (\varepsilon \mu_{n_\varepsilon}, \lambda_{n_\varepsilon}),$$

and by taking into account inequality (4), it follows that

$$\int_{S_j} |f_n| \, \mathrm{d}\mathbf{m} \leqslant \varepsilon \, z_0, \quad (n \in \mathbb{N}),$$

as soon as

$$\mathbf{m}(S_j) \leqslant \eta(\varepsilon) y_0. \tag{5}$$

Since $\mathbf{m}(S_j) \overset{o}{\to} \mathbf{0}$, by taking into account the fact that $\mathscr{Y}$ is a regular space, we infer that there exists a number j_ε such that inequality (5) be satisfied for any $j \geqslant j_\varepsilon$.

The lemma is proved.

Lemma 2. *If f is a $(\mathscr{T}, \mathbf{m})$-integrable function, then there exists an $l \in \mathscr{Z}$, such that*

$$l = (o)\text{-}\lim_n \int_T P(f_n(t), \mathrm{d}\mathbf{m}(t)), \tag{6}$$

for any integrating sequence $\{f_n\}_{n \in \mathbb{N}}$.

Proof. If $\{f_n\}_{n\in\mathbb{N}}$ is an integrating sequence for f, then the (o)-limit in (6) exists as a consequence of a proposition from Section 3.2.6 (proposition 1).

Let $\{f_n^1\}_{n\in\mathbb{N}}$ and $\{f_n^2\}_{n\in\mathbb{N}}$ be two integrating sequences for f. For any $A\in\mathcal{T}$ we have

$$\left|\int_T f_n^1 \mathrm{d}\mathbf{m} - \int_T f_n^2 \mathrm{d}\mathbf{m}\right| \leqslant \int_A |f_n^1 - f_n^2|\, \mathrm{d}\mathbf{m} + \int_{\complement A} |f_n^1|\, \mathrm{d}\mathbf{m} + \int_{\complement A} |f_n^2|\, \mathrm{d}\mathbf{m}. \tag{7}$$

Let $\{A_j\}_{j\in\mathbb{N}}$ be a sequence of sets from $\mathcal{T}$, such that $\mathbf{m}(\complement A_j) \overset{o}{\to} \mathbf{0}$ and such that the sequence $\{f_n^1\}_{n\in\mathbb{N}}$, $\{f_n^2\}_{n\in\mathbb{N}}$ be uniformly (o)-convergent to f on each A_j. By virtue of lemma 1, there exists a $z_0\in\mathcal{Z}$, and for any number $\varepsilon>0$ there exists a number $j_\varepsilon\in N$, such that

$$\int_{\complement A_j} |f_n^i|\, \mathrm{d}\mathbf{m} \leqslant \varepsilon\, z_0, \quad (i=1, 2; j \geqslant j_\varepsilon), \tag{8}$$

for any $n\in\mathbb{N}$.

On the other hand, one can immediately check that for any $j\in\mathbb{N}$ there exists an $r_j\in\mathcal{X}$ and a sequence $\{\delta_{jn}\}_{n\in\mathbb{N}}$ of numbers, which converges to zero, such that

$$|f_n^1(t) - f_n^2(t)| \leqslant \delta_{jn} r_j, \quad t\in A_j,$$

whence

$$\int_{A_j} |f_n^1 - f_n^2|\, \mathrm{d}\mathbf{m} \leqslant \delta_{jn}\, P(r_j, \mathbf{m}(A_j)). \tag{9}$$

By making $A = A_{j_\varepsilon}$ in inequality (7) and taking into account inequalities (8) and (9) we obtain

$$\left|\int_T f_n^1 \mathrm{d}\mathbf{m} - f_n^2 \mathrm{d}\mathbf{m}\right| \leqslant \delta_{j_\varepsilon n}\, P(r_{j_\varepsilon}, \mathbf{m}(A_{j_\varepsilon})) + 2\varepsilon z_0.$$

By tending to the (o)-limit with respect to n and taking into account the fact that $\varepsilon>0$ was arbitrary , the uniqueness of the limit immediately follows.

The lemma is proved.

If f is a $(\mathcal{T}, \mathbf{m})$-integrable function, defined on T, we shall define the *integral of f with respect to* $\mathbf{m}$, *on* T, by the formula

$$\int_T P(f(t), \mathrm{d}\mathbf{m}(t)) = (o)\text{-}\lim_n \int_T P(f_n(t), \mathrm{d}\mathbf{m}(t)),$$

where $\{f_n\}_{n\in\mathbb{N}}$ is an arbitrary integrating sequence for f.

The integral on an arbitrary set from $\mathcal{T}$ can be defined in an obvious manner.

It is immediately verified that if f and g are integrable then $\alpha f + \beta g$ is integrable too (for any $\alpha, \beta \in \mathbb{R}$) and

$$\int_T P(\alpha f(t) + \beta g(t), \mathbf{dm}(t)) = \alpha \int_T P(f(t), \mathbf{dm}(t)) + \beta \int_T P(g(t), \mathbf{dm}(t)).$$

If f and g are integrable and if $f(t) \leqslant g(t)$, $(\mathcal{T}, \mathbf{m})$-almost everywhere, then

$$\int_T P(f(t), \mathbf{dm}(t)) \leqslant \int_T P(g(t), \mathbf{dm}(t)).$$

Lemma 3. *If f is a $(\mathcal{T}, \mathbf{m})$-measurable function and if g is a $(\mathcal{T}, \mathbf{m})$-integrable function such that $|f(t)| \leqslant g(t)$, $(\mathcal{T}, \mathbf{m})$-almost everywhere, then f is $(\mathcal{T}, \mathbf{m})$-integrable.*

Proof. Let $\{f_n\}_{n \in \mathbb{N}}$ be an approximating sequence for f and $\{g_n\}_{n \in \mathbb{N}}$ an integrating sequence for g.

Let us consider the functions defined by

$$f_n^1(t) = |f_n(t)| \wedge g_n(t), \; n \in \mathbb{N}.$$

The sequence $\{f_n^1\}_{n \in \mathbb{N}}$ obviously is an approximating sequence for $|f(.)|$. Let us show that it is an integrating sequence. In order to do this, let us remark that for any $A \in \mathcal{T}$ we have

$$\int_T |f_i^1 - f_k^1| \mathbf{dm} \leqslant \int_A |f_i^1 - f_k^1| \, \mathbf{dm} + \int_{\complement A} g_i \mathbf{dm} + \int_{\complement A} g_k \mathbf{dm}. \tag{10}$$

Let $\{A_j\}_{j \in \mathbb{N}}$ be a sequence of sets from $\mathcal{T}$, such that $\mathbf{m}(\complement A_j) \overset{o}{\to} \mathbf{0}$, whereas the sequence $\{f_n^1\}_{n \in \mathbb{N}}$ be uniformly (o)-convergent to f on any set A_j. By virtue of lemma 1, there exists a $z_0 \in \mathcal{Z}$, such that for any number $\varepsilon > 0$ there exists a $j_\varepsilon \in \mathbb{N}$ having the property that

$$\int_{\complement A_j} g_n \, \mathbf{dm} \leqslant \varepsilon \, z_0, \; j \geqslant j_\varepsilon, \; \forall n \in \mathbb{N}. \tag{11}$$

On the other hand, one can observe that

$$(o)\text{-}\lim_n \left(\sup_{i,k \geqslant n} \int_{A_j} |f_i^1 - f_k^1| \, \mathbf{dm} \right) = \mathbf{0}. \tag{12}$$

From relations (10), (11) and (12) one infers that $\{f_n^1\}_{n \in \mathbb{N}}$ is an integrating sequence for $|f(.)|$. From $\mathbf{0} \leqslant (f(t))_+ \leqslant |f(t)|$ and $\mathbf{0} \leqslant (f(t))_- \leqslant |f(t)|$ it follows that f is integrable [7]).

[7]) It is easily checked that if f is measurable, then f is integrable iff $|f|$ is integrable.

Theorem. *Let* $\{f_n\}_{n\in\mathbb{N}}$ *be a sequence of* $(\mathscr{T}, \mathbf{m})$*-measurable functions and* g *a* $(\mathscr{T}, \mathbf{m})$*-integrable function, such that* $|f_n(t)| \leqslant g(t)$, $(\mathscr{T}, \mathbf{m})$*-almost everywhere. If the sequence* $\{f_n\}_{n\in\mathbb{N}}$ $(\mathscr{T}, \mathbf{m})$*-almost uniformly (o)-converges to a function* f*, then* f *is* $(\mathscr{T}, \mathbf{m})$*-integrable and*

$$\int_T P(f(t), \mathbf{dm}(t)) = (o)\text{-}\lim_n \int_T P(f_n(t), \mathbf{dm}(t)).$$

Proof. By virtue of lemma 2, the functions f_n are integrable and from the fact that $|f(t)| \leqslant g(t)$ holds $(\mathscr{T}, \mathbf{m})$-almost everywhere, it follows that f is $(\mathscr{T}, \mathbf{m})$-integrable.

Let us now observe that for any $A \in \mathscr{T}$ we have

$$\left| \int_{\complement A} f_n\,\mathbf{dm} - \int_{\complement A} f\,\mathbf{dm} \right| \leqslant \int_{\complement A} g\,\mathbf{dm} + \int_{\complement A} |f|\,\mathbf{dm}. \tag{13}$$

Let $\{A_j\}_{j\in\mathbb{N}}$ be a sequence of sets from $\mathscr{T}$, such that $\mathbf{m}(\complement A_j) \xrightarrow{o} 0$, and the sequence $\{f_n\}_{n\in\mathbb{N}}$ be uniformly (o)-convergent to f on any A_j. By virtue or lemma 1, there exists a $z_0 \in \mathscr{Z}$, and for any number $\varepsilon > 0$ there exists a $j_\varepsilon \in \mathbb{N}$, such that

$$\int_{\complement A_{j_\varepsilon}} |f|\,\mathbf{dm} \leqslant \varepsilon z_0, \int_{\complement A_{j_\varepsilon}} g\,\mathbf{dm} \leqslant \varepsilon\, z_0$$

By making $A = A_{j_\varepsilon}$ in inequality (13), we obtain

$$\left| \int_{\complement A_{j_\varepsilon}} f_n\,\mathbf{dm} - \int_{\complement A_{j_\varepsilon}} f\,\mathbf{dm} \right| \leqslant 2\varepsilon z_0. \tag{14}$$

On the other hand

$$\int_{A_{j_\varepsilon}} f_n\,\mathbf{dm} \xrightarrow{o} \int_{A_{j_\varepsilon}} f\,\mathbf{dm},$$

as one can easily prove. Hence, by taking into account inequality (14), we get

$$\overline{\lim_n} \left| \int_T f_n\,\mathbf{dm} - \int_T f\,\mathbf{dm} \right| \leqslant 2\varepsilon\, z_0;$$

since the number $\varepsilon > 0$ was arbitrary, the left-hand member vanishes

The theorem is proved.

Remark. One can also prove that if $\{S_n\}_{n\in\mathbb{N}}$ is a sequence of mutually disjoint sets from $\mathscr{T}$, and if f is a $(\mathscr{T}, \mathbf{m})$-integrable function on $\bigcup_{n\in\mathbb{N}} S_n$, then

$$\int_{\bigcup_{n\in\mathbb{N}} S_n} P(f(t), \mathrm{d}\mathbf{m}(t)) = (o)\text{-}\sum_{n=1}^{\infty} \int_{S_n} P(f(t), \mathrm{d}\mathbf{m}(t))$$

7.2.3 The integral representation of some linear operators

In this section we shall consider the integral we have introduced in the preceding section for the case of real functions. In this case, the operator P will be the multiplication of the elements of $\mathscr{Y}$ by real numbers.

Let T and $\mathscr{T}$ be as in the preceding sections. Let μ be a real measure defined on $\mathscr{T}$.

We shall consider the space $S(T, \mathscr{T}, \mu)$ consisting of all classes of $(\mathscr{T}, \mu)$-measurable functions, defined similarly to the space $S([\alpha, \beta])$ from Section 6.4.1. The unit element will be the element corresponding to the constant function, equal to 1.

We shall denote by $S_0(T, \mathscr{T}, \mu)$ an arbitrary normal subspace of $S(T, \mathscr{T}, \mu)$, which contains the unit element.

Let $\mathscr{Y}$ be a σ-regular space.

If $\mathbf{m}$ is a measure on $\mathscr{T}$, taking values in the space $\mathscr{Y}$, then $\mathbf{m}$ is, by definition, *absolutely* (o^*)*-continuous* with respect to μ, if $\mu(A_n) \to 0$ implies that $\mathbf{m}(A_n) \xrightarrow{o^*} 0$. We shall denote $\mathbf{m} \ll \mu$.

Theorem. *The general form of the* (o)*-continuous, positive, linear operators, defined on* $S_0(T, \mathscr{T}, \mu)$ *and taking values in* $\mathscr{Y}$*, is given by the formula*

$$U(f) = \int_T f(t)\,\mathrm{d}\mathbf{m}(t), \tag{1}$$

where $\mathbf{m}$ *is a measure on* $\mathscr{T}$*, taking values in* $\mathscr{Y}$*, such that* $\mathbf{m} \ll \mu$*, and such that any* $f \in S_0(T, \mathscr{T}, \mu)$ *be* $(\mathscr{T}, \mathbf{m})$*-integrable.*

Proof. Let U be of the form (1), where $\mathbf{m}$ has the mentioned properties. By taking into account the properties of the integral, we infer that U is linear and positive.

Let $f_n \xrightarrow{o} f$ in $S_0(T, \mathscr{T}, \mu)$. There then exists a $g \in S_0(T, \mathscr{T}, \mu)$, such that $|f_n| \leqslant g$, i.e., $|f_n(t)| \leqslant g(t)$, $(\mathscr{T}, \mu)$-almost everywhere, and therefore, $(\mathscr{T}, \mathbf{m})$-almost everywhere. Since $\mathbf{m} \ll \mu$, the sequence $\{f_n\}_{n\in\mathbb{N}}$ $(\mathscr{T}, \mathbf{m})$-almost uniformly (o)-converges to f. By taking into account the theorem from the preceding section, we infer that $U(f_n) \xrightarrow{o} U(f)$. It follows that U is (o)-continuous.

Conversely, let U be an (o)-continuous, positive, linear operator defined on $S_0(T, \mathscr{T}, \mu)$. For any $A \in \mathscr{T}$ let us define $\mathbf{m}(A) = U(\chi_A)$, where χ_A is the characteristic function of the set A. It is not difficult to check that $\mathbf{m}$ is a measure on $\mathscr{T}$(taking values in $\mathscr{Y}$). If $\mu(A_n) \to 0$, then there exists a subsequence of indices $\{j_n\}_{n\in\mathbb{N}}$, such that $\chi_{A_{j_n}} \overset{o}{\to} 0$ in $S_0(T, \mathscr{T}, \mu)$, whence $\mathbf{m}(A_{j_n}) \overset{o}{\to} 0$. Consequently, $\mathbf{m} \ll \mu$.

Formula (1) obviously holds for the simple elements of the space $S_0(T, \mathscr{T}, \mu)$. Let f be arbitrary in $S_0(T, \mathscr{T}, \mu)$. There then exists a sequence $\{f_n\}_{n\in\mathbb{N}}$ of $\mathscr{T}$-simple functions, which $(\mathscr{T}, \mu)$-almost uniformly converges to f. Since $\mathbf{m} \ll \mu$, it follows that the sequence $\{f_n\}_{n\in\mathbb{N}}(\mathscr{T}, \mathbf{m})$-almost uniformly (o)-converges to f. On the other hand, $f_n(t) \to f(t)$ $(\mathscr{T}, \mu)$-almost everywhere and one can assume that $|f_n(t)| \leqslant |f(t)|$. It follows that $f_n \overset{o}{\to} f$, whence $U(f_n) \overset{o}{\to} U(f)$.

Since $\{f_n\}_{n\in\mathbb{N}}$ is an integrating sequence for f, equality (1) now immediately follows.

Bibliographical comments

The results in this chapter, excepting Sections 7.1.1 and 7.1.2, are the author's.

Spaces with vector norm (especially the spaces of type (B_k)) were considered by L. V. Kantorovitch in connection with the study of operator equations. The linear operators in spaces of type (B_k) were also studied by L. V. Kantorovitch. (See L. V. Kantorovitch, B. Z. Vulich, A. G. Pinskner [1]).

The importance of developing a theory of functions taking values in a vector lattice, was stressed by L. V. Kantorovitch [5]. In connection with the study of the continuous or semi-continuous functions of a real variable, taking values in a complete or a σ-complete vector lattice, see L. V. Kantorovitch [5], B. P. Frumkin [1], R. Cristescu [7], etc. The notion of derivative for functions taking values in a complete vector lattice has been studied by A. O'Neille [1].

The Stieltjes type integral studied in Section 7.1, and the representation theorem from the Section 7.1.4, have been given by R. Cristescu [17]. Special types of vector Stieltjes integrals were previously introduced by B. Z. Vulich [4] and L. V. Kantorovitch [17]. For the case of the functions taking values in a normed vector lattice, results analogous to those given in Section 7.1.4, were established by R. Cristescu [13].

If $\mathscr{X}$ is a vector lattice, T an arbitrary set and $\mathscr{T}$ an algebra of subsets of T, a function $f\colon T \to \mathscr{X}$ is said to be *totally $\mathscr{T}$-measurable* if there exists a sequence $\{f_n\}_{n\in\mathbb{N}}$ of $\mathscr{T}$-simple function which uniformly converges to f. In connection with the vector Radon integrals and the representation of the linear operators on spaces of totally measurable vector functions, see R. Cristescu [14], [20].

The Hellinger integral from Section 7.1.5, as well as the integral representation of the bounded linear operators defined on the space $L(T, \mathscr{X})$, belong to the author (see R. Cristescu [22], [25]).

For other representations by Hellinger vector integrals, see R. Cristescu [35] and Gh. Grigore [4].

In connection with the results from Section 7.2, see R. Cristescu [1], [5]. The integral from Section 7.2 is a direct generalization of the Bochner integral, defined for vector functions, with respect to a real measure.

Various results about the positive vector measure were obtained by R. L. Gomes [1] and by other authors.

In the work of R. L. Gomes [1] one considers a locally compact space T and the space $C_0(T)$ of the continuous real functions, defined on T and having compact supports. One also considers an ordered vector space $\mathscr{X}$ in which a locally convex topology is given, subject to a compatibility condition with respect to the order relation.

By a positive measure one means a positive linear operator, defined on $C_0(T)$ and taking values in $\mathscr{X}$. With the help of some extensions of the measure, one then defines some vector integrals.

M. Zamansky [1] shows that the theorem on the completion of a lattice ordered group leads to a rapid construction of the Daniell integral.

A general theory of the real integral on lattices is developed by E. M. Alfsen [1]. In particular, in the cited paper the Daniell integral on a vector lattice is also considered.

8

Some applications of the theory of ordered vector spaces

This chapter gives a brief exposition of the manner in which the theory of ordered vector spaces can be used in various branches of mathematics [1])

8.1 Operator equations

8.1.1 The method of the successive approximations

Let $\mathscr{X}$ be a σ-complete vector lattice and $\mathscr{F}_{\mathscr{X}}$ a (v)-complete vector space, with a vector norm (Section 7.1.1).

Let z be a positive element of $\mathscr{X}$ and $E = [0, z]$.

An operator V, defined on a subset A of $\mathscr{X}$, and taking values in $\mathscr{X}$ is said to be isotone if $x_1 \leqslant x_2 (x_1, x_2 \in A)$ implies $V(x_1) \leqslant V(x_2)$.

Let V be an isotone and (o)-continuous operator, defined on E, taking values in E, and let us consider the equation

$$x = V(x). \tag{1}$$

If we put $x_0 = \mathbf{0}$ and if we recurrently define

$$x_n = V(x_{n-1}), \quad n \in \mathbb{N}, \tag{2}$$

then $x_n \in E$ and $x_n \uparrow$. There then exists the limit $a = (o)\text{-lim}\, x_n$ and we obviously have $a = V(a)$. Analogously, if we define $x'_0 = z$ and if we define x'_u by a formula analogous to formula (2), then the element $b = (o)\text{-lim}\ x'_u$ exists and $b = V(b)$. Obviously, $a \leqslant b$. If x is an element which satisfies condition (1), then $a \leqslant x \leqslant b$, as one can easily prove.

Let us now define

$$D = \{f \in \mathscr{F}_{\mathscr{X}};\ |f| \leqslant z\},$$

[1]) We do not intend to develop fully, in this chapter, the applications we give here.

and let U be an operator defined on D, taking values in $\mathscr{F}_{\mathfrak{X}}$ and satisfying the following conditions

(i) $|U(\mathbf{0})| \leqslant V(\mathbf{0})$;

(ii) $|U(f_1 + f_2) - U(f_1)| \leqslant V(x_1 + x_2) - V(x_1)$

as soon as $|f_1| \leqslant x_1$, $|f_2| \leqslant x_2$ and $x_1 + x_2 \in E$.

In this case we shall say that U is *dominated* by V. One can observe that the values of U belong to the set D.

Let us consider the equation

$$f = U(f). \tag{3}$$

Let us define $f_0 = \mathbf{0}$ and, recurrently,

$$f_n = U(f_{n-1}) \quad n \in \mathbb{N}. \tag{4}$$

By considering the sequence of the elements x_n, defined by formula (2), where $x_0 = \mathbf{0}$, one can prove by induction that

$$|f_{n+1} - f_n| \leqslant x_{n+1} - x_n,$$

for any n. It follows that

$$|f_m - f_n| \leqslant x_m - x_n$$

for $m > n$, and, since

$$x_m - x_n \leqslant a - x_n \downarrow \mathbf{0},$$

it follows that the element $f^* = (v)\text{-}\lim f_n$ exists.

From

$$|f_n| \leqslant x_n, \quad n \in \mathbb{N},$$

it follows that

$$|f^*| \leqslant a.$$

Since

$$|U(f^*) - U(f_n)| \leqslant V(a) - V(x_n) = a - x_{n+1},$$

it follows that $f^* = U(f^*)$.

If $a = b$, then the solution of equation (3) is unique.

Indeed, let $f_0 \in D$ and define

$$f'_n = U(f'_{n-1}), \quad n \in \mathbb{N}.$$

One can prove by induction that

$$|f'_n - f_n| \leqslant x'_n - x_n,$$

and, since $x'_n - x_n \xrightarrow{o} \mathbf{0}$, it follows that $f'_n \xrightarrow{v} f^*$.

In particular, if f_0' is a solution of equation (3), then $f_n' = f_0'$, for any $n \in \mathbb{N}$ and, therefore $f^* = f_0'$.

8.1.2 Linear operator equations

Let $\mathscr{X}$ and $\mathscr{F}_x$ be as in the preceding section.

Let V be an (o)-continuous, positive, linear operator, which maps $\mathscr{X}$ into $\mathscr{X}$. Let us consider the equation

$$x = V(x) + v, \tag{1}$$

where v is a given element of $\mathscr{X}$.

Let us assume that there exists an element $z > \mathbf{0}$, such that

$$w = z - V(z) \geqslant \mathbf{0}. \tag{2}$$

If $|v| \leqslant w$, then equation (1) has a solution x^* which can be obtained by the method of successive approximations, beginning with $x_0 = \mathbf{0}$.

Indeed, let us first assume that $\mathbf{0} \leqslant v \leqslant w$ and define $E = [\mathbf{0}, z]$ and

$$V_0(x) = V(x) + v, \quad x \in E.$$

The operator V_0 maps E into E, is isotone and (o)-continuous. From what we showed in the preceding section, it follows that there exists a solution $x^* \in E$ of the equation

$$x = V_0(x),$$

which can be obtained by the method of successive approximations, beginning with $x_0 - \mathbf{0}$.

If $|v| \leqslant w$, then $\mathbf{0} \leqslant v_+ \leqslant w$ and $\mathbf{0} \leqslant v_- \leqslant w$. Replacing in equation (1) the element v by v_+, and then by v_-, we obtain the solutions x_1^* and x_2^*, respectively; by defining $x^* = x_1^* - x_2^*$, we obtain a solution of equation (1). Obviously, $|x^*| \leqslant z$.

Let us now consider the operator equation

$$f = U(f) + g \tag{3}$$

in the space $\mathscr{F}_{\mathscr{X}}$, where U is a linear operator which maps $\mathscr{F}_{\mathscr{X}}$ into itself, g is an element of $\mathscr{F}_{\mathscr{X}}$, such that $|g| \leqslant w$ (where w satisfies condition (2)), whereas

$$|U(f)| \leqslant V(|f|),$$

for any $f \in \mathscr{F}_{\mathscr{X}}$. Under these conditions, equation (3) has a solution f^*, satisfying the condition $|f^*| \leqslant z$, and it can be obtained by the method of successive approximations, beginning with $f_0 = 0$.

Indeed, by putting

$$U_1(f) = U(f) + g \quad (|f| \leqslant z),$$

we obtain an operator which is dominated by the operator V_1 given by the formula

$$V_1(x) = V(x) + w.$$

8.1.3 Equations in topological ordered vector spaces

Let $\mathscr{X}$ be a sequentially (τ)-complete topological ordered vector space, in which the set $\mathscr{X}_+$ is (τ)-closed. We shall assume that the following condition is also satisfied: if $x_n \uparrow_{n \in \mathbb{N}}$ and $x_n \leqslant y$ ($n \in \mathbb{N}$), then $x = (\tau)\text{-}\lim x_n$ exists.

Let U and V be sequentially (τ)-continuous linear operators, which map $\mathscr{X}$ into $\mathscr{X}$ and are such that

$$-V \leqslant U \leqslant V.$$

We note that $-y \leqslant x \leqslant y$ implies in this case

$$-V(y) \leqslant U(x) \leqslant V(y). \tag{1}$$

Indeed, since $V - U \geqslant \mathbf{0}$ and $V + U \geqslant \mathbf{0}$, it follows that

$$-(V - U)(y) \leqslant (V - U)(x) \leqslant (V - U)(y), \tag{2}$$

$$-(V + U)(y) \leqslant (V + U)(x) \leqslant (V + U)(y). \tag{3}$$

From inequalities (2) we infer the following inequalities

$$-(V - U)(y) \leqslant (U - V)(x) \leqslant (V - U)(y),$$

whereas, by taking into account inequalities (3), we get inequalities (1).

Let $a, b \in \mathscr{X}$ be such that

$$-b \leqslant a \leqslant b$$

and let us consider the equations

$$x = U(x) + a, \tag{4}$$

$$y = V(y) + b. \tag{5}$$

Let us assume that equation (5) has a positive solution y^*. We shall show that, in this case, equation (4) has solution x^* which can be obtained as the (τ)-limit of the sequence $\{x_n\}_{n\in\mathbb{N}}$, defined by

$$x_n = U(x_{n-1}) + a, \quad n \in \mathbb{N};\ x_0 = \mathbf{0}. \tag{6}$$

Let us put $y_0 = \mathbf{0}$ and

$$y_n = V(y_{n-1}) + b, \quad n \in \mathbb{N}. \tag{7}$$

Obviously, $y_n \uparrow$ and since $y_0 \leqslant y^*$ one immediately infers, by induction, that $y_n \leqslant y^*$. By hypothesis, the limit $z = (\tau)\text{-lim } y_n$ exists. From (7) we infer that z is a solution of equation (5) and, obviously, $z \leqslant y^*$.

Let us now consider the sequence $\{x_n\}_{n\in\mathbb{N}}$ given by (6). By induction, we can show that

$$-(y_n - y_{n-1}) \leqslant x_n - x_{n-1} \leqslant y_n - y_{n-1}. \tag{8}$$

Indeed, for $n = 1$, the inequalities are obviously true. Let us assume that they hold for a given n. With a preceding remark, from $-V \leqslant U \leqslant V$ and from inequalities (8) it follows that

$$-V(y_n - y_{n-1}) \leqslant U(x_n - x_{n-1}) \leqslant V(y_n - y_{n-1}),$$

and we have only to take into account the fact that

$$U(x_n - x_{n-1}) = (U(x_n) + a) - (U(x_{n-1}) + a) = x_{n+1} - x_n,$$

$$V(y_n - y_{n-1}) = y_{n+1} - y_n.$$

Consequently, inequalities (8) hold for any $n \in \mathbb{N}$. From these inequalities we infer the inequalities

$$-(y_{n+m} - y_n) \leqslant x_{n+m} - x_n \leqslant y_{n+m} - y_n, \tag{9}$$

for any $n, m \in \mathbb{N}$.

Since the topology τ is locally full, there exists in $\mathscr{X}$ a fundamental system $\mathscr{W}$ consisting of equilibrated and full neighbourhoods of the origin (if W is equilibrated, then $\text{fl}(W)$ is equilibrated and full). For any $W \in \mathscr{W}$ there exists an $n_0 \in \mathbb{N}$, such that $y_{n+m} - y_n \in W$, for any $n \geqslant n_0$. From (9) we infer that $x_{n+m} - x_n \in W$, for any $n \geqslant n_0$; consequently, $\{x_n\}_{n\in\mathbb{N}}$ is a Cauchy sequence for the topology τ. Since the space $\mathscr{X}$ is sequentially (τ)-complete the limit $x^* = (\tau)\text{-lim } x_n$ exists and, since U is sequentially (τ)-continuous, from (6) it follows that x^* is a solution of equation (4).

Since from (8) we obtain

$$-y_n \leqslant x_n \leqslant y_n, \quad n \in \mathbb{N},$$

and since the set $\mathscr{X}_+$ is (τ)-closed, it follows that $-z \leqslant x^* \leqslant z$, and this implies that $-y^* \leqslant x^* \leqslant y$, because $z \leqslant y^*$.

8.1.4 A problem in mathematical programming

Let $\mathscr{X}$ be a sequentially (τ)-complete, Hausdorff, topological vector lattice, having the property that $x_n \underset{n \in \mathbb{N}}{\uparrow}$ and $x_n \leqslant y$, $(n \in \mathbb{N})$ implies the existence of the limit $x = (\tau)\text{-lim } x_n$.

Let us consider an operator $U:\mathscr{X}_+ \to \mathscr{X}$, which is isotone and sequentially (τ)-continuous. Let us define

$$E = \{x \in \mathscr{X}_+; \quad U(x) \leqslant x\}.$$

If the set E has the smallest element, then for any isotone functional f, defined on $\mathscr{X}_+$, the problem of minimizing f on the set E is solved.

If we define

$$U_0(x) = (U(x))_+, \quad x \in \mathscr{X}_+,$$

then U_0 maps $\mathscr{X}_+$ into $\mathscr{X}_+$, is isotone and sequentially (τ)-continuous.

By defining

$$G = \{x \in \mathscr{X}_+; \quad U_0(x) = x\},$$

we obviously have $G \subset E$. If $E \neq \varnothing$ and $x \in E$, then by defining

$$x_n = U_0(x_{n-1}), \quad n \in \mathbb{N}; \quad x_0 = \mathbf{0},$$

we have $\mathbf{0} \leqslant x_n \uparrow$ and $x_n \leqslant x$.

By hypothesis, the limit $a = (\tau)\text{-lim } x_n$ exists and we have $a \leqslant x$, $a \in G$. It follows that a is the smallest element of E.

8.1.5 Applications to equations in concrete spaces

(1) Let us consider an infinite system of linear equations

$$\xi_i = \sum_{j=1}^{\infty} \alpha_{ij}\, \xi_j + \gamma_i, \qquad i \in \mathbb{N}, \tag{1}$$

where α_{ij} and γ_i are given real numbers.

Let us consider also the system

$$\xi_i = \sum_{j=1}^{\infty} \beta_{ij}\xi_j + \delta_i, \quad i \in \mathbb{N}, \tag{2}$$

(where $\beta_{ij} \geqslant 0$, $\delta_i \geqslant 0$) and assume that it has a solution $\{\xi_i'\}_{i \in \mathbb{N}}$, such that $\xi_i' \geqslant 0$.

Let us now assume that $|\alpha_{ij}| \leqslant \beta_{ij}$ and $|\gamma_i| \leqslant \delta_i$, for any i, j.

Let (s) be the space of all sequences of real numbers, endowed with the usual structure of complete vector lattice (Section 3.4.2).

Let us define

$$A(x) = \left\{ \sum_{j=1}^{\infty} \alpha_{ij}\xi_j + \gamma_i \right\}_{i \in \mathbb{N}} \tag{3}$$

for any $x = \{\xi_j\}_{j \in \mathbb{N}}$, such that the series $\sum_j \alpha_{ij}\xi_j$ be convergent.

If we denote $z = \{\xi_i'\}_{i \in \mathbb{N}}$ and $D = \{x \in (s);\ |x| \leqslant z\}$, with formula (3) we define an operator, at least on the set D.

If we now define

$$B(x) = \left\{ \sum_{j=1}^{\infty} \beta_{ij}\xi_j + \delta_i \right\}_{i \in \mathbb{N}},$$

for any $x \in [\mathbf{0}, z]$, we obtain an isotone and (o)-continuous operator on $[\mathbf{0}, z]$, taking values in $[\mathbf{0}, z]$, whereas A is dominated by B on the set D.

From what we have shown in Section 8.1.1, there exists a solution $\{\xi_j^*\}_{j \in \mathbb{N}}$ of the system (1), such that $|\xi_j^*| \leqslant \xi_j'$; $j \in \mathbb{N}$, and it can be obtained by the method of successive approximations.

If, for any $i \in \mathbb{N}$, the series $\sum_{j=1}^{\infty} \alpha_{ij}$ is convergent and if

$$\sum_{j=1}^{\infty} |\alpha_{ij}| < 1,$$

for any $i \in \mathbb{N}$, then the system (1) is said to be regular.

Let us assume that the system (1) is regular and that the sequence $c = \{\gamma_i\}_{i \in \mathbb{N}}$ is bounded.

Let us consider the complete vector lattice (m) of all bounded sequences of real numbers and the following linear operators (which map (m) into (m))

$$U(x) = \left\{ \sum_{j=1}^{\infty} \alpha_{ij}\,\xi_j \right\}_{i \in \mathbb{N}},$$

$$V(x) = \left\{ \sum_{j=1}^{\infty} |\alpha_{ij}|\,\xi_j \right\}_{i \in \mathbb{N}}.$$

The operator V is positive and (o)-continuous, whereas U is majorized by V. The system (1) can be written in the following form

$$x = U(x) + c.$$

Let us now denote

$$\lambda_i = 1 - \sum_{j=1}^{\infty} |\alpha_{ij}|,$$

and assume that there exists a number $\zeta > 0$, such that $|\gamma_i| \leqslant \zeta\lambda_i$, for any $i \in \mathbb{N}$. If z is the sequence whose members are all equal to ζ and if we define

$$w = z - V(z),$$

we have $w = \{\zeta\lambda_i\}_{i\in\mathbb{N}}$, hence $w \geqslant \mathbf{0}$. By virtue of what we have already proved in Section 8.1.2, the regular system has a bounded solution $x = \{\xi_n\}_{n\in\mathbb{N}}$ which can be obtained by the method of successive approximations.

(2) Let us consider the integral equation

$$x(s) = \int_0^1 k(s, t)\, x(t)\mathrm{d}t + y(s), \tag{4}$$

where k is a Lebesgue integrable function defined on $[0,1] \times [0,1]$ and y is an essentially bounded measurable function, defined on $[0,1]$.

Let us denote

$$h(s) = \int_0^1 |k(s, t)|\, \mathrm{d}t, \tag{5}$$

and assume that $h(s) < 1$ and that a real number α exists, such that

$$|y(s)| \leqslant \alpha(1 - h(s)).$$

Under these assumptions, the integral operator U, corresponding to the kernel k, i.e.,

$$g = U(x) \Leftrightarrow g(s) = \int_0^1 k(s, t)\, x(t)\, \mathrm{d}t,$$

maps the complete vector lattice $M([0,1])$ into itself [2]). The integral operator V, whose kernel is $|k(s, t)|$, majorizes U, as well as $-U$, and V is an (o)-continuous, positive, linear operator.

On the other hand, the equation

$$x = V(x) + w,$$

where

$$w(s) = \alpha(1 - h(s)),$$

has the positive solution $x(s) \equiv \alpha$.

---

[2]) The space $M(T)$ is defined in Section 6.4.1.

From what we have shown in Section 8.1.2, equation (4) has a bounded, measurable solution which can be obtained by the method of successive approximations.

Let us now assume that in equation (4) the functions k and y are continuous functions and that a continuous solution of the equation is sought.

Let us consider the vector space $C([0,1])$ of all (real) continuous functions, defined on [0,1], on which the following vector norm, taking values in $\mathbb{R}^n$, is considered

$$|x| = (\max_{t_0 \leqslant t \leqslant t_1} |x(t)|, \ldots, \max_{t_{n-1} \leqslant t \leqslant t_n} |x(t)|);$$

here t_i ($i = 0, 1, \ldots, n$) are the points of a given partition of the interval [0,1] (such that $t_0 = 0 < t_1 < \ldots < t_n = 1$).

Let us denote $\Delta_{ij} = [t_{i-1}, t_i] \times [t_{j-1}, t_j]$ and

$$\alpha_{ij} = (t_j - t_{j-1}) \max_{(s,\, t) \in \Delta_{ij}} |k(s, t)|.$$

Let A be the operator defined on $\mathbb{R}^n$ by the matrix whose members are α_{ij} ($i, j = 1, \ldots, n$)

It is easy to check that

$$|U(x)| \leqslant A(|x|);$$

consequently, if the equation

$$z = A(z) + |y| \tag{6}$$

has a positive solution in $\mathbb{R}^n$, then the results from Section 8.1.2 can be applied.

Equation (6) obviously is a finite system of algebraic equations.

8.2 Various applications

8.2.1 Order relations in normed vector spaces

Let $\mathscr{X}$ be a (real) normed vector space. Let us choose a non zero element v of the space $\mathscr{X}$. We shall assume that $\|v\| = 4$, in order to achieve some simplifications.

Let us denote by $\mathscr{P}$ the set of all elements of the form $x = \alpha(v + a)$, where $\alpha \geqslant 0$ and $\|a\| \leqslant 1$.

It is obvious that if $x \in \mathscr{P}$ and $0 \leqslant \lambda \in \mathbb{R}$, then $\lambda x \in \mathscr{P}$ and one can immediately prove that if $x, y \in \mathscr{P}$, then $x + y \in \mathscr{P}$. Also, the following equality holds

$$\mathscr{P} \cap (-\mathscr{P}) = \{0\}. \tag{1}$$

Indeed, let

$$x = \alpha(v + a), \tag{2}$$

$$-x = \beta(v + b), \tag{3}$$

where $\alpha, \beta \geqslant 0$, and $\|a\| \leqslant 1$, $\|b\| \leqslant 1$.

If $\alpha = 0$ or $\beta = 0$, then $x = \mathbf{0}$. Let us then assume that α and β do not vanish. Then

$$(\alpha + \beta)v + \alpha a + \beta b = \mathbf{0}, \tag{4}$$

whence

$$4(\alpha + \beta) = \|\alpha a + \beta b\| \leqslant \alpha + \beta,$$

which is a contradiction.

By defining $x \leqslant y$ to be equivalent to $y - x \in \mathscr{P}$, the vector space $\mathscr{X}$ becomes an ordered vector space.

It is easy to prove that for any element $x \in \mathscr{X}$ the following inequalities hold

$$-\|x\|v \leqslant x \leqslant \|x\|v. \tag{5}$$

In particular, $\mathscr{X}$ is a directed space.

Let $x_1, x_2 \geqslant \mathbf{0}$, namely

$$x_i = \alpha_i(v + a_i), \quad i = 1, 2$$

where $\alpha_i \geqslant 0$ and $\|a_i\| \leqslant 1$, $i = 1, 2$.

We have

$$x_1 + x_2 = (\alpha_1 + \alpha_2)(v + a_1) + \alpha_2(a_2 - a_1)$$

and, therefore,

$$\|x_1 + x_2\| \geqslant (\alpha_1 + \alpha_2)\|v + a_1\| - \alpha_2 \|a_2 - a_1\|$$

$$= \|x_1\| + \alpha_2(\|v + a_1\| - \|a_2 - a_1\|).$$

From

$$\|v + a_1\| \geqslant 3,$$

and since

$$\|a_2 - a_1\| \leqslant 2$$

we obtain

$$\|x_1 + x_2\| \geqslant \|x_1\| + \alpha_2. \tag{6}$$

On the other hand it is obvious that $\|x_2\| \leqslant 5\alpha_2$, and this, by taking into account (6), implies that

$$\|x_1 + x_2\| \geqslant \|x_1\| + \frac{1}{5}\|x_2\|.$$

From these inequalities it follows that if $\mathbf{0} \leqslant x \leqslant y$, then

$$\frac{1}{5} \|y - x\| \leqslant \|y\| - \|x\|. \tag{7}$$

In particular, we have $\|x\| \leqslant \|y\|$, i.e., the norm is monotonous with respect to the order relation we have just defined.

If $\{x_\delta\}_{\delta \in \Delta}$ is a generalized sequence of elements of $\mathscr{X}$, such that $x_\delta \underset{\delta \in \Delta}{\downarrow} 0$, then $\|x_\delta\| \to 0$.

Indeed, let

$$\lambda = \inf_{\delta \in \Delta} \|x_\delta\|.$$

Let us assume that $\lambda > 0$. Let $\delta_0 \in \Delta$ be such that

$$\|x_{\delta_0}\| \leqslant \frac{26}{25} \lambda.$$

If $\delta \geqslant \delta_0$, then from (7) we have

$$\frac{1}{5} \|x_{\delta_0} - x_\delta\| \leqslant \|x_{\delta_0}\| - \|x_\delta\| \leqslant \frac{\lambda}{25},$$

whence

$$\|x_{\delta_0} - x_\delta\| \leqslant \frac{\lambda}{5}.$$

If we now define

$$x = x_{\delta_0} - \frac{\lambda}{5} v,$$

then one can immediately check that $x \leqslant x_\delta$ for any $\delta \geqslant \delta_0$; it follows that $x \leqslant \mathbf{0}$. From $-x \geqslant \mathbf{0}$ we then obtain

$$\|x_{\delta_0}\| \leqslant \|x_{\delta_0} - x\| = \frac{4}{5} \lambda,$$

and this contradicts the definition of λ.

Consequently, $\lambda = 0$, i.e., $\|x_\delta\| \to 0$.

With the help of this result one can show that a generalized sequence $\{x_\delta\}_{\delta \in \Delta}$ of elements of $\mathscr{X}$ converges in norm to an element x iff there exists a generalized sequence $\{c_\xi\}_{\xi \in \Xi}$ of elements of $\mathscr{X}$, such that: (i) $c_\xi \underset{\xi \in \Xi}{\downarrow} \mathbf{0}$, (ii) for any $\xi \in \Xi$ there exists a $\delta' \in \Delta$, such that $-c_\xi \leqslant x_\delta - x \leqslant c_\xi$, for any $\delta \geqslant \delta'$.

★

Let us now assume that $\mathscr{X}$ is a normed vector space in which the closed ball $\overline{S}(0; 1)$ has an extremal point [3]) and such that the set of its balls has the binary intersection property.

Let u be an extremal point of $\overline{S}(0; 1)$.

Let us denote by $\mathscr{P}$ the set of all elements of the form $x = \alpha(u + a)$, where $\alpha \geqslant 0$ and $\|a\| \leqslant 1$.

---

[3]) We recall that an element $u \in \overline{S}(\mathbf{0}; 1)$ is an extremal point if it cannot be written as $u = \alpha x + (1 - \alpha) y$, where $x, y \in \overline{S}(\mathbf{0}; 1)$ $x \neq y$, $0 < \alpha < 1$.

As in the first part of this section, we can show that if $x \in \mathscr{P}$ and $0 \leqslant \lambda \in \mathbb{R}$, then $\lambda x \in \mathscr{P}$, whereas $x, y \in \mathscr{P}$ implies $x + y \in \mathscr{P}$.

Equality (1) is now a consequence of the fact that u is an extremal point of $\overline{S}(0; 1)$.

Indeed, let x and $-x$ be given by formulae (2) and (3) where $v = u$ (but α, β, a and b are submitted to the same hypotheses).

We must consider equality (4), in which $v = u$ and $\alpha + \beta > 0$. In this case

$$-u = \frac{1}{\alpha + \beta} (\alpha a + \beta b),$$

and since $-u$ is an extremal point of $\overline{S}(0; 1)$ too, it follows that $a = -u$ or $b = -u$. Both results imply that $x = \mathbf{0}$.

Consequently, $\mathscr{X}$ is an ordered vector space, in which the order relation is defined by

$$x \leqslant y \Leftrightarrow y - x \in \mathscr{P}.$$

Inequalities (5), in which $v = u$, continue to hold. In particular, $\mathscr{X}$ is a directed space. One can show that

$$\overline{S}(\mathbf{0}; 1) = [-u; u],$$

whence

$$\overline{S}(x; \rho) = [x - \rho u; x + \rho u].$$

Consequently, any closed ball is an interval. On the other hand, one can show that any interval is an intersection of two closed balls. One can then infer that the family of all intervals has the binary intersection property. From these facts one can show that the ordered set $\mathscr{X}$ has the interpolation property.

Since $\mathscr{X}$ is an ordered set, which is directed on both sides, and since it has the interpolation property, whereas the set of its intervals has the binary intersection property, the set $\mathscr{X}$ is a relatively complete lattice.

Consequently, $\mathscr{X}$ is a complete vector lattice with a strong unit (the element u). The norm which corresponds to u, coincides with the original norm of $\mathscr{X}$, because the interval $[u, u]$, is the closed ball of radius 1, centered at 0 for both norms.

The order relation introduced in $\mathscr{X}$ is the only one for which $\mathscr{X}$ becomes a normed complete vector lattice, with u as a strong unit, and such that the original norm coincide with the norm associated to u.

8.2.2 Applications to the problem of the extension of the continuous linear operators in normed spaces

We shall say that a normed vector space $\mathscr{X}_0$ has property (P) if for any normed vector space $\mathscr{X}$, which contains $\mathscr{X}_0$ as a normed subspace, and any continuous linear operator U_0, which maps $\mathscr{X}_0$ into an arbitrary normed vector space $\mathscr{Y}$, there exists an extension of U_0 to $\mathscr{X}$, which is linear, continuous and has the same norm.

We shall say that a normed vector space $\mathscr{Y}$ has property (E), if for any normed vector space $\mathscr{X}$, any vector subspace $\mathscr{X}_0$, and any continuous linear operator U_0 which maps $\mathscr{X}_0$ into $\mathscr{Y}$, there exists an extension of U_0 to $\mathscr{X}$, which is linear, continuous, and has the same norm.

If $\mathscr{X}$ is a normed vector space and $\mathscr{X}_0$ a vector subspace of $\mathscr{X}$, a mapping P of $\mathscr{X}$ onto $\mathscr{X}_0$ is said to be a *projection* (of $\mathscr{X}$ onto $\mathscr{X}_0$) if P is a continuous, linear operator, such that $P(x) = x$ for $x \in \mathscr{X}_0$, and $\|P\| = 1$.

Remark. Let $\mathscr{X}$ be a normed vector space and $\mathscr{X}_0$ a vector subspace of $\mathscr{X}$. Then there exists a projection P of $\mathscr{X}$ onto $\mathscr{X}_0$ iff any continuous linear operator, defined on $\mathscr{X}_0$ and taking values in a normed vector space $\mathscr{Y}$, can be extended to $\mathscr{X}$ as a continuous, linear operator, having the same norm.

Indeed, the existence of the projection implies the existence of the extension, given by the formula

$$U(x) = (U_0P)(x), \quad x \in \mathscr{X}.$$

Conversely, from the possibility of extending the operator $U_0(x) = x$, $x \in \mathscr{X}_0$, by preserving the properties mentioned, one can infer the existence of the projection.

Let now $\mathscr{X}$ be a complete vector lattice, which, at the same time, is a normed vector lattice. Let $S = \{x;\ \|x\| \leqslant 1\}$. If the set S is (o)-bounded and if $\|\sup S\| = 1$, then $\mathscr{X}$ will be called a *space of type* $\mathscr{M}$.

Lemma. *Any space of type* $\mathscr{M}$ *has property* (E).

Indeed, let $\mathscr{Y}$ be a space of type $\mathscr{M}$ and U_0 a (τ)-continuous (i.e. norm continuous) linear operator, defined on a normed linear space $\mathscr{X}_0$ and taking values in $\mathscr{Y}$. By defining

$$y_0 = \sup \{y;\ \|y\| \leqslant 1\},$$

we obviously have

$$|U_0(x)| \leqslant \|U_0\| \cdot \|x\| y_0.$$

Consequently, U is a (τo)-bounded operator and

$$|U| \leqslant \|U_0\|\, y_0.$$

If $\mathscr{X}_0$ is a vector subspace of a normed vector space $\mathscr{X}$, then U_0 can be extended to $\mathscr{X}$, such that the extension be linear and with the same vector norm (Section 6.3.3, proposition 1). If U is such an extension of U_0, then

$$|U| \leqslant \|U_0\|\, y_0,$$

whence we get

$$\|U\| \leqslant \||U|\| \leqslant \|U_0\|,$$

and this implies that $\|U\| = \|U_0\|$.

Proposition 1. *Any normed vector space is isomorphic to a normed vector subspace of a space having property (E).*

Indeed, let $\mathscr{Y}$ be an arbitrary normed vector space and

$$T = \{f \in \mathscr{Y}^*_\tau ; \|f\| \leqslant 1\}.$$

Let $B(T)$ be the set of all bounded real functions defined on T. Endowed with the usual operations, the natural order relation ($\varphi \leqslant \psi \Leftrightarrow \varphi(f) \leqslant \psi(f)$, $\forall f \in T$) and with the norm

$$\|\varphi\| = \sup_{f \in T} |\varphi(f)|,$$

the set $B(T)$ becomes a space of type $\mathscr{M}$.

In virtue of the lemma, $B(T)$ has property (E).

Let us now consider the mapping $y \mapsto \varphi_y$ of $\mathscr{Y}$ into $B(T)$, given by the formula $\varphi_y(f) = f(y)$. It is obvious that this mapping is one-to-one (between $\mathscr{Y}$ and a subset of $B(T)$) and it respects the linear operations. Moreover,

$$\|\varphi_y\| = \sup_{f \in T} |\varphi_y(f)\| = \sup_{\|f\| \leqslant 1} |f(y)| = \|y\|,$$

and the proposition is proved.

Proposition 2. *A normed vector space has property (P) iff it has property (E).*

Proof. Let $\mathscr{Y}$ be a normed vector space which has property (P). In virtue of proposition 1, $\mathscr{Y}$ can be considered to be a normed vector subspace of a normed vector space $\mathscr{Z}$, having property (E). In virtue of a preceding remark, there exists a projection P of $\mathscr{Z}$ onto $\mathscr{Y}$.

Let now $\mathscr{X}$ be an arbitrary normed vector space, $\mathscr{X}_0$ a vector subspace of $\mathscr{X}$ and U_0 an arbitrary continuous, linear operator defined on $\mathscr{X}_0$ and taking values in $\mathscr{Y}$. Since U_0 maps $\mathscr{X}_0$ into $\mathscr{Z}$, and since $\mathscr{Z}$ has property (E), there exists an extension U_1 of U_0 to $\mathscr{X}$, which is linear, continuous and has the same norm. By defining

$$U(x) = (PU_1)(x), \; x \in \mathscr{X},$$

we obtain a continuous, linear operator, defined on $\mathscr{X}$ and taking values in $\mathscr{Y}$, and such that $\|U\| = \|U_0\|$.

Conversely let $\mathscr{X}_0$ be a normed vector space which has property (E) and let $\mathscr{X}$ be a normed vector space, such that $\mathscr{X}_0$ be a normed vector subspace of $\mathscr{X}$. Let us consider the operator $U_0(x) = x$, $x \in \mathscr{X}_0$. The extension of U_0 to $\mathscr{X}$, which is linear, continuous and has the same norm, is a projection of $\mathscr{X}$ onto $\mathscr{X}_0$. From the existence of this projection if follows that $\mathscr{X}_0$ has property (P).

One can prove that: a normed vector space $\mathscr{X}$ has property (E) iff it is isomorphic (as a normed vector space) to a space $\mathscr{Y}$, which is a complete vector lattice with a strong unit, in which the norm is that associated with the unit (L. Nachbin [5], J. L. Kelley, [1]).

8.2.3 Applications to self-adjoint operators in Hilbert spaces [4]

Let $\mathscr{X}$ be a Hilbert space and $\mathscr{A}(\mathscr{X})$ the set of all self-adjoint operators which map $\mathscr{X}$ into $\mathscr{X}$.

An operator $U \in \mathscr{A}(\mathscr{X})$ is said to be *positive* and one writes $U \geqslant \mathbf{0}$, if $< U(x), x > \geqslant 0$, for any $x \in \mathscr{X}$.

One writes $U_1 \leqslant U_2$ $(U_1, U_2 \in \mathscr{A}(\mathscr{X}))$ if $U_2 - U_1 \geqslant \mathbf{0}$.

If $U_1 \leqslant U_2$, then $U_1 + U \leqslant U_2 + U$, for any $U \in \mathscr{A}(\mathscr{X})$, whereas if $0 \leqslant \lambda \in \mathbb{R}$, then $\lambda U_1 \leqslant \lambda U_2$.

One also knows that if $\{U_\delta\}_{\delta \in \Delta}$ is an increasing, majorized, generalized sequence of elements of $\mathscr{A}(\mathscr{X})$, then it admits of a least upper bound U in $\mathscr{A}(\mathscr{X})$ namely

$$U(x) = \lim_{\delta \in \Delta} U_\delta(x), \quad x \in \mathscr{X},$$

i.e., U is the point-wise limit of the generalized sequence $\{U_\delta\}_{\delta \in \Delta}$.

Let now $\mathscr{A}_0$ be a subset of $\mathscr{A}(\mathscr{X})$ having the following properties

(i) $U_1, U_2 \in \mathscr{A}_0 \Rightarrow U_1 + U_2 \in \mathscr{A}_0$;

(ii) $U \in \mathscr{A}_0, \lambda \in \mathbb{R} \Rightarrow \lambda U \in \mathscr{A}_0$;

(iii) $U_1, U_2 \in \mathscr{A}_0 \Rightarrow U_1 U_2 \in \mathscr{A}_0$;

(iv) if $\{U_\delta\}_{\delta \in \Delta}$ is a generalized sequence of elements of $\mathscr{A}_0$ and if $\{U_\delta\}_{\delta \in \Delta}$ is point-wise convergent to an operator U, then $U \in \mathscr{A}_0$.

We shall show that $\mathscr{A}_0$ is a *complete vector lattice*.

From what we have already recalled, (and by taking into account (iv)), it is enough to show that if $U \in \mathscr{A}_0$, then $U \vee \mathbf{0}$ exists in $\mathscr{A}_0$.

In order to do this, let us observe that $U \in \mathscr{A}_0$ implies $U^2 \in \mathscr{A}_0$ and $U^2 \geqslant \mathbf{0}$. By denoting by V the positive square root of U^2, from (iv) we infer that [5] $V \in \mathscr{A}_0$.

Let us now define

$$U' = \frac{1}{2}(V + U), \; U'' = \frac{1}{2}(V - U).$$

Obviously, we have $U', U'' \in \mathscr{A}_0$. Let

$$\mathscr{X}_1 = \{x; U''(x) = \mathbf{0}\},$$

and denote by $\mathscr{X}_2$ the orthogonal complement of $\mathscr{X}_1$. One immediately observes that $\mathscr{X}_1$ is invariant for any operator which belongs to $\mathscr{A}_0$. It follows that $\mathscr{X}_2$ is invariant for any operator of $\mathscr{A}_0$ too. In particular, $x \in \mathscr{X}_2$ implies that $U'(x) \in \mathscr{X}_2$.

---

[4]) For the notions used in this section, see R. Cristescu [31].

[5]) For the positive square root of a positive operator see R. Cristescu [31], Section 7.1.5, theorem 1.5.3.

Since U and V are commuting, from $V^2 = U^2$ we infer that $U''U' = \mathbf{0}$, hence $U'(x) \in \mathscr{X}_1$, for any $x \in \mathscr{X}$. It follows that $U'(x) = \mathbf{0}$, for any $x \in \mathscr{X}_2$. Since $U'' = V - U'$ we also infer that $U''(x) = V(x)$, for any $x \in \mathscr{X}_2$.

Let now x be arbitrary in $\mathscr{X}$ and let $x = x_1 + x_2$, where $x_1 \in \mathscr{X}_1$, $x_2 \in \mathscr{X}_2$. We have

$$\langle U'(x), x \rangle = \langle U'(x_1) + U'(x_2), x_1 + x_2 \rangle$$

$$= \langle U'(x_1), x_1 + x_2 \rangle = \langle U'(x_1), x_1 \rangle = \langle V(x_1), x_1 \rangle \geqslant 0,$$

where we have taken into account the fact that $U'(x_2) = \mathbf{0}$ and $\langle U'(x_1), x_2 \rangle = 0$. Consequently, $U' \geqslant \mathbf{0}$.

Analogously, one can prove that $U'' \geqslant \mathbf{0}$. Since $U'' = U' - U$, it follows that $U' \geqslant U$.

If $W \in \mathscr{A}_0$ and $W \geqslant \mathbf{0}$, $W \geqslant U$, then, as before, let $x \in \mathscr{X}$ and $x = x_1 + x_2$, where $x_1 \in \mathscr{X}_1$, $x_2 \in \mathscr{X}_2$. We have

$$\langle U'(x), x \rangle = \langle V(x_1), x_1 \rangle = \langle U(x_1), x_1 \rangle \leqslant \langle W(x_1), x_1 \rangle. \tag{1}$$

On the other hand since the subspaces $\mathscr{X}_1$ and $\mathscr{X}_2$ are invariant for W, we have

$$\langle W(x_1), x_1 \rangle + \langle W(x_2), x_2 \rangle = \langle W(x), x \rangle. \tag{2}$$

From relations (1) and (2) it follows that

$$\langle U'(x), x \rangle \leqslant \langle W(x), x \rangle,$$

i.e., $U' \leqslant W$.

We have thus shown that $U' = U \vee \mathbf{0}$.

Consequently, $\mathscr{A}_0$ is a complete vector lattice.

If $\mathscr{A}_0$ contains the identity operator I, then I is a total element in $\mathscr{A}_0$; by choosing I as a unit element, $\mathscr{A}_0$ becomes a space with a strong unit and the basis of $\mathscr{A}_0$ is the set of all projectors belonging to $\mathscr{A}_0$.

In order to prove this assertion, let us first observe that if $U_1, U_2 \in \mathscr{A}_0$, and if $U = U_1 \wedge U_2$, then there exists two complementary closed vector subspaces $\mathscr{E}_1, \mathscr{E}_2 \subset \mathscr{X}$, such that $U(x) = U_1(x)$, if $x \in \mathscr{E}_1$, and $U(x) = U_2(x)$, if $x \in \mathscr{E}_2$.

Indeed, from what we have already shown, there exist two complementary, closed, vector subspaces $\mathscr{X}_1, \mathscr{X}_2$, such that

$$(U_1 - U_2)(x) = (U_1 - U_2)_+(x) = |U_1 - U_2|(x),$$

for any $x \in \mathscr{X}_1$, and

$$(U_1 - U_2)(x) = -(U_1 - U_2)_-(x) = -|U_1 - U_2|(x),$$

for any $x \in \mathscr{X}_2$.

Since

$$U_1 \wedge U_2 = U_1 - (U_1 - U_2)_+,$$

we can take $\mathscr{E}_1 = \mathscr{X}_2$, $\mathscr{E}_2 = \mathscr{X}_1$.

From what we have already shown it is easy to show that if $U_1, U_2 \in \mathscr{A}_0$, then $U_1 \wedge U_2 = \mathbf{0}$ iff there exist two complementary closed vector subspaces $\mathscr{E}_1, \mathscr{E}_2$ of $\mathscr{X}$, such that $U_1(x) = 0$, for any $x \in \mathscr{E}_1$, and $U_2(x) = \mathbf{0}$, for any $x \in \mathscr{E}_2$.

Consequently, if $U \in \mathscr{A}_0$ and $U \wedge I = \mathbf{0}$, then one can now easily infer that $U = \mathbf{0}$. It follows that I is total in $\mathscr{A}_0$.

If $U \in \mathscr{A}_0$ and if we denote

$$\omega_U = \inf_{||x||=1} \langle U(x), x \rangle, \ \Omega_U = \sup_{||x||=1} \langle U(x), x \rangle,$$

then

$$\omega_U I \leqslant U \leqslant \Omega_U I,$$

and this implies that $\mathscr{A}_0$ is a subspace with a strong unit.

If $P \in \mathscr{A}_0$, and if

$$P \wedge (I - P) = \mathbf{0}. \tag{3}$$

then, by denoting by $\mathscr{E}_0$ the subset of $\mathscr{X}$ on which P vanishes, and by putting $\mathscr{E}_1 = \mathscr{E}_0^{\perp}$, from equality (3) we infer that $P(x) = I(x) = x$, for any $x \in \mathscr{E}_1$. Consequently, P is the projector on $\mathscr{E}_1$.

Conversely, if $P \in \mathscr{A}_0$ is the projection on a closed vector subspace $\mathscr{E}_1 \subset \mathscr{X}$, then $I - P$ is the projection on $\mathscr{X}_1^{\perp}$. Consequently, $P(x) = \mathbf{0}$, for any $x \in \mathscr{E}_1^{\perp}$ and $(I - P)(x) = \mathbf{0}$, for any $x \in \mathscr{E}_1$. Hence we infer equality (3) and, therefore, P belongs to the basis of $\mathscr{A}_0$.

Let now U_0 be an arbitrary self-adjoint operator (on the Hilbert space $\mathscr{X}$).

Let

$$\mathscr{A}_1 = \{U \in \mathscr{A}(\mathscr{X});\ UU_0 = U_0 U\},$$

and define

$$\mathscr{A}_0 = \{U \in \mathscr{A}_1;\ UV = VU, \forall V \in \mathscr{A}_1\}.$$

The set $\mathscr{A}_0$ has all of properties (i) — (iv), hence it is a complete vector lattice. It is obvious that $I \in \mathscr{A}_0$, hence $\mathscr{A}_0$ is a space with a strong unit (I is such a unit).

In particular, one can apply to U_0 the theorem of the integral representation for the elements of a space with unit (Section 4.1.3). One retrieves the known integral representation of the self-adjoint operators.

8.2.4 Positive contractions

Let T be a set, $\mathscr{T}$ a σ-algebra of subsets of T and μ a measure on T, such that $\mu(T) = 1$.

Let us consider the space $L(T, \mathscr{T}, \mu)$ (see Section 6.4.2), in which the element **1** corresponding to the constant function equal to 1 has been chosen as a unit element. Let us also consider the subspace $M(T, \mathscr{T}, \mu)$ of all bounded elements of $L(T, \mathscr{T}, \mu)$.

Let V be a (τ)-continuous, positive, linear operator, defined on $L(T, \mathscr{T}, \mu)$, taking values in the same space, such that $\|V(x)\| \leqslant \|x\|$. By using the canonical embedding of a normed vector space in its second dual space, and by identifying with $M(T, \mathscr{T}, \mu)$ the dual of the space $L(T, \mathscr{T}, \mu)$, let us denote by φ_n the (τ)-continuous, linear functional, defined on $M(T, \mathscr{T}, \mu)$ and corresponding to the element $V^n(\mathbf{1})$ of $L(T, \mathscr{T}, \mu)$.

In order that an element $x \in L(T, \mathscr{T}, \mu)$ should exist, such that $x(t) > 0$, μ-almost everywhere, and $V(x) = x$, it is necessary and sufficient that, for any $y \in M(T, \mathscr{T}, \mu)$ such that $y(t) > 0$, μ-almost everywhere, we have $\lim \varphi_n(y) > 0$.

One can prove a more general result, by considering spaces of type (KB). We shall first prove some lemmas.

Lemma 1. *Let $\mathscr{X}$ be a σ-complete vector lattice and f, g two positive linear functionals on $\mathscr{X}$, such that g be (o)-continuous. If $\mathbf{0} \leqslant b \in \mathscr{X}$ and $(f \wedge g)(b) = 0$, then there exists a sequence $\{a_{n-1}\}_{n \in \mathbb{N}}$ of positive elements, such that $b = (o)\text{-}\sum_{n=0}^{\infty} a_n$ and $g(a_0) = f(a_i) = 0$, $i \in \mathbb{N}$.*

Proof. We shall first prove that for any number $\varepsilon > 0$ there exists a $b_1 \in [\mathbf{0}, b]$, such that $g(b_1) < \varepsilon$ and $f(b_1) = f(b)$. Let us denote $c_0 = \mathbf{0}$ and observe that

$$(f \wedge g)(b) = \inf\{f(b_1) + g(b_2);\ b_1 + b_2 = b,\ \ b_1, b_2 \geqslant \mathbf{0}\}.$$

We can recurrently construct a sequence $\{c_n\}_{n \in \mathbb{N}}$ of elements of $\mathscr{X}$ such that $c_n \in \left[\mathbf{0}, b - \sum_{i=0}^{n-1} c_i\right]$ and

$$g(c_n) < \frac{\varepsilon}{2^n},\ f\left(b - \sum_{i=1}^{n} c_i\right) < \frac{\varepsilon}{2^n}.$$

It is then enough to take

$$b_1 = (o)\text{-} \sum_{n=1}^{\infty} c_n.$$

It follows that for any $n \in \mathbb{N}$, there exists a $b_n \in [\mathbf{0}, b]$, such that

$$g(b_n) < \frac{1}{2^n},\quad f(b_n) = f(b).$$

If we now define

$$a_0 = \overline{\lim}\, b_n = (o)\text{-}\lim_n \bigvee_{j=n}^{\infty} b_j,$$

we have

$$g\left(\bigvee_{j=n}^{\infty} b_j\right) \leqslant \sum_{j=n}^{\infty} g(b_j) < \frac{1}{2^{n-1}}.$$

It follows that $g(a_0) = 0$. If we now define

$$z_n = \bigwedge_{j=1}^{n} (b - b_j),$$

then $z_n \uparrow b - a_0$. Consequently, if we define $a_1 = z_1$, $a_{n+1} = z_{n+1} - z_n$, $n \in \mathbb{N}$, it follows that $a_n \geqslant \mathbf{0}$ ($\forall_n \in \mathbb{N}$) and

$$b - a_0 = (o)\text{-}\sum_{n=1}^{\infty} a_n,$$

whereas $f(a_n) = 0$, ($\forall n \in \mathbb{N}$), because $\mathbf{0} \leqslant z_n \leqslant b - b_n$ and $f(b - b_n) = 0$.

Remark. From the preceding lemma it follows that if f and g are (o)-continuous, positive, linear functionals on a σ-complete vector lattice and if $(f \wedge g)(x_0) = 0$ for a positive element x_0, then there exist positive elements x_1, x_2, such that $x_0 = x_1 + x_2$ and $f(x_1) = g(x_2) = 0$.

Lemma 2. *If $\mathscr{X}$ is a space of type KB with the unit u, if $\mathbf{0} < f \in \mathscr{X}_\tau^*$ and $f(a) > 0$ for a given element $a > \mathbf{0}$ of $\mathscr{X}$, then there exists a positive element $b \in \mathscr{X}$, such that $f(b) > 0$ and $([f]g)(a) \geqslant g(b)$, as soon as $\mathbf{0} \leqslant g \in \mathscr{X}_\tau^*$.*

Proof. We shall first remark that if we define

$$E = \{x \in \mathscr{X};\ f(|x|) = 0\},$$

then E is a component. Since $\mathscr{X}$ is a complete vector lattice with unit, the projector $[E]$ is principal (Section 4.1.2). Let e_0 be the unitary element such that $[E] = [e_0]$ and denote $a_0 = u - e_0$, $b = [a_0]a$.

From $a = [e_0]a + b$ and $f(a) > 0$ it follows that $f(b) > 0$. Let $\mathbf{0} \leqslant g \in \mathscr{X}_\tau^*$ and define $g_1 = [f]g$, $g_2 = f - g_1$. Since $g_2 \wedge f = \mathbf{0}$, we can write $b = b_1 + b_2$, where b_1, b_2 are such that $b_1, b_2 \geqslant \mathbf{0}$, and $g_2(b_1) = f(b_2) = 0$ (see the preceding remark). From $\mathbf{0} \leqslant b_2 \leqslant b$ and $b \perp e_0$ it follows that $b_2 \perp e_0$. On the other hand, from $f(b_2) = 0$ we infer that $b_2 \in E$. Since $[E] = [e_0]$, we infer that $b_2 = 0$ and, therefore, $g_2(b) = g_2(b_1) = 0$.

It follows that

$$g(b) = g_1(b) = ([f]g)([a_0]a) \leqslant ([f]g)(a),$$

and the lemma is proved.

In the following lemma we shall denote by F_x the functional defined on the dual $\mathscr{X}^*$ of a normed vector space $\mathscr{X}$, by the formula $F_x(f) = f(x)$. We shall take into account the fact that if $\mathscr{X}$ is a normed vector lattice, then $\mathscr{X}_\tau^*$ is a Banach vector lattice and, at the same time, a complete vector lattice with respect to the usual norm and the usual order relation. We shall denote by $\mathscr{X}^{**}$ the space $(\mathscr{X}_\tau^*)_\tau^*$.

Lemma 3. *If $\mathscr{X}$ is a space of type KB with a unit u and if F is an (o)-continuous, positive, linear functional on $\mathscr{X}_\tau^*$, then there exists an $x \in \mathscr{X}$, such that $F = F_x$.*

Proof. We shall assume that $F \neq \mathbf{0}$. From the relations

$$\mathbf{0} \leqslant F - \left(F - \frac{1}{n} F_u\right) \vee \mathbf{0} = \frac{1}{n} F_u \wedge F \leqslant \frac{1}{n} F_u,$$

it follows that $\left(F - \frac{1}{n} F_u\right)_+ \uparrow F$ in $\mathscr{X}^{**}$. Since $F \neq \mathbf{0}$, there exists an $m \in \mathbb{N}$, such that $\left(F - \frac{1}{m} F_u\right)_+ > \mathbf{0}$. Let us denote

$$G = F - \frac{1}{m} F_u.$$

Let f_0 be such that $G_+(f_0) > 0$. Since G is (o)-continuous, G_+ and G_- possess the same property, and from $(G_+ \wedge G_-)(f_0) = 0$ it follows that there exist $f_1, f_2 \geqslant \mathbf{0}$, such that $f_0 = f_1 + f_2$ and $G_+(f_1) = G_-(f_2) = 0$. It follows that

$$\inf \{G(g);\ \mathbf{0} \leqslant g \leqslant f_2\} = 0,$$

and, therefore, $G(g) \geqslant 0$ as soon as $\mathbf{0} \leqslant g \leqslant n f_2$ $(n \in \mathbb{N})$. From

$$[f_2]g = \bigvee_{n \in \mathbb{N}} g \wedge n f_2, \qquad (\mathbf{0} \leqslant g \in \mathscr{X}_\tau^*),$$

it follows that $G([f_2]g) \geqslant 0$ for any positive $g \in \mathscr{X}_\tau^*$. There then exists an $f \geqslant \mathbf{0}$ in $\mathscr{X}_\tau^*$, such that

$$\left(F - \frac{1}{m} F_u\right)([f]g) \geqslant 0, \ (\mathbf{0} \leqslant g \in \mathscr{X}_\tau^*),$$

whence

$$F(g) \geqslant ([f]g)\left(\frac{1}{m}\, u\right), \ (\mathbf{0} \leqslant g \in \mathscr{X}_\tau^*).$$

From lemma 2, in which we make $a = \frac{1}{m} u$, it follows that there exists a $b \in \mathscr{X}$, $b > \mathbf{0}$, such that

$$([f]g)\left(\frac{1}{m} u\right) \geqslant g(b), \ (\mathbf{0} \leqslant g \in \mathscr{X}_\tau^*).$$

Consequently, we have

$$F(g) \geqslant g(b), \quad (\mathbf{0} \leqslant g \in \mathscr{X}_\tau^*),$$

.e., $F_b \leqslant F$.

Let us now define

$$c = \sup\{x \in \mathscr{X}; \quad \mathbf{0} \leqslant F_x \leqslant F\}.$$

If $F - F_c > \mathbf{0}$, then from what we have shown, there would exist an $x_0 \in \mathscr{X}$, such that $x_0 > \mathbf{0}$ and $F_{x_0} \leqslant F - F_c$; consequently, $F_{c+x_0} \leqslant F$, and this contradicts the definition of c.

This shows that $F = F_c$ and the lemma is proved.

Theorem. *Let $\mathscr{X}$ be a space of type KB with the unit u, and V a positive, linear operator which maps $\mathscr{X}$ into $\mathscr{X}$, such that $\|V(x)\| \leqslant \|x\|$, $x \in \mathscr{X}$. Then there exists a positive, total element $a > \mathbf{0}$, such that $V(a) = a$, iff $\underline{\lim} f(V^n(u)) > 0$, for any non-zero positive $f \in \mathscr{X}^*_\tau$.*

Proof. Let us assume that the element a from the statement of the theorem exists, and let $\mathbf{0} \leqslant f_0 \in \mathscr{X}^*_\tau$ be such that $\underline{\lim}\, f_0(V^n(u)) = 0$. Let then $\{n_i\}_{i \in \mathbb{N}}$ be an increasing sequence of natural numbers, such that $\lim_i f_0(V^{n_i}(u) = 0$.

If $\mathbf{0} \leqslant x \leqslant mu$, then obviously $\lim_i f_0(V^{n_i}(x)) = 0$. The last inequality holds for an arbitrary x, too. Indeed, it is enough to assume $x > \mathbf{0}$. In this case, $x \wedge mu \uparrow x$ and, therefore, $x \wedge mu \xrightarrow{\tau} x$. By denoting $x_m = x \wedge (mu)$, we have

$$|f_0(V^{n_i}(x_m - x)| \leqslant \|f_0\| \cdot \|x_m - x\|,$$

whence we infer that

$$\overline{\lim_i} f_0(V^{n_i}(x))| \leqslant \|f_0\| \cdot \|x_m - x\| \qquad m \in \mathbb{N},$$

and this shows that $f_0(V^{n_i}(x)) \to 0$. In particular, by taking $x = a$, we get $f_0(a) = 0$. But in this case $f_0 = \mathbf{0}$, because the element a is total.

Conversely, let us assume that $\lim f(V^n(u)) > 0$, for any non-zero positive $f \in \mathscr{X}^*_\tau$. We shall consider the Banach limit [6] $\mathscr{L}$.

If $f \in \mathscr{X}^*_\tau$, we shall put

$$F(f) = \mathscr{L}(\{f(V^n(u))\}_{n \in \mathbb{N}}),$$

and we shall obtain a positive linear functional F on $\mathscr{X}^*_\tau$. From the definition of $\mathscr{L}$ and from the condition we imposed on V, it follows that $F(f) > 0$ for any $f > \mathbf{0}$.

From [7]

$$\mathscr{L}(\{f(V^n(u))\}_{n \in \mathbb{N}}) = \mathscr{L}(\{f(V^{n+1}(u))\}_{n \in \mathbb{N}}) = \mathscr{L}(\{V^*(f)(V^n(u))\}_{n \in \mathbb{N}})$$

[6]) The Banach limit is a positive, linear operator $\mathscr{L}$, defined on the space (m) of the bounded sequences of real numbers, and satisfying the conditions $\mathscr{L}(\{x_{n+1}\}_{n \in \mathbb{N}}) = \mathscr{L}(\{x_n\}_{n \in \mathbb{N}})$; $\mathscr{L}((1, 1, \ldots, 1)) = 1$.

[7]) We shall consider the adjoint V^* of V, i.e., the operator, which maps $\mathscr{X}^*_\tau$ into $\mathscr{X}^*_\tau$, given by the formula

$$(V^*(f))(x) = f(V(x)).$$

it follows that

$$F(f) = F(V^*(f)) = (V^{**}(F))(f),$$

and, therefore, $V^{**}(F) = F$. By identifying $\mathscr{X}$ with its canonical image in $\mathscr{X}^{**}$ and by defining

$$b = \sup \{x \in \mathscr{X}; \quad 0 \leqslant x \leqslant F\},$$

it follows that $b \in \mathscr{X}$ (see proposition 5 from Section 6.2.9).

From the equality $V^{**}(F) = F$ and since $\mathbf{0} \leqslant b \leqslant F$, it follows that $V(b) \leqslant b$, if we take into account the definition of b. The element b is total. Indeed, from

$$\mathbf{0} \leqslant (F - F_b) \wedge F_u \leqslant F_u$$

it follows that $(F - F_b) \wedge F_u$ is an (o)-continuous functional. With lemma 3, we can write

$$(F - b) \wedge u \in \mathscr{X},$$

and, by taking into account the definition of b, it follows that

$$(F - b) \wedge u = \mathbf{0}.$$

If $\mathbf{0} \leqslant f \in \mathscr{X}_\tau^*$ and $f(b) = 0$, then (with lemma 1) there exists a sequence $\{f_{n-1}\}_{n \in \mathbb{N}}$ of positive elements of $\mathscr{X}_\tau^*$, such that $f = \sum_{n=0}^{\infty} f_n$, $f_0(u) = 0$ and

$$(F - b)(f_n) = 0, \; n \in \mathbb{N}.$$

From $f_n \leqslant f$ it follows that $F(f_n) = 0$ $(n \in \mathbb{N})$. Since $F(g) > 0$, as soon as $g > \mathbf{0}$, it follows that $f_n = \mathbf{0}$, $(\forall n \in \mathbb{N})$. From $f_0(u) = 0$, it follows that $f_0 = \mathbf{0}$, because u is a unit element. Hence, $f = \mathbf{0}$. We have thus shown that $\mathbf{0} < f \in \mathscr{X}_\tau^*$ implies $f(b) \neq 0$. Consequently, b is total. Indeed, let us assume that there exists an element $x_0 > \mathbf{0}$ in $\mathscr{X}$, such that $x_0 \wedge b = \mathbf{0}$. There then exists an $f_0 \in \mathscr{X}_\tau^*$, such that $f_0 > \mathbf{0}$ and $f_0(x_0) \neq 0$. By defining

$$f_1(x) = \sup f_0(x \wedge nx_0), \qquad x \in \mathscr{X}_+,$$

$$f_1(x) = f_1(x_+) - f_1(x_-), \quad x \in \mathscr{X},$$

the functional f obviously is linear and $\mathbf{0} \leqslant f_1 \leqslant f_0$; hence $\mathbf{0} \leqslant f_1 \in \mathscr{X}_\tau^*$. We have $f_1(x_0) = f_0(x_0) \neq \mathbf{0}$, hence $f_1 > \mathbf{0}$, whereas $f_1(b) = 0$, and this is a contradiction. Therefore, b is total.

From $V(b) \leqslant b$, it follows that $V^n(b) \downarrow$; therefore there exists an $a \in \mathscr{X}$, such that $V^n(b) \downarrow a$. It follows that $a = (\tau)\text{-}\lim V^n(b)$, and, therefore, $V(a) = a$. If $\mathbf{0} \leqslant f \in \mathscr{X}_\tau^*$ and $f(a) = 0$, then $\lim f(V^n(a)) = 0$, and this shows that $f = \mathbf{0}$. Consequently, the element a is total.

The theorem is proved.

8.2.5 Sum-functions[8])

Let $\mathscr{B}$ be a σ-complete Boolean lattice. We shall denote by $\mathbf{0}$ the smallest element of $\mathscr{B}$ and by u its greatest element. If $e \in \mathscr{B}$, we shall denote by e^* the complement of e.

Let P be a probability on $\mathscr{B}$ i.e., a real function having the following properties: (i) $P(e) \geqslant 0$, for any $e \in \mathscr{B}$; (ii) $P(u) = 1$; (iii) if $\{e_n\}_{n\in\mathbb{N}}$ is a sequence of elements of $\mathscr{B}$, such that $e_i \wedge e_j = \mathbf{0}$, for $i \neq j$, then

$$P(\bigvee_{n\in\mathbb{N}} e_n) = \sum_{n=1}^{\infty} P(e_n).$$

In particular, it follows that P is an increasing function on $\mathscr{B}$ (i.e., $e_1 \leqslant e_2$ implies $P(e_1) \leqslant \leqslant P(e_2)$), and (o)-continuous (i.e., $e_n \xrightarrow{o} e \Rightarrow P(e_n) \to P(e)$).

A real function F, defined on $\mathscr{B}$, is called a *sum-function* if it has the following properties:

(a) if $e_1 \wedge e_2 = \mathbf{0}$, then

$$F(e_1 \vee e_2) = F(e_1) + F(e_2);$$

(b) $\lim\limits_{P(a)\to 0} F(a) = 0.$

If a probability space $(\Omega, \mathscr{A}, P)$ is considered, whose elements are subsets of a set of elementary events, then a sum-function F on $\mathscr{A}$ can be represented as

$$F(A) = \int_A f(\omega)\, \mathrm{d}P(\omega),$$

where $A \in \mathscr{A}$, and f is a random variable.

In the probability theory, based on the notion of sum-function, one can work directly on the algebra of events, without any reference to the elementary events.

One can prove that any sum-function F has the following properties:

(1) if $P(e) = 0$, then $F(e) = 0$;

(2) if $e_n \xrightarrow{o} e$, then $F(e_n) \to F(e)$;

(3) if $\{e_n\}_{n\in\mathbb{N}}$ is a sequence of elements of $\mathscr{B}$, such that $e_i \wedge e_j = \mathbf{0}$, for $i \neq j$, then

$$F(\bigvee_{n\in\mathbb{N}} e_n) = \sum_{n=1}^{\infty} F(e_n);$$

(4) The set $F(\mathscr{B})$ is bounded [9]).

We shall now denote by $\mathscr{F}(\mathscr{B}, P)$ the set of all sum-functions defined on $\mathscr{B}$. We shall consider in this set the natural linear operations and the order relation given by: $F_1 \leqslant F_2$ iff $F_1(e) \leqslant F_2(e)$, $\forall e \in \mathscr{B}$.

---

[8]) In the classical theory of probability one of the fundamental notions is that of random variable. In the theory started by O. Onicescu, the field of events is a σ-complete Boolean lattice, endowed with a probability, whereas the role of the random variable is played by the "sum-function".

[9]) Property (4) is implied by (3), as I. Cuculescu has shown; see the work *Sum-functions*, by O. Onicescu and his co-workers, p. 45 (in Romanian).

It is easy to check that $\mathcal{F}(\mathcal{B}, P)$ is an ordered vector space.

We shall define on $\mathcal{F}(\mathcal{B}, P)$ the following norm

$$\|F\| = \sup_{e\in\mathcal{B}} |F(e) - F(e^*)|.$$

The set $\mathcal{F}(\mathcal{B}, P)$ is a space of type KB in which P is a total element.

Indeed, if $F \in \mathcal{F}(\mathcal{B}, P)$, and if we define

$$G(e) = \sup_{e'\leqslant e} F(e'),$$

then G is a sum-function on $\mathcal{B}$, and $G = F \vee \mathbf{0}$, as one can easily prove.

Consequently, $\mathcal{F}(\mathcal{B}, P)$ is a vector lattice, in which the positive part of an element is given by the formula

$$F_+(e) = \sup_{e'\leqslant e} F(e').$$

The following equalities also hold

$$F_-(e) = -\inf_{e'\leqslant e} F(e'),$$

$$|F|(e) = \sup_{e'\leqslant e} |F(e') - F(e \wedge e'^*)|. \tag{1}$$

If $\{F_j\}_{j\in J}$ is a bounded from above system of elements of $\mathcal{F}(\mathcal{B}, P)$, and if we define

$$F(e) = \sup\left\{\sum_{i=1}^{n} F_{j_i}(e_i),\ e_i \wedge e_h = 0 \quad (i \neq h);\ \bigvee_{i=1}^{n} e_i = e,\ j_i \in J,\ n \in \mathbb{N}\right\},$$

then it is not difficult to show that F is a sum-function and that $F = \bigvee_{j\in J} F_j$.

Consequently, $\mathcal{F}(\mathcal{B}, P)$ is a complete vector lattice.

From (1) it follows that $\|F\| = |F|\,(u)$. Hence we immediately infer that if $|F_1| \leqslant |F_2|$, then $\|F_1\| \leqslant \|F_2\|$, and that $F_n \downarrow \mathbf{0}$ in $\mathcal{F}(\mathcal{B}, P)$ implies $\|F_n\| \to 0$.

If $\mathbf{0} \leqslant F_n \uparrow$ and $\|F_n\| \leqslant \alpha$, then $F_n(e) \leqslant \alpha$, for any $e \in \mathcal{B}$; by defining

$$F(e) = \lim_n F_n(e),$$

it is easy to check that F is a sum-function and $F_n \uparrow F$.

Consequently, $\mathcal{F}(\mathcal{B}, P)$ is a space of type KB.

In order to show that P is a total element, we shall use a proposition of O. Onicescu[10]): *for any sum-function G, there exists an element $e_0 \in \mathcal{B}$ such that $G(e) \geqslant 0$ if $e \leqslant e_0$, and $G(e) \leqslant 0$, if $e \leqslant e_0^*$.*

Let $F > \mathbf{0}$ and $e_1 \in \mathcal{B}$ be such that $P(e_1) \neq 0$, and $F(e_1) > 0$. Let us consider a real number $\alpha \in (0,1)$, such that

$$F(e_1) > \alpha P(e_1),$$

and let us define

$$G(e) = F(e) - \alpha P(e), \quad e \in \mathcal{B}.$$

---

[10]) For the proof of this proposition see the work *Sum-functions* by O. Onicescu and co-workers p. 47 (In Romanian).

Then G is a sum-function. Let $e_0 \in \mathscr{B}$ be such that $G(e) \geqslant 0$ for any $e \leqslant e_0$, and $G(e) \leqslant 0$, for any $e \leqslant e_0^*$. Let us denote $e_2 = e_0 \wedge e_1$. From $e \leqslant e_2$ it follows that $e \leqslant e_0$ and, therefore, $G(e) \geqslant 0$. On the other hand, we have

$$G(e_1) = G(e_0 \wedge e_1) + G(e_0^* \wedge e_1),$$

where $G(e_0 \wedge e_1) \geqslant 0$ and $G(e_0^* \wedge e_1) \leqslant 0$. It follows that $P(e_2) \neq 0$ and $F(e) \geqslant \alpha P(e)$, for any $e \leqslant e_2$.

If we now define

$$H(e) = \alpha P(e \wedge e_2),$$

then $0 < H \in \mathscr{F}(\mathscr{B}, P)$ and $F(e) \geqslant H(e)$, $P(e) \geqslant H(e)$. Consequently, $F \wedge P > \mathbf{0}$.

This shows that P is a total element in $\mathscr{F}(\mathscr{B}, P)$.

One can show that if P is chosen to be the unit of the space $\mathscr{F}(\mathscr{B}, P)$, then an element F of this space is a unitary element iff it can be represented as $F(e) = P(e \wedge e_0)$, for a certain element $e_0 \in \mathscr{B}$.

In particular, it follows that any sum-function on $\mathscr{B}$ is the uniform limit of a sequence of simple sum-functions. By a simple sum-function one obviously means a sum-function of the form

$$F(e) = \sum_{i=1}^{n} \alpha_i P(e \wedge e_i), \quad e \in \mathscr{B},$$

where $e_i (i = 1, 2, \ldots, n)$ are given elements of $\mathscr{B}$, whereas α_i $(i = 1, 2, \ldots, n)$ are given real numbers.

With this representation of the sum-functions one introduces in a natural manner various notions in the theory of sum-functions, e.g., the entropy, the correlation, the quantity of information.

Bibliographical comments

Operator equations in complete vector lattice or in spaces with vector norm were studied by L. V. Kantorovitch [8], [14], [16], to whom the results from Sections 8.1.1 and 8.1.2 belong (see also L. V. Kantorovitch, B. Z. Vulich, A. G. Pinsker [1]). For the operator equations in an ordered group, in which any bounded from above sequence of elements has a least upper bound, some results of L. V. Kantorovitch were generalized by R. Cristescu [18], [19].

M. A. Krasnoselski [1] studies the existence, the uniqueness and the properties of the positive solutions of linear as well as the non-linear equations in a Banach space endowed with an order relation, submitted to various conditions.

For other results concerning operator equations in ordered vector spaces, see S. N. Slugin [1] and A. R. Esajan [1] (in Banach ordered spaces). The problem of mathematical programming from Section 8.1 was considered by V. Z. Belenkin [1].

The use of the theory of ordered vector spaces in the study of the extension of continuous linear operators between normed spaces is due to G. P. Akilov [1].

For the applications of the theory of ordered vector spaces to the theory of self-adjoint operators in Hilbert spaces, see V. I. Sobolev [2], B. Z. Vulich [15], A. G. Pinsker [11].

The theory of ordered vector spaces was used by J. Neveu [1] in the problem of the existence of invariant measures in ergodic theory. His results were generalized by P. C. Shields [1], to whom the theorem and the lemmas from Section 8.2.4 belong.

A theory of probability based on the notion of sum-function was started by O. Onicescu [1]. In connection with various results on the sum-functions see the work: *Sum-functions,* by O. Onicescu and co-workers [1] (in Romanian). The use of the theory of ordered vector spaces in the study of sum-functions is due to R. Cristescu [23], [34].

For various applications of the theory of ordered vector spaces to probability theory, see G. I. Kalmykov [1], V. V. Sazonov [1], M. Driml [1], etc.

The introduction of an order relation in arbitrary normed vector spaces (Section 8.2.4), was studied by R. E. de Marr [1] and B. Z. Vulich and J. F. Danilenko [1].

The ordered vector spaces which are normed, such that the order convergence coincides with the convergence in norm, were studied by B. Z. Vulich and O. S. Korsakova [1]. The introduction of a structure of complete vector lattice with strong unit in some normed spaces (see Section 8.2.1) belongs to L. Nachbin [5].

Some methods for ordering a normed vector space, such that its norm or topology be determined by the order relation, were given also by A. J. Ellis [2].

The ordering of a Hilbert space was studied by H. Nakano [7].

In connection with the ordering of some groups endowed with a function more general than the norm, see M. H. Stone [4].

Applications of the theory of ordered vector spaces to the positive moment problem were given by L. V. Kantorovitch [3].

Applications to the basis problem in Banach spaces were given by H. H. Schaefer [2].

For applications to the theory of Banach algebras, see M. Nakamura [3], R. V. Kadison [1], S. Sherman [1], J. L. Kelley and R. L. Vaught [1], M. Fukamiya, J. Misonou, Z. Takeda [1], J. P. C. Curtis [1], etc.

The linear problem of optimal control in a Banach space endowed with an order relation was studied by Z. I. Halilov [1].

For linear operator equations, which are defined on some types of Banach spaces, which are also vector lattices, A. L. Kalasnikov and S. N. Slugin [1] gave some criteria for the existence of optimal positive solutions.

References

AKILOV, G.P.

[1] On the extension of linear operations. *Dokl. Akad. Nauk*, **57**, 643–646 (1947) (in Russian).

ALFSEN, E. M.

[1] Order theoretic foundations of integration. *Math. Ann.*, **149**, 419–461 (1963).

ALIPRANTINS, C. D.; LANGFORD, ERIC

[1] Almost σ-Dedekind complete Riesz spaces and the main inclusion theorem. *Proc. Amer. Mat. Soc.*, **44**, 421–426 (1974).

AMEMIYA, ICHIRO

[1] On a topological method in semiordered linear spaces. *Proc. Jap. Acad.*, **27**, 138—140 (1951).

[2] A generalization of Riesz-Fischer's theorem. *Journ. Math. Soc. Japan*, **5**, 353–354 (1953).

AMEMIYA, I.; MORI, T.

[1] Topological structures in ordered linear spaces. *Journ. Math. Soc. Japan*, **9**, 131–142 (1957).

AMEMIYA, ICHIRO; ANDÔ, TSUYOSHI; SASAKI, MASAHUMI

[1] Monotony and flatness of the norms by modulars, *J. Fac. Sci. Hokkaido Univ.*, Ser. 1, **14**, 96–113 (1959).

ANDERSON, B.; NAKANO, H.

[1] Semi-continuous linear lattices. *Studia Math.*, **37**, 191–195 (1971).

ANDÔ, TSUYOSHI

[1] Positive linear operators in semi-ordered linear spaces. *J. Fac. Sci. Hokkaido Univ.*, Ser. I, **13**, 214–228 (1957).

[2] On the structure of the associated modular. *Proc. Japan Acad.*, **34**, 587–588 (1958).

[3] Convexity and evenness in modulared semi-ordered linear spaces. *J. Fac. Sci. Hokkaido Univ.*, Ser. I, **14**, 59–95 (1959).

[4] On fundamental properties of a Banach space with a cone. *Pacific J. Math.*, **12**, 1163–1169 (1962).

BANACH, S.

[1] *Théorie des opérations linéaires*, Warsaw, 1932.

BAUER, H.

[1] Sur le prolongement des formes linéaires positives dans un espace vectoriel ordonné. *C. R. Acad. Paris*, **244**, 289–292 (1957).

[2] Uber die Forsetzung positiver Linearformen. *Bayer. Akad. Wiss. Math. Nat.* Kl. S. B., **1957**, 177–190 (1958).

BELENKII, V. Z.

[1] Problems of mathematical programming which have a minimal point. *Dokl. Akad. Nauk SSSR*, **183**, 15–17 (1968) (in Russian).

BERNSTEIN, B. A.

[1] Operations with respect to which the elements of a Boolean algebra form a group. *Trans. Amer. Math. Soc.*, **26**, 171–174 (1924).

BERTOLINI, FERNANDO

[1] Le funzioni misurabili di ultrafiltro come elementi di un reticolo lineare numerabilmente completo. *Annali della Scuola norm. sup. di Pisa*, serie III, XV, 115–227 (1961).

[2] Le misure su un reticolo di Boole, come elementi di un reticolo lineare. *Annali della Scuola norm. sup. di Pisa*, serie III, **XV**, 129–144 (1961).

BIEBINGER, G.

[1] *Topologische tensorprodukte geordneter lokalkonvexe raume.* Diplomarbeit, Frankfurt, 1967.

BIRKHOFF, GARRETT

[1] On the combination of subalgebras. *Proc. Cam. Phil. Soc.*, **29**, 441–464 (1933).

[2] On the structure of abstract algebras. *Proc. Cam. Phil. Soc.*, **31**, 433–454 (1935).

[3] Moore-Smith convergence in general topology. *Ann. of Math.*, **38**, 39–56 (1937).

[4] Partially ordered linear spaces. *Bull. Amer. Mat. Soc.*, **44**, 186 (1938).

[5] Dependent probabilities and spaces (L). *Proc. Nat. Acad. Sci.*, **24**, 155–159 (1938).

[6] Lattice-Ordered Groups. *Ann. of Math.*, **43**, 298–331 (1942).

[7] *Lattice Theory*, 3rd ed., Amer. Math. Soc. Coll. Publ. 25, New York, 1973.

[8] Uniformly semi-primitive multiplicative processes. *Transactions Amer. Math. Soc.*, **104**, 37–51 (1962).

BIRKHOFF, GARRETT; FRINK, ORRIN

[1] Representations of lattices by sets. *Trans. Am. Math. Soc.*, **64**, 229–316 (1948).

BIRKHOFF, G. D.

[1] Proof of the ergodic theorem. *Proc. Nat. Acad. Sci.*, **17**, 650–655 (1931).

BOCHNER, S.

[1] Integration and differentiation in partially ordered spaces. *Proc. Nat. Acad. Sci.*, **26** (1940).

[2] Finitely additive integral. *Ann. of Math.*, **41**, 495 (1940).

BOCHNER, S.; FAN, KY.

[1] Distributive order preserving operations in a partially ordered vector set. *Ann. of Math.*, **48** (1947).

BOCHNER, S.; PHILIPS, R. S.

[1] Additive set functions and vector lattices. *Ann. of Math.*, **42**, 316–324 (1941).

BONSALL, F. F.

[1] Sublinear functionals and ideals in partially ordered vector spaces. *Proc. London Math. Soc.*, **4**, 402–418 (1954).

[2] Extreme maximal ideals of a partially ordered vector space. *Proc. Amer. Math. Soc.*, **7**, 831–837 (1956).

[3] The decomposition of continuous linear functionals into non-negative components. *Proc. Univ. Durham Philos. Soc.* Ser. A, **13**, 6–11 (1957).

[4] Linear operators in complete positive cones. *Proc. London Math. Soc.*, **(3) 8**, 53–75 (1958).

[5] Positive operators compact in an auxiliary topology. *Pacific. J. Math.*, **10**, 1131–1138 (1960).

BONSALL, F. F.; REUTER, G. E. H.

[1] A fixed point theorem for transition operators in an (L)-space. *Quart. J. Math. Oxford*, Ser. (2), 244–248 (1956).

BOOLE, G.

[1] *The Mathematical Analysis of Logic*, Cambridge, 1847.

BOURBAKI, N.

[1] *Topologie générale*, Hermann & Cie, Paris, 1951.

[2] *Algèbre*, Hermann & Cie, Paris, 1951.

[3] *Espaces vectoriels topologiques*, Hermann & Cie, Paris, 1953.

[4] *Intégration*, Hermann & Cie, Paris, 1956.

BRAINERD, BARRON

[1] On a class of lattice-ordered rings. II. *Nederl. Akad. Wettensch.*, Proc. Ser A 60, Indag. Math., **19**, 541–547 (1957).

[2] On a class of lattice-ordered rings. *Proc. of the Amer. Math. Soc.*, **8**, 673–684 (1957).

[3] On the embedding of vector lattices in F-rings. *Trans. Amer. Math. Soc.*, **93**, 132–144 (1959).

BRAUNSCHWEIGER, CHRIS, C.

[1] A geometric construction of the M-space conjugate to an L-space. *Proc.Amer. Mat. Soc.*, **10**, 77–82 (1959).

BRODSKII, V. M.

[1] On the eigenvectors of the completely continuous operators, defined on normed semiordered spaces. *Sibirski matem. jurnal*, v. 468–471 (1962).

BUCK, R. C.

[1] Multiplication operators. *Pacific J. Math.*, **11**, 95–103 (1961).

BUSULINI, BRUNO

[1] Su la relazione triangulare in un l-gruppo, *Rend. Sem. Maz. Univ. Padova*, **28**, 68–70 (1958).

CARTAN, H.

[1] Un théorème sur les groupes ordonnés. *Bull. Sci. Mat.*, **63**, 201–205 (1939).

CIORĂNESCU, IOANA

[1] Asupra distribuţiilor vectoriale pozitiv definite. *St. cerc. mat.*, **22**, 993–996 (1970).

CLIFFORD, A. H.

[1] Partially ordered Abelian groups. *Ann. of Math.*, **41**, 465–473 (1940).

COHEN, L. W.

[1] Transformation of spaces infinitely many-dimensional. *Ann. of Math.*, **37**, 326–335 (1936).

COHEN, L. W.; GOFFMAN, C.

[1] The topology of ordered Abelian groups. *Trans. Amer. Math. Soc.*, **67**, 310–320 (1949).

CRISTESCU, ROMULUS

[1] Integrarea în spaţii semiordonate. *Bul. Ştiinţ. Acad. R.P.R.* Secţiunea de ştiinţe matematice şi fizice, **IV**, 291–310 (1952).

[2] Spaţii parţial-ordonate pseudonormate. *Com. Acad. R.P.R.*, **IV**, 1–2, 15–19 (1954).

[3] Asupra unei noţiuni de convergenţă. *Bul. Ştiinţ. Acad. R.P.R.*, Secţiunea de ştiinţe matematice şi fizice, **VI**, 2, 297–304 (1954).

[4] Operatori produs. *Revista Universităţii "C. I. Parhon" şi a Politehnicii Bucureşti*, **8**, 23–24 (1955).

[5] O teoremă asupra reprezentării operaţiilor lineare. *Comunicările Acad. R.P.R.*, **V**, 4, 655–660 (1955).

[6] Cu privire la definiţia produsului în K-spaţii. *Bul. Ştiinţ. Acad. R.P.R.*, Secţiunea de ştiinţe matematice şi fizice, **VII**, 3, 553–564 (1955).

[7] Operatorul produs în spaţiile liniare semiordonate şi aplicaţiile sale. *St. cerc. mat.*, **VI**, 3–4, 357–493 (1955).

[8] Clase de spaţii liniare semiordonate pseudonormate. *St. cerc. mat.*, **VII**, 3–4, 291–305 (1956).

[9] Sopra gli operatori biadditivi regolari. *Rendic. di Matem. e delle sue appl.*, Roma serie 5, **15**, 271–290 (1956).

[10] Teorema lui Radon-Nikodym în K-spaţii. *Analele Universităţii "C. I. Parhon"*, **14**, 25–27 (1957).

[11] Cu privire la teoremele ergodice individuale ale lui H. Nakano. *Analele Universităţii "C. I. Parhon" (şt. nat.)*, **20**, 23–25 (1958).

[12] La notion de composante dans un groupe dirigé. *C.R. Acad. Sci. Paris*, **246**, 1700–1702 (1958).

[13] Integrali vettoriali di Stieltjes ed operatori lineari. *Rendic. Acad. Naz. Lincei, Roma*, **XXVII**, Ferie, 31–34 (1959).

[14] Intégrales vectorielles et représentation de certains opérateurs linéaires. *Bull. math. soc. sci. math. phys. R.P.R.*, **3** (51), 7–15 (1959).

[15] *Spaţii liniare ordonate*, Editura Acad. R.P.R., Bucureşti, 1959.

[16] Sur les groupes dirigés. *Czechoslovak math. journal*, **10** (85), 17–26 (1960).

[17] The Stieltjes Integral in K-spaces. *Doklady Akad. Nauk*, **133**, 519–522 (1960) (in Russian).

[18] Asupra metodei aproximaţiilor succesive pentru ecuaţii operatoriale. *St. cerc. de matem. (Cluj)*, **XII**, 41–44 (1961).

[19] Il metodo delle approssimazioni successivi nei gruppi ordinati. *Boll. dell'Un. mat. ital.*, **16**, 39–43 (1961).

[20] Asupra unor integrale vectoriale. *Comunicările Acad. R.P.R.*, **11**, 1169–1173 (1961).

[21] Asupra unor operatori liniari pe spaţii reticulate pseudonormate. *Analele Univ. Bucureşti*, Mat. Fiz., **XII**, 40, 93–98 (1963).

[22] Sur la représentation des opérateurs linéaires par des intégrales vectorielles de Hellinger. *Rev. roum. math. pures et appl.*, **10**, 437–441 (1965).

[23] Sur les fonctions-somme de O. Onicescu. *Bull. Sc. math.* Paris, 2e série, **89**, 49–63 (1965).

[24] Integrali vettoriali su uno spazio localemente compatto. *Rendic. Accad. Naz. Lincei*, Roma ser, **VIII**, 40, 792—795 (1966).

[25] Cu privire la integralele vectoriale de tip Hellinger. *Analele Univ. Bucureşti*, Seria şt. nat. mat. mec., **XV**, *I*, 61—64 (1966).

[26] Observaţii privind suporturile unor operatori liniari. *St. cerc. mat.*, **19**, 493—495 (1967).

[27] Asupra unor operatori liniari mărginiţi. *Analele Univ. Bucureşti (Seria şt. nat. mat. mec.)*, **XVI**, I, 39—43 (1967).

[28] Operatori liniari mărginiţi pe spaţii liniare reticulate topologice. *St. cerc. mat.*, **20**, 1313—1316 (1968).

[29] Sur l'extension de type Dedekind d'un espace réticulé localement convexe bornologique. *Rev. roum. math. pures et appl.*, **13**, 1293—1296 (1968).

[30] Spaţii liniare ordonate topologice. *St. cerc. mat.*, **21**, 5, 685—709 (1969).

[31] *Analiză funcţională*, ed. II, Edit. didactică şi pedagogică, Bucureşti, 1970.

[32] Sur le dual d'un espace dirigé localement convexe. *Rev. Roum. Math. Pures et Appl.*, **17**, 17—20 (1967).

[33] Spaţii de şiruri vectoriale. *Analele Univ. Bucureşti (mat. mec.)*, **17**, 9—14, (1968).

[34] Remarques sur les fonctions-somme. *Analele Univ. Bucureşti (mat-mec.)*, **20**, 43—47 (1971).

[35] Intégrales de Hellinger dans les espaces linéaires ordonnés. *Rev. Roum. math. pures et appl.*, **16**, 483—486 (1971).

[36] Sur les espaces ordonnés de suites vectorielles. *Rev. roum. math. pures et appl.*, **18**, 19—24 (1973).

[37] Sur les opérateurs réguliers et continus définis sur un espace ordonné de suites vectorielles bornées. *Rev. Roum. math. pures et appl.*, **18**, 515—518 (1973).

[38] Asupra unor spaţii de operatori liniari. *St. cerc. mat.*, **26**, 521—524 (1974).

[39] Elemente difuze în spaţii ordonate local convexe. *St. cerc. mat.*, 26 1289—1292 (1974).

[40] Opérateurs linéaires sur les espaces ordonnés de suites vectorielles. *Bull. math. soc. sci. math. R. S. Roumanie*, **18** 31—36 (1974).

CURTIS, JR. P. C.

[1] Order and commutativity in Banach algebras. *Proc. Amer. Math. Soc.*, **9**, 643—646 (1958).

DANIELL, P. J.

[1] The modular difference of classes. *Bul. Amer. Math. Soc.*, **23**, 446—450 (1966).

DAVIES, E. B.

[1] The structure and ideal theory of the predual of a Banach lattice. *Trans. Amer. Math. Soc.*, **131**, 544—555 (1968).

DE MARR, R. E.

[1] Order convergence in linear topological spaces. *Pacific J. Math.*, **14**, 17—20 (1964).

DIEUDONNÉ, J.

[1] Sur le théorème de Lebesgue-Nikodym. *Ann. of math.*, **42**, 547—555 (1941).

DIKANOVA, Z. T.

[1] On some properties of the semi-ordered space of the continuous functions on a compact space. *Uspehi mat. nauk.*, **14**, *6*, 165—172 (1959) (in Russian).

DINCULEANU, N.

[1] *Teoria măsurii și funcții reale*. Edit. didactică și pedagogică. București, 1964.

DRIML, M.

[1] Les fonctions de répartition sur les espaces semiordonnés, *Bull. Soc. math.* 2e série, **90**, 129–140 (1966).

DUHOUX, MICHEL

[1] Topologies pour un sur-espace d'un espace vectoriel topologique ordonné. *Acad. Royale Belgique. Bull.* Sc. 5-serie, **57**, 296–313 (1971).

[2] Le complété d'un treillis localement solide. *Bull. Soc. Math. Belgique*, **24**, 133–151 (1972).

DUSHNIK, B.; MILLER, E.

[1] Partially ordered sets. *Amer. J. Math.*, **63** (1941).

EISENSTADT, B. J.; LORENTZ, G. G.

[1] Boolean rings and Banach lattices. *Illinois J. Math.* **3**, 524–531 (1959).

ELLIS, A. J.

[1] The duality of partially ordered normed linear spaces. *Journ. London Math. Soc.*, 39, 730–734 (1964).

[2] On partial ordering of normed spaces. *Math. Scand.* **23**, 123–132 (1968).

ELLIS, H. W.; NAKANO, HIDEGORO

[1] Monotone functions on linear lattices. *Canad. J. Math.*, **15**, 226–236 (1963).

ERDŐS, J.

[1] On the structure of ordered real vector spaces. *Publ. Math. Debrecen*, **4**, 334–343 (1956).

ERMOLIN, M. A.

[1] *On the theory of regular operators in K-spaces*. Dissertation. Len. Pedag. Inst. (1953) (in Russian).

ESAJAN, A. R.

[1] A method of successive approximation in spaces with a regular cone (Russian) *Dokl. Akad. Nauk. Tadszik.S'S.R.*, **8**, *11*, 7–11 (1965).

FAN, KY.

[1] Le prolongement des fonctionnelles continues sur un espace semiordonné. *Revue Sci. (Rev. Rose)*, 82, 131–139 (1944).

[2] Partially ordered additive groups of continuous functions. *Ann. of Math.*,**51**, 409–437 (1950).

FET, A. I.

[1] The space of analytic functions and its application to the Cauchy– Kovalewski problems. *Uspehi mat. nauk.*, **XI**, *2*, 215–222 (1956) (in Russian).

FREUDENTHAL, H.

[1] Teilweise geordnete Moduln. Proc. Acad. Amsterdam, **39**, 641–651 (1936).

FRINK, O.

[1] Topology in Lattices. *Trans. Amer. Math. Soc.*, **51**, 569–582 (1942).

FRUMKIN, B. P.

[1] On the Egorov theorem about measurable functions. *Doklady Akad. Nauk SSSR* (N.S), **60**, 973–975 (1948) (in Russian).

FUKAMIYA, M.; MISONOU, Y.; TAKEDA, Z.

[1] On ordered and commutativity of B-algebra. *Tohoku Math. Journal*, **6**, 89–93 (1954).

GEBA K.; SEMADENI, Z.

[1] On the embedding of one space of continuous functions into another. *Bull.Acad. Polon. Sci.* Ser. Math. Astr. Phys., **7**, 187–189 (1959).

[2] On the M-subspaces of the Banach spaces of continuous functions. *Zeszyty Nauk*, Univ. Michiewicza, **25**, 53–68 (1960).

GHIKA AL.

[1] Prelungirea funcţionalelor generale lineare în module seminormate. *St. cerc. mat.*, **1**, 2, 251–281 (1960).

[2] Asupra inelelor comutative ordonate. *Bul. Ştiinţ. Acad. R.P.R.*, Seria matematică fizică, chimie II, 6, 509–518 (1960).

[3] Două criterii de complet reticulare a mulţimilor ordonate. *Comunic. Acad. R.P.R.*, **XI**, 151–154 (1961).

[4] *Analiză funcţională.* Editura Academiei, Bucureşti, 1967.

GLIVENKO, V. I.

[1] *Théorie générale des structures*, Paris, 1938.

GOFFMAN, CASPER

[1] Compatible seminorms in a vector lattice. *Proc. Nat. Acad. Sci.*, **42**, 536–538 (1956).

[2] Remarks on lattice ordered groups and vector lattices I. Carathéodory functions. *Trans. Amer. Math. Soc.*, **88**, 107–120 (1958).

[3] A class of lattice ordered algebras. *Bull. Amer. Math. Soc.*, **64**, 170–173 (1958).

[4] Completeness in topological vector lattices. *Amer. Math. Monthly*, **66**, 87–92 (1959).

GOMES, RUY LUIS

[1] Mesures vectorielles positives. *Portugal. Math.*, **22**, 101–116 (1963).

GORDON, HUGH

[1] Topologies and projection on Riesz spaces. *Trans. Amer. Math. Soc.*, **84**, 529–551 (1960).

[2] Measures defined by abstract L_p spaces. *Pacific J. Math.*, **10**, 557–562 (1960).

[3] Decomposition of linear functionals on Riesz spaces. *Duke Math. J.*, **27**, 597–606 (1960).

[4] Relative uniform convergence. *Math. Ann.*, **153**, 418–427 (1964).

GRIGORE, G.

[1] Reprezentarea unor spaţii reticulate local convexe. *Analele Univ. Bucureşti (mat. mec.)*, **18**, 45–61 (1969).

[2] Reprezentarea spaţiilor reticulate local convexe de tip L. *St. cerc. mat.*, **22**, 1183–1188 (1970).

[3] Sur les espaces de type (S). *Bull. Math. Soc.Sci. Math. R. S. Roumanie*, **15**, 29–31 (1971).

[4] Reprezentarea unor operatori liniari (no)-continui prin integrala Hellinger. *St. cerc. mat.*, **24**, 837–840 (1972).

[5] Spaţii liniare dirijate topologice. *St. cerc. mat.*, **25**, 993–996 (1973).

GROTHENDIECK, A.

[1] *Espaces vectoriels topologiques*, 2nd ed., Sociedade de Mat. de São Paulo, 1958.

GUILLAUME, MARCEL

[1] Sur les topologies définies à partir d'une relation. *Acad. Roy Belg.*, Cl. Sci. Mem. Coll. in **8**, 29 (1956).

HALILOV, Z. I.

[1] The linear control problem in a Banach space. *Doklady Akad. Nauk. SSSR*, **155**, 767–770 (1964) (in Russian).

HALMOS, PAUL

[1] *Measure Theory*, Van Nostrand, New York, 1950.

HALPERIN, I.; NAKANO, H.

[1] Discrete semi-ordered linear spaces. *Canadian Journ. Math.*, **3**, 293–298 (1951).

HALPERIN, I.; LUXEMBURG, W. A. J.

[1] The Riesz-Fischer completeness theorem for function spaces and vector lattices. *Trans. Royal Soc. Canada*, **50**, 33–39 (1956).

HAUSNER, M.; WENDEL, J. G.

[1] Ordered vector spaces. *Proc. Amer. Math. Soc.*, **3**, 977–982 (1952).

HEIDER, L. J.,

[1] Lattice orderings on Banach spaces. *Proc. Amer. Math. Soc.*, **3**, 823–838 (1952).

[2] T-Sets and abstract L-spaces. *Pacific Journ. of Math.*, **7**, 1611–1618 (1957).

HENRIKSEN, M.; JOHNSON, D. H.

[1] On the structure of a class of Archimedean lattice-ordered algebra. *Fund. Math.*, **50**, 73–94 (1961, 1962).

HEWITT, E.

[1] Rings of real valued continuous functions. *Trans. Amer. Math. Soc.*, **64**, 45–99 (1948).

HONDA, KOJI

[1] A characteristic property of Lp-spaces ($p > 1$) II. *Proc. Japan. Acad.*, **36**, 123–127 (1960).

HUSTAD, OTTO

[1] On positive and continuous extension of positive functionals defined over dense subspaces. *Math. Scand.*, **7**, 392–404 (1959).

[2] Linear inequalities and positive extension of linear functionals. *Math. Scand.*, **8**, 333–338 (1960).

[3] Extension of positive linear functionals. *Math. Scand.* **11**, 63–78 (1962).

[4] Convex cones with properties related to weak local compacteness. *Math. Scand.*, **11**, 78–90 (1962).

IONESCU TULCEA, C. T.

[1] Integrale în spaţii vectoriale ordonate. *Rev. Univ. C. I. Parhon şi a Politehnicii Bucureşti*, **2**, 9–14 (1953).

[2] *Spaţii Hilbert*, Ed. Acad. R.P.R., Bucureşti, 1956.

ISEKI, KIYOSHI

[1] Une condition pour qu'un lattice soit distributif. *C.R. Acad. Sci.* Paris, **230**, 1726–1727 (1950).

ISHIL, JYUN

[1] On equivalence of modular function spaces. *Proc. Japan. Acad.*, **35**, 551–556 (1959).

ITO, TAKASHI

[1] On conjugately similar transformations. *J. Fac. Sci. Hokkaido Univ.* Ser. I, **15**, 221–232 (1961).

IUDIN, A. I.

[1] Solution de deux problemes de la théorie des espaces semi-ordonnés. *Doklady Akad Nauk.*, **23**, 418–422 (1939) (in Russian).

[2] On the extension of semi-ordered vector spaces. *Ucenie Zapiski Len. univ.* Ser. mat. **12**, 57–61 (1941) (in Russian).

IWASAWA, K.

[1] On the structure of conditionally complete lattice groups. *Jap. J. Math.*, **18**, 777–789 (1943).

JAFFARD, P.

[1] Contribution à l'étude des groupes ordonnées. *J. Math. Pures Appl.*, **32**, 203–280 (1953).

JAMES, R. C.; MICHAEL, A. G.; WYMAN, MAX

[1] Topological abelian groups with ordered norms. *Bull. Amer. Math. Soc.*, **53** (1947).

JAMESON, G. J. O.

[1] Topological M-spaces. *Math. Z.*, **103**, 139–150 (1968).

JOHNSON, D. G.; KIST, J. E.

[1] Complemented ideals and extremally disconnected spaces. *Arch. Math.*, **12**, 349–359 (1961).

[2] Prime ideals in vector lattices. *Canad. J. Math.*, **14**, 517–528 (1962).

KADISON, R. V.,

[1] A representation theory for commutative topological algebra. *Memoirs. Amer. Math. Soc.*, **7** (1951).

KAKUTANI, S.

[1] Mean ergodic theorem in abstract L-spaces. *Proc. Imp. Acad. Tokyo*, **15**, 121–123 (1939).

[2] Concrete representation of abstract L-spaces and the mean ergodic theorem. *Ann of Math.*, **42**, 523–537 (1941).

[3] Concrete representation of abstract M-spaces. *Ann. of Math.*, **42**, 994–1024 (1941).

KALAŠNIKOV, A. L.; SLUGIN, S. N.

[1] Pointwise semi-ordered Sobolev spaces and the positiveness of the optimal solution. *Dokl. Akad. Nauk SSSR*, **207**, 281–283 (1972) (in Russian).

KALMAN, J. A.

[1] An identity for l-groups. *Proc. Amer. Math. Soc.*, **7**, 931–932 (1956).

KALMYKOV, G. I.

[1] On the semi-ordering of Markov processes on real line. *Teoria veroiatnych primen*, **VIII**, 466–469 (1962) (in Russian).

KANTOROVITCH, L. V.

[1] Sur les espaces semi-ordonnées linéaires et leur applications à la théorie des opérations linéaires. *C. R. Acad. Sci. de l'U.R.S.S.* (DAN), **4**, 11–14 (1935).

[2] Sur les propriétés des espaces semi-ordonnés linéaires *C.R. Acad. Sci. de L'U.R.S.S. (DAN)*, **202**, 813–816 (1936).

[3] On the general theory of operations in semi-ordered spaces. *Dokl. Akad. Nauk*, **1**, 271–274 (1936) (in Russian).

[4] Einige Sätze uber halbgeordnete Rāume allgemeiner Art. *C.R. Acad. Sci de l'URSS (DAN)*, **2**, 7–10 (1936).

[5] The element of the theory of functions of a real variable with values belonging to a semi ordered linear space. *C.R. Acad. Sci de l'U.R.S.S. (DAN)*, **2**, 365–369 (1936).

[6] On some clases of linear operators. *Dokl. Akad. nauk*, **3**, 9–114 (1936)(in Russian).

[7] General forms of some classes of linear operators. *Doklady Akad. Nauk*, **3**, 101–106 (1936) (in Russian).

[8] On a class of functional equations. *C.R. Acad. Sci de l'URSS (DAN)*, **4**, 219–224 (1936).

[9] On sequences of linear operations. *Doklady Akad. Nauk*, **14**, 255–260 (1937) (in Russian).

[10] On the problem of moments for a finite interval. *Doklady Akad. Nauk.*, **14**, 531–536 (1937) (in Russian).

[11] Some theorems on the convergence almost everywhere. *Doklady Akad. Nauk*, **14**, 537–540 (1937) (in Russian).

[12] On semi-ordered spaces. *Izvestia Akad. Nauk*, Ser. Mat. 91–110 (1937) (in Russian).

[13] Linear halbgeordnete Rāume. *Recueil Mat. (Mat. sbornik)*, **2**, 121–168 (1937).

[14] On functional equations. *Uchenie zap. Len. univ.*, **3**, 17–33 (1937) (in Russian).

[15] Sur la continuité et sur le prolongement des opérations linéaires. *C.R. Acad. Sci de l'URSS (DAN)*, **206**, 833–835 (1938).

[16] The method of successive approximations for functional equations. *Acta Math.*, **71**, 63–97 (1939).

[17] Linear operations in semi-ordered spaces. *Matem. sbornik*, **7**, 209–284 (1940) (in Russian).

[18] The principle of majorants and Newton's method. *Dokl. Akad. Nauk*, **76**, 17–20 (1951) (in Russian).

[19] Some further applications of the method of majorants. *Doklady Akad. Nauk*, **80**, 849–852 (1951) (in Russian).

KANTOROVITCH, L. V.; VULICH B. Z.; PINSKER, A. G.

[1] *Functional analysis in semi-ordered spaces*, Gostehizdat, Moscow–Leningrad, 1950 (in Russian).

[2] Semi-ordered groups and semi-ordered linear spaces. *Uspehi mat. nauk.*, **6**, 41–98 (1951) (in Russian).

KAPLANSKI, IRVING

[1] Lattices of continuous functions. I. *Bull. Amer. Math. Soc.*, **53**, 617–623 (1957).

[2] Lattices of continuous functions. II. *Amer. Journ. of Mat.*, **70**, 626–634 (1958).

KAPPOS, D. A.

[1] *Die Theorie der Boole'schen Algebra*, Teubner, Berlin, 1960.

[2] Remarks on the structure of spaces of generalized random variables. *Bull. Soc. Math. de Grèce*, N.S., **6**, 129–142 (1965).

KAWAI, ITIZO

[1] Locally convex lattices. *Journ. Math. Soc. Japan*, **9**, 281–314 (1957).

KELLEY, J. L.; VAUGHT, R. L.

[1] The positive cone in Banach algebras. *Trans. Amer. Math. Soc.*, **74**, 44–55 (1953).

KENDALL, DAVID G.

[1] Simplexes and vector lattices. *J. London Math. Soc.*, **37**, 365–371 (1962).

KIST, JOSEPH

[1] Locally *o*-convex spaces. *Duke Math. J.*, **25**, 569–582 (1958).

[2] Indecomposable maximal ideals of a partially ordered vector space. *J. London Math. Soc.*, **36**, 436–438 (1961).

KOMM, HORACE

[1] On the dimension of partially ordered sets. *Amer. Journ. Math.*, **70**, 507 (1948).

KOSHI, SHÔZÔ

[1] Modulars on semi-ordered linear spaces II. Aproximately additive modulars, *J. Fac. Sci. Hokkaido Univ. Ser.* **I**, **13**, 166–200 (1957).

[2] On some type of the modulared linear spaces. *J. Fac. Sci. Hokkaido Univ.* ser I, **14**, 16–28 (1958).

KOSHI, SHÔZÔ; SHIMOGAKI, TETSUYA,

[1] On quasimodular spaces. *Studia Math.*, **21**, 15–35 (1961–62).

KŐTHE G.

[1] *Topologische Lineare Raume*, 1. Springer-Verlag, Berlin Gottingen-Heidelberg, 1960.

KRASNOSELSKI, M. A.

[1] *Positive solutions of operator equations*, Gosudarstvenoe izsdatelstvo, Moscow, 1962 (in Russian).

KREIN, M. G.

[1] On linear operators which leave invariant some conic sets. *Dokl. Akad. Nauk*, **23**, 749–752 (1939) (in Russian).

[2] Fundamental properties of normal conic sets in Banach spaces. *Dokl. Akad. Nauk*, **28**, 13–17 (1940) (in Russian).

[3] On the minimal decomposition of a linear functional into positive parts. *Dokl. Akad. Nauk*, **28**, 18–22 (1940) (in Russian).

KREIN, M. G.; KREIN, S. G.

[1] On an inner characteristic of the set of all continuous functions defined on a bicompact Hausdorff space. *C.R. Acad. Sci. URSS (DAN)*, **27**, 427–430 (1940).

[2] Sur les espaces des fonctions continues definies sur un bicompact de Hausdorff et ses sousespaces semiordonnés *Rec. Math. (Mat. Sbornik)* NS, **13** (55) 1–38 (1943).

KREIN, M. G.; RUTMAN, M. A.

[1] Linear operators which leave invariant a cone in a Banach space. *Uspehi mat. nauk*, **3**, 1, 3–95 (1948) (in Russian).

KULLER, R. G.

[1] Locally convex topological vector lattices and their representations. *Michigan Math. J.*, **5**, 83–90 (1958).

LEADER, SOLOMON

[1] Separation and approximation in topological vector lattices. *Canad. J. Math.*, **11**, 286–296 (1959).

LEENDERS, J. H.

[1] On a partially ordered matrix-algebra. *Simon Stevin*, **32**, 102–141 (1958).

LORCH, EDGAR

[1] L'integrazione ed i funtionali lineari. *Rendiconti dei sem mat. e fis di Milano*, **XXV**, 113–120 (1953–54).

LOWENSLAGER, D. B.

[1] Duality in vector lattices. *Proc. Amer. Math. Soc.*, **8**, 476–483 (1957).

LOZANOVSKII, G. Y.

[1] On topological reflexive KB-spaces. *Dokl. Acad. Nauk*, **158**, 516–519 (1964) (in Russian).

[2] On almost integral operators in KB-spaces. *Vestnik Leningrad. Univ.*, **21**, 7, 35–44 (1966) (in Russian).

LUXEMBURG, W. A. J.

[1] Notes on Banach functions spaces. *Proc. Koninklijke Nederlands Akad. van Wetens*, Series A, Note XVI, A **68**, 646–647 (1965).

MAC NEIL, H. M.

[1] Extensions of partially ordered sets. *Proc. Nat. Acad. Sci.*, **22** (1936).

[2] Partially ordered sets. *Trans. Amer. Math. Soc.*, **42**, 416–460 (1937).

MAEDA, F.

[1] Partially ordered linear spaces. *J. Hiroshima Univ.* Ser. A, **10**, 137–150 (1940).

MAEDA, F.; OGASAWARA, T.

[1] Representation of vector lattices. *J. Hiroshima Univ.* Ser. A, **12**, 17–35 (1942).

MALLIOS, ANASTASSIOS

[1] On approximation in topological vector lattices. *Bull. Soc. Math. Grèce* (N.S.), **1**, 48–54 (1960).

MAMEDOV, R. G.

[1] Asymptotic values of approximation of differentiable functions by positive linear operators. *Dokl. Akad. Nauk*, **128**, 471–474 (1959) (in Russian).

[2] On the order of approximation of functions by positive linear operators. *Dokl. Akad. Nauk.*, **128**, 674–676 (1959) (in Russian).

MAŘIK, J.

[1] Vrcholy jednotkové koule v prostoru functional na daném polousporadanem prostoru. *Časop. pestov. mat.*, **79**, 3–40 (1954).

MARINESCU, G.

[1] Metoda aproximaţiilor succesive în grupuri cu normă abstractă. *Com. Acad. RPR*, **I**, 547–550 (1951).

[2] Spaţii vectoriale parţial-ordonate cu coeficienţi într-o algebră. *Rev. Univ. C. I. Parhon şi A Politehnicii Bucureşti*, **1**, 17–19 (1952).

[3] *Spaţii vectoriale normate*. Ed. Acad., Bucureşti, 1956.

[4] *Spaţii vectoriale topologice şi pseudo-topologice*. Ed. Acad., Bucureşti, 1959.

MATHES, KLAUS

[1] Uber eine Schar von Regularitätsbedingungen fur Verbände II. *Math. Nachr.*, **23**, 149–159 (1961).

MATUSZEVSKA, W.; ORLICZ, W.

[1] A note on modular spaces. V. *Bull. Acad. Polon Sci. Ser. Sci. Math. Astron, Phys.*, **11**, 51–54 (1963).

MC ARTHUR, C. W.
[1] Convergence of monotone nets. *Studia math.*, **34**, 1–16 (1970).
[2] In what spaces is every closed normal cone regular? *Proc. Edinburgh Math. Soc.*, **17**, 121–125 (1970).

MILLES, P. E.
[1] Order isomorphisms of B*-algebras. *Trans. Amer. Math., Soc.*, **107**, 217–236 (1963).

MIYAKAWA, MICHIYO; NAKANO, HIDEGORO
[1] Modulars on semiordered linear spaces. I. *J. Fac. Sci. Hokkaido Univ. Ser.*, I, **13**, 41–53 (1956).

MUSIELAK, J.; ORLICZ, W.
[1] Some remarks on modular spaces. *Bull. Acad. Polon. Sci. Sér. Math. Astr. Phys.*, **7**, 661–668 (1959).

NACHBIN, LEOPOLDO
[1] Sur les espaces topologiques ordonnés. *C.R. Acad. Sci. Paris*, **226**, 381–382, (1948).
[2] Sur les espaces uniformisables ordonnés. *C. R. Acad. Sci. Paris*, **226**, 547 (1948).
[3] Sur les espaces uniformes ordonés. *C. R. Acad. Sci. Paris*, **226**, 774 (1948).
[4] A characterization of the normed vector ordered spaces of continuous functions over a compact space. *Amer. Journ. of Math.*, **71**, 701–705 (1949).
[5] A theorem of the Hahn-Banach type for linear transformations. *Trans. Amer. Math. Soc.*, **68**, 28 (1950).
[6] *Topology and order*, 1965.

NAKAMURA, M.
[1] Notes on Banach space (9). The Vitali-Hahn-Saks theorem and K-spaces. *Tohoku Math. J.*, **1**, 100–108 (1949).
[2] Notes on Banach spaces (11). Banach lattices with positive bases, *Tohoku Math. J.*, **2**, 135–141 (1950).
[3] Notes on Banach space (12). A remark on a theorem of Gelfand and Neumark. *Tohoku Math. J.*, **2**, 182–187 (1950).

NAKANO, H.
[1] Teilweise Geordnete Algebra. *Japanese Journ. Math.*, **17**, 425–511 (1941).
[2] Eine Spektraltheorie. *Proc. Phys. Math. Soc. Japan*, **23**, 285–511 (1941).
[3] Uber normierte teilweise geordnete Modular. *Proc. Imp. Acad. Tokyo*, **17**, 311–317 (1941).
[4] Stetige lineare Funktionale auf dem teilweise geordneten Moduln. *Journ. Fac. Sci. Imp. Univ. Tokyo*, **4**, 201–382 (1942).
[5] Riesz-Fischer Satz in normierten Teilweise geordneten Moduln. *Proc. Imp. Acad. Tokyo*, **18**, 350–353 (1942).
[6] Uber die Stetigkeit des normierten telweise geordneten Moduls. *Proc. Imp. Acad. Tokyo*, **19**, 10–11 (1943).
[7] Über Einführung der teilweisen Ordnung in reellen Hilbertschen Raum. *Proc. Phys-Math. Soc. Japan*, **26**, 1–8 (1941).
[8] On the product of relative spectra. *Ann. of Math.*, **49**, 281–315 (1948).
[9] Ergodic theorems in semiordered linear spaces. *Annals of Math.*, **49**, 538 (1948).

[10] *Modulared semiordered linear spaces*, Maruzen Co. Ltd., Tokyo, 1950.
[11] *Modern spectral theory*, Maruzen Co. Ltd., Tokyo, 1950.
[12] Modulared sequence spaces. *Proc. Japan. Acad.*, **27**, 508–512 (1951).
[13] On transcendental points in proper spaces of discrete semiordered linear spaces. *Journ. Fac. Sci. Hokkaido Univ.*, **12**, 105–110 (1953).
[14] Linear topology on semiordered linear spaces. *Journ. Fac. Sci. Hokkaido Univ.* Ser. 1 Math., **12**, 87–104 (1953).
[15] Concave modulars. *Journ. of Math. Soc. Jap.* **5**, 29–49 (1953).
[16] Product spaces of semiordered linear spaces. *Journ. Fac. Sci. Hokkaido Univ.* ser I. Math., **12**, 163 (1953).
[17] Convergence concepts in semiordered linear spaces. II .*Proc. Japan, Acad.*, **35**, 83–88 (1959).
[18] On semicontinuous linear lattices. *Proc. Am. Math. Soc.*, **34**, 115–117 (1972).

NAKANO, H.; SASAKI, MASAHUMI
[1] Convergence concepts in semiordered linear spaces. *Proc. Japan Acad.*, **35**, 25–30 (1959).

NAMIOKA, ISAAC
[1] Partially ordered linear topological spaces. *Memoirs of the Amer. Math. Soc.*, **24** (1957).

NEVEU, J.
[1] Sur l'existence des mesures invariantes en théorie ergodique. *C. R. Acad. Sci. Paris*, **260**, 393–396 (1965).

NG, KUNG-FU
[1] Solid sets in ordered topological vector spaces. *J. London Math. Soc.*, **22**, 106–120 (1971).
[2] A duality theorem on partially ordered normed spaces. *J. London Math. Soc.*, **22**, 403–404 (1971).

NICOLESCU, MIRON
[1] *Analiza matematică*, vol. I, II, III, Ed. tehnică, București, 1957, 1958, 1960.

NICULESCU, CONSTANTIN P.
[1] Operateurs absolument continus. *Rev. Roum. math. pures et appl.*, **19**, 225–236 (1974).

OGASAWARA, TOZIRO
[1] Theory of vector lattices. *J. Sci. Hiroshima Univ.*, Ser. A, **12**, 37–100 (1942).
[2] Some general theorems and convergence theorems in vector lattices. *J. Sci. Hiroshima Univ.*, Ser. A, **14**, 14–25 (1949).
[3] Remarks on representation of vector lattices. *J. Sci. Hiroshima Univ.*, Ser. A, **12**, 217–234 (1943).
[4] On Frechét lattices. *J. Sci. Hiroshima Univ.*, Ser. A, **12**, 235–248 (1943).
[5] *Theory of vector lattices.* II. *J. Sci. Hiroshima Univ.*, Ser. A, **13**, 41–161 (1944).

OLMSTED, J.
[1] Lebesgue theory on a Boolean algebra. *Trans. Amer. Math. Soc.*, **51**, 164–194 (1942).

OLUBUMMO, A.
[1] The Laplace-Stieltjes transform of an increasing vector-valued function. *Quart. J. Math. Oxford Aer*, **(2)**, **8**, 97–107 (1957).

[2] Monotone semi-groups of bounded operators. *J. Math. Mech.*, **12**, 385–390 (1963).

O'NEILL, ANNE

[1] Contributions to the theory of derivates. *Duke Math. Journ.*, **12**, 89 (1945).

ONICESCU, O.

[1] Notes sur les b-Algèbres. *Rev. Math. pures et appl.*, **4**, 345–350 (1959).

[2] Un théorème ergodique. *C. R. Acad. Sci. Paris*, **265**, 2761–2764 (1963).

ONICESCU, O.; CUCULESCU, I.; SÎMBOAN, G.; MARINESCI, G.; THEODORESCU, R.

[1] *Funcţii sumă.* Ed. Acad., Bucureşti 1963.

ORLICZ, W.

[1] A note on modular spaces. IV. *Bull. Acad. Polon. Ser. Sci. Math. Astronom. Phys.*, **10**, 479–484 (1962).

PAPANGELOU, FREDOS

[1] Note on the functional analogue of the Jordan decompositions of measures. *Arch. Math. (Basel)*, **17**, 253–256 (1966).

PAPERT, D.

[1] A note on vector lattices and interration theory. *J. London Math. Soc.*, **38**, 477–485 (1963).

PASHENKOV, V. V.

[1] Some remarks on the completion of partially ordered sets. *Uspehi matem. nauk*, **17**, 6, 155–162 (1962) (in Russian).

PEIRCE, C. S.

[1] *Mem. Am. Acad. Sci.*, **9**, 317–378 (1870).

PERESSINI, A. L.

[1] On topologies in ordered linear spaces. *Math. Ann.*, 1144, 199–223 (1961).

[2] Concerning the order structure of Köthe sequence spaces. *Michigan Math. J.*, **10**, 409–415 (1963).

[3] A note on abstract (M)-spaces. *Illinois J. Math.*, **7**, 118–120 (1963).

[4] A note on the lattice properties of continuous linear mappings. *Math. Z.*, **115**, 18–22 (1970).

PHILLIPS, R. S.

[1] Semi-groups of positive contraction operators. *Czechoslovak Math. J.*, **12**, 294–313 (1962).

PINSKER, A. G.

[1] Analytic representation of some partially additive functionals. *Dokl. Akad. Nauk*, **18**, 397–401 (1938) (in Russian).

[2] Sur l'extension des espaces semi-ordonnés. *C. R. Acad. Sci. d'URSS*, **21**, 6–10 (1938).

[3] Sur certaines propriétés des K-espaces études. *C. R. Acad. Sci. de l'URSS*, **22**, 220–224 (1939).

[4] Universal K-spaces. *Dokl. Akad. Nauk*, **49**, 8–11 (1945) (in Russian).

[5] Decomposition of K-spaces into elementary spaces. *Dokl. Akad. Nauk*, **49**, 169–172 (1945) (in Russian).

[6] On separable K-spaces. *Dokl. Akad. Nauk*, **49**, 327–328 (1945) (in Russian).

[7] Completely linear functionals in K-Spaces. *Dokl. Akad. Nauk*, **55**, 303–306 (1947) (in Russian).

[8] On concrete representations of semi-ordered linear spaces. *Dokl. Akad. Nauk*, **55**, 383–386 (1947) (in Russian).

[9] Lattices equivalent to K-spaces. *Dokl. Akad. Nauk*, **99**, 503–506 (1954) (in Russian).

[10] On the conditions for the equivalent of a Banach space to an L-space. *Dokl. Akad. Nauk*, **99**, 677–679 (1954) (in Russian).

[11] On the representation of K-spaces as rings of self-adjoint operators. *Dokl. Akad. Nauk*, **106**, 195–198 (1956) (in Russian).

POPA, NICOLAE

[1] Obiecte injective în categoria spaţiilor ordonate archimediene cu elemente axiale. *St. cerc. mat.*, **17**, 1333–1336 (1965).

[2] Produits tensoriels ordonnés. *Rev. Roum. math. pures et appl.*, **13**, *2*, 235–246 (1968).

[3] Les duals des espaces de Fréchet ordonnés. *Rev. Roum. math. pures et appl.*, **14**, *2*, 223–230 (1969).

[4] Une condition suffisante pour que $E \otimes F$ soit un treillis localement-convexe. *C. R. Acad. Sc. Paris*, **268**, 487–489 (1969).

[5] Produse tensoriale topologice ordonate. *St. cerc. mat.*, **23**, 201–286 (1971).

POTEPUN, A. V.

[1] Generalized regularity of K-lineals and properties of the order topology. *Dokl. Akad. Nauk, SSSR*, **207**, 541–543 (1972) (in Russian).

RIBENBOIM, P.

[1] *Théorie des groupes ordonés*, Bahia Blanca, 1964.

RIEDL, JOHN

[1] Partially ordered locally convex vector spaces and extensions of positive continuous linear mappings. *Math. Annalen*, **157**, 95–124 (1964).

RIESZ, F.

[1] Sur la décomposition des opérations functionnelles linéaires. *Atti. Cong. Int. dei Mat. (Bologna)*, **(VI)**, *3*, 143–148 (1928).

[2] Sur quelques notions fondamentales dans la théorie générale des opérations linéaires. *Ann. of. Math.*, **41**, 174–206 (1940).

[3] Sur la théorie ergodique des espaces abstraits. *Acta Univ. Szeged, Acta Sci. Math.* **10**, 1—20 (1941–1943)

ROBERTS, G. T.

[1] Topologies in vector lattices. *Proc. Cambridge Phil. Soc.*, **48**, 533–546 (1952).

ROBERTSON, A. P.; ROBERTSON, W.

[1] *Topological vector spaces*, Cambridge University Press, Cambridge, 1964.

SASAKI, MASAHUMI

[1] On semi-lower bounded modulars. *J. Fac. Sci. Hokkaido Univ.* Ser. 1, **14**, 114–124 (1959).

SASONOV, V. V.

[1] On distributions in spaces dual to ordered linear spaces. *Trudy Mat. Inst. Steklova*, **71**, 102–103 (1964) (in Russian).

SCHAEFER, HELMUT

[1] Positive transformationen in lokalkonvexen halbgeordneten Vektorräumen. *Math. Ann.*, **129**, 323–329 (1955).

[2] Halbgeordnete lokalkonvexe Vektorräume. *Math. An.*, **135**, 115–141, (1958).
[3] Halbgeordnete lokalkonvexe Vektorräume. II. *Math. An.*, **138**, 259–286 (1959).
[4] On the completeness of topological vector lattices. *Michigan Math. Journ.*, **7**, 303–309 (1960).
[5] Halbgeordnete lokalkonvexe Veltorräume. III. *Math. Ann.*, 113–142 (1960).
[6] A generalized moment problem. *Math. Ann.*, **146**, 326–330 (1962).
[7] Spectral measures in locally convex algebras. *Acta Math.*, **107**, 125–173 (1962).
[8] *Topological vector spaces*, 1966.

SCHÄFFER, JUAN, JORGE
[1] Sur une classe de treillis vectoriels topologiques. *Univ. Rep. Fac. Ingen. Agrimen. Montivideo Publ. Inst. Mat. Estadist.*, **3**, 135–146 (1961).

SEMADENI, Z.
[1] Sur les fonctionnelles linéaires dans des espaces vectoriels semiordonnés avec application à la représentation des fonctionnelles par des intégrales. *Bull. de l'Acad. Polon des sciences*, serie des sci. math. astr. et phys., **IV**, 457–462 (1958).
[2] Sur la convergence faible dans certains espaces vectoriels semiordonnés, *Zeszyty Nauk. Univ. Mickiewicza*, **25**, 69–77 (1960).

ŞERBAN, DOINA
[1] Asupra unor operatori liniari pe spaţii de şiruri vectoriale. *St. cerc. Mat.*, **26**, 123–127 (1974).

SHERMAN, S.
[1] Order in operator algebras. *Amer. Math.*, **73**, 227–232 (1951).

SHIELDS, P. C.
[1] Invariant elements for positive contractions on a Banach Lattice. *Math. Z.*, **96**, 189–195 (1967).

SHIMBIREVA, E. P.
[1] On the theory of partially ordered groups. Recueil mathématique. *Math. sbornik*, **20 (62)**, 145–178 (1947).

SHIMOGAKI, TETSUYA
[1] On the norms by uniformly finite modulars. *Proc. Japan. Acad.*, **33**, 304–309 (1957).
[2] On certain property of the norms by modulars. *J. Fac. Sci. Hokkaido Univ.* ser I, **13**, 201–213 (1957).
[3] On this continuity and the monotonousness of norms. *J. Fac. Sci. Hokkaido Univ.* Ser. I, **16**, 225–237 (1962).

SHIROKHOV, M. F.
[1] Functions of elements of semi-ordered spaces. *Dokl. Akad. Nauk*, **74**, 1057–1060 (1950) (in Russian).
[2] Applications of functions of decompositions to the theory of semi-ordered spaces. *Vestnik Leningrad. Univ.*, **19**, 29–36 (1960) (in Russian).

ŠIK, FRANTISEK
[1] On the theory of lattice ordered groups. *Czechoslovak, Math. J.*, **6 (181)**, 1–25 (1956) (in Russian).

SHIKORSKI, ROMAN
[1] *Boolean algebras.* Ergebnisse der Matematik. Berlin–Göttingen–Heidelberg, 1960.

SIMEONOV, S. V.

[1] On a process of successive approximations and its applications to the solutions of functional equations with non-linear operators of monotonous type. *Dokl. Akad. Nauk*, **138**, 1033–1034 (1961) (in Russian).

SIMONS, S.

[1] Partial orderings on real linear topological spaces. *Proc. Cambridge Phil. Soc. (Math. Phys. Sci.)*, **59** part 2, 323–328 (1963).

SINGER, I.

[1] Contribuţii la teoria KB-linealelor. *St. cerc. mat.*, **6**, 465–475 (1955).

SKORDEV, D.

[1] On some semi-ordered spaces. *Dokl. Akad. Nauk*, **138**, 553–555 (1961) (in Russian).

SLUGIN, S. N.

[1] Bilateral valuations of the solution of a linear functional equation. *Dokl. Akad. Nauk*, **127**, 34–36 (1959) (in Russian).

[2] A linear semi-ordered topological space. *Dokl. Akad. Nauk*, **131**, 1261–1263 (1960) (in Russian).

[3] Partially ordered topological groups. *Dokl. Akad. Nauk*, **136**, 36–38 (1961) (in Russian).

[4] Monotonous processes of bilateral approximation in a partially ordered convergence group. *Dokl. Akad. Nauk*, **147**, 34–37 (1962) (in Russian).

[5] Complex Hilbert lattice. *Izvestia Akad. Nauk* (Ser. Mat.), **29**, 215–226 (1965) (in Russian).

SOBOLEV, V. I.

[1] On inverse elements in semi-ordered rings. *Dokl. Akad. Nauk*, **56**, 237–239 (1947) (in Russian).

[2] On a property of self-adjoint operators in Hilbert space. *Uspehi Mat. Nauk*, **7**, 169–172 (1952) (in Russian).

[3] On the semi-ordered measure of sets of measurable functions and some abstract integrals. *Dokl. Akad. Nauk*, **91**, 23–26 (1953) (in Russian).

SOLOVEV, V.

[1] On the extension of a monotonous norm from a normed lattice to its Dedekind completion. *Syrbirkii Mat. Jur.*, **VII**, 1360–1369 (1966) (in Russian).

STANCOVIČ, B.

[1] Complétion d'un groupe réticulé. *Acad. Serbe Ser-Publ. Inst. Math.*, **14**, 115–122 (1960).

STEEN, S. W. P.

[1] An introduction to the theory of operators. *Proc. London Math. Soc.*, **41**, 361–392 (1936).

STONE, M. H.

[1] Subsumption of Boolean algebras under the theory of rings. *Proc. Nat. Acad. Sci.*, **21**, 103–105 (1935).

[2] The theory of representation for Boolean algebras. *Trans. Amer. Math. Soc.*, **40**, 37–111 (1936).

[3] Application of the theory of Boolean rings to general topology. *Trans. Amer. Math. Soc.*, **41**, 375–481 (1937).

[4] Pseudo-norms and partial ordering in Abelian groups. *Ann. of Math.*, **48**, 851 (1947).

SYNNAČKE, J.

[1] The almost integral operators in K-spaces. *Vestnik Leningrad. Univ.*, no. 13, 81–89 (1971) (in Russian).

SZPILRAJN, E.

[1] Sur l'extension de l'ordre partiel. *Fundam. Math.* **16**, 386–389 (1930).

TAGAMILITZKI, Y.

[1] Sur quelques applications de la théorie générale des espaces vectoriells partiellement ordonnés. *Annuaire Univ. Sofia*, Fac. Sci. Livre 1, **45**, 260–286 (1949).

TANAKA, TOSHIO

[1] On Archimedean vector lattices with a basis. *Bull. Fukuoka Gakugei Univ.*, **114**, 8–12 (1954).

TELEMAN, SILVIU

[1] La répresentation des automorphismes de réseaux vectoriels. *Rev. Roum. math. pures appl.*, **13**, 95–105 (1968).

THOMPSON, A. C.

[1] On certain contraction mappings in a partially ordered vector space. *Proc. Amer. Math. Soc.*, **14**, 438–443 (1963).

TSITSAS, LEONIDAS N.

[1] Integral representations of linear maps. *Bull. Soc. Math. Grèce* (N.S.), **6II**, Fasc. 2, 298–355 (1965).

TURBATU-BILCIURESCU, S.

[1] Functionale liniare pe anumite spaţii ordonate pseudonormate. *Comunic. Acad. R.P.R.*, **XII**, 265–268 (1962).

VEKSLER, A. I.

[1] On the principle of Archimedes in semi-ordered quotient lineals. *Dokl. Akad. Nauk.*, **121**, 775–777 (1958) (in Russian).

[2] On the homomorphism between the classes of regular operators in K-linears and in their completions. *Izvestia vis. uceb. zav. Matem.*, **1**, 48–47 (1960) (in Russian).

[3] On the completeness and the σ-completeness of normed and topological linear lattices. *Izvestia vis uceb. zaved. Mat.*, **3**, 22–30 (1962) (in Russian).

[4] On the conditions for the principle of Archimedes in semi-ordered quotient groups and quotient lineals. *Matem. Sbornik*, **57**, *4*, 477–492 (1962) (in Russian).

[5] On the topological and order completeness of normed and topological linear lattices. *Dokl. Akad. Nauk.*, **143**, 262–264 (1962) (in Russian).

[6] On the linear lattices with a sufficient set of maximal l-ideals. *Dokl. Akad. Nauk.*, **150**, 715–718 (1963).

VULICH, B. Z.

[1] On a generalized notion of convergence in a Banach space. *Ann. of Math.*, **38**, 156–174 (1937).

[2] Une définition du produit dans les espaces semiordonnés linéaires. *C.R. Sci. de L'URSS (DAN)*, **26**, 850 854 (1940).

[3] Les propriétés du produit et de l'élement inverse dans les espaces semi-ordonés linéaires. *C.R. Acad. Sci. de L'URSS (DAN)*, **26**, 855–859 (1940).

[4] The Stieltjes integral for functions with values in semi-ordered spaces. *Ucenie Zap. Leningrad. Univ.* **12**, 3–29 (1941) (in Russian).

[5] On linear multiplicative operations. *Dokl. Akad. Nauk*, **41**, 148–151 (1943).

[6] Sur la représentation analytique d'opérations linéaires multiplicatives. *C.R. Acad. Sci. de l'URSS (DAN)*, **41**, 187–190 (1943).

[7] On the linear functionals in semi-ordered linear spaces. *Dokl. Akad. Nauk*, **52**, 95–98 (1946) (in Russian).

[8] Sur les opérations linéaires multiplicatives. *C.R. Acad. Sci. de l'URSS (DAN)*, **52**, 383–386 (1946).

[9] Sur quelques opérations non-linéaires dans les espaces semi-ordonnés linéaires *C.R. Acad. Sci. de l'URSS (DAN)*, **52**, 475–478 (1946).

[10] Concrete representations of semi-ordered linear spaces. *Dokl. Akad. Nauk*, **58**, 733–736 (1947) (in Russian).

[11] Multiplication in semi-ordered linear spaces and its application to the theory of operations. *Math. Sbornik*, **22**, 27–78; 267–317 (1948) (in Russian).

[12] On the concrete representation of semi-ordered lineals. *Dokl. Akad. Nauk*, **78**, 189–192 (1951) (in Russian).

[13] Generalized semi-ordered rings. *Mathem. Sbornik*, **33**, *2*, 343–358 (1953) (in Russian).

[14] Partial ordering of rings of bounded self-adjoint operators. *Vestnik Leningrad. Univ.*, **13**, 13–21 (1957) (in Russian).

[15] Application of the theory of semi-ordered rings and the study of self-adjoint operators in Hilbert space, *Uspehi mat. nauk*, **12**, *1*, 169–172 (1957) (in Russian).

[16] *Introduction to the theory of semi-ordered spaces.* Gosudarstvennoe izdatelstvo, Moscow, 1961 (in Russian).

[17] On linear lattices equivalent to lattice with monotonous norm. *Dokl. Akad. Nauk*, **147**, 271–274 (1962) (in Russian).

VULICH, B. Z.; DANILENKO, I. F.

[1] A certain method of partial ordering of a normed space. *Vestnik Leningrad. Univ.*, **25**, 19, 18–22 (1970) (in Russian).

VULICH, B. Z.; KORSAKOVA, O. S.

[1] Spaces in which convergence in norm coincides with order convergence *Mat. Zametki*, **13**, 259–268 (1973) (in Russian).

VULICH, B-Z.; LOZANOVSKI, G. I.

[1] The representation of completely linear and of regular functionals in partially ordered spaces. *Math. Sb. (N.S.)*, **84 (126)**, 331–352 (1971).

WADA, JUNZO

[1] Positive linear functionals on ideals of continuous functions. *Osaka Math. J.*, **11**, 173–185 (1959).

WALLMAN, H.

[1] Lattices and topological spaces. *Ann. of Math.*, **39**, 112–126 (1938).

WALSH, B.

[1] On characterizing Köthe sequence spaces as vector lattice. *Math. Ann.*, **175**, 253–256 (1968).

WASSILKOFF, D.
[1] On the theory of partially ordered linear systems and linear spaces. *Ann. of Math.*, **44**, 580–609 (1943).

WESTON, J. D.
[1] Convergence of monotonic sequences in vector spaces. *Journ. of the London Math. Soc.*, **32**, 476–477 (1957).
[2] Relations between order and topology in vector spaces. *Quart. J. Math. Oxford Ser.*, **(2) 10**, 1–8 (1959).
[3] The decomposition of a continuous linear functional into non-negative components. *Math. Scand.*, **5**, 54–56 (1957).

WINKELBAUER, KAREL
[1] The ergodic theorem in semi-ordered linear spaces with abstract norm. *Czechoslovak. Math. J.*, **8 (83)**, 1–21 (1958).

WONG, YAU CHUEN
[1] Locally o-convex Riesz spaces. *Proc. London Math. Soc.*, **19**, 289–309 (1969).
[2] Reflexivity of locally convex Riesz spaces. *Journ. London Math. Soc.*, **1**, 725–732 (1969).
[3] The order bound topology. *Proc. Camb. Phil. Soc.*, **71**, 321–327 (1972).

YAMAMURO, SADAYUKI
[1] Monotone completeness of normed semi-ordered linear spaces. *Pacif. J. Math.*, **7**, 1715–1725 (1957).
[2] On the theory of some non-linear operators. *Yokohama Math. J.*, **10**, 11–17 (1962).

ZAMANSKI, MARK
[1] Groupes de Riesz et théorie de l'intégration. *C.R. Acad. Sci. Paris.*, **248**, 3393–3395 (1959).

ZORN, M.
[1] A remark on method in transfinite algebra. *Bull. A M S*, **41**, 667–670 (1935).

Subject index